INVERSE PROBLEMS OF ACOUSTIC AND ELASTIC WAVES

Inverse Problems of Acoustic and Elastic Waves

edited by

Fadil Santosa, Yih-Hsing Pao,
Cornell University

William W. Symes,
Rice University

and

Charles Holland
Office of Naval Research

siam Philadelphia/1984

Library of Congress Catalog Card Number: 84-52372
ISBN: 0-89871-050-2

CONTENTS

PREFACE

An International Conference on Inverse Problems of Acoustic and Elastic Waves was held at the campus of Cornell University on June 4–6, 1984. Collected here are the proceedings of that meeting.

Inverse problems in wave propagation are measurement problems in continuum mechanics. They differ from other (traditional) measurement problems in that the data set and the set of parameters to be inferred from it are both very large. In fact, ideally both sets are *continua*, i.e., functions of several variables. In contrast, most conventional measurement theory concerns the inference of one parameter or a few parameters from a measured data set.

The data for the problem considered in this volume consist (in principle) of the output of instruments which respond to acoustic or elastic wave motion. The parameters to be inferred from this data are the (distributed) mechanical parameters of the fluid or solid body supporting the wave motion. The wave character of the motion implies that information about mechanical parameters is carried to remote locations by the wave field, where it may be measured. This possibility of inference of complex mechanical structures from remote measurements gives the subject its characteristic flavor.

The best-known example of inverse problems in wave propagation is the problem of seismic prospecting for petroleum. Seismic crews stimulate motion in the earth using energy sources (dynamite, vibrator trucks, etc.) near the earth's surface, and record the signals from motion-sensitive sensors also distributed on or near the surface. From this (very large volume of) information, seismic data processors attempt to infer the sedimentary lithology of the upper crust, hence the presence of petroleum-bearing rock formations. This inference is presumed to be possible, at least to some extent, because the motion of the earth seems to consist (approximately) of linearly elastic waves, at least in the regimes used in seismic prospecting. Because of the wave nature of the motion, the motion near the surface is sensitive to (at least some aspects of) the distribution of mechanical parameters at depths of thousands of meters. The mechanical parameters in turn are related (in a complicated way) to the lithology.

Similar measurement problems occur in ocean acoustics, ultrasonic nondestructive materials evaluation, biomedical ultrasonics, and other fields. Applications range from location of submarines to the identification of flaws in manufactured articles, to the characterization of tumors and other tissue anomalies.

Since all of these problems share the features of very large data and parameter sets, the availability of larger and faster computers has encouraged a great deal of work along formerly impractical lines. The ability to process large amounts of data also has brought to the fore the key question of *information content*; for very large data and parameter sets, the significance of the inference can be difficult to discern. For instance, seismic reflection data is sensitive to, and therefore carries information about, changes in mechanical properties of rock along interfaces with some orientations, but not about changes along interfaces with other orientations.

The task of assessing the information content of data and devising methods to extract it, may be divided into three parts:

(i) modeling—deciding which physical description of the problem is appropriate;
(ii) mathematical analysis of the model;
(iii) numerical analysis of algorithms for inference of parameters from data, in context of the model.

These processes often are treated as distinct and ordered as above. The careful reader of the following pages will note, however, that all three activities often are carried on simultaneously, and that information flows in all ways between them. Mathematical analysis and numerical (and laboratory) experiments lead to refinements in models, analysis of mathematical structure leads to improved efficiency of algorithms, numerical experiments suggest refinements in mathematical analysis, and so on.

Most of the work on inverse problems in wave propagation has taken as its basic physical model either linear acoustics or linear elastodynamics. Many modeling concerns remain, however, such as appropriate models of sensor/medium interaction, useful classes of coefficients (that is distribution of mechanical parameters), various approximations to the relation between coefficients and solution to the equations of motion (high-frequency asymptotics, linearization), and so on.

The mathematical analysis of these problems usually reveals that they are ill-posed in some sense—that is, the coefficients (parameters) are extremely sensitive to data error. In other words, for most of these problems, the data do not determine the parameters unambiguously in the presence of noise. For example, energy sources and receiving instrumentation commonly are modeled by bandlimited transfer functions, which assumption (roughly speaking) imposes a related bandlimit on the parameter estimates.

This oversensitivity to data error, or inadequacy of the data set, forces the analyst to impose, either explicitly or implicitly, additional constraints, usually called *"a priori"* to distinguish them from the primary data of the problem, to render the parameter estimate unambiguous. These additional constraints on "admissible" models generally prevent an exact match of parameter distribution to noisy data. Thus many inverse problems must be formulated as "best-fit" or optimization problems.

These themes—the necessity of additional information, inverse problems as optimization problems—will be found throughout the volume, muted in some places, loud in others.

To some extent, the papers fall into rough subject groups: one-dimensional transient problems (Sondhi, Bube, Oldenburg, Fawcett), inverse problems in the frequency domain (McLaughlin, Gladwell, Bates, Colton), multiparameter estimates for layered media (Stickler, Levy), inverse problems as optimization problems (Sabatier, Lailly, Tarantola, Chen, Lee et al.), algorithms based on linearization and/or high-frequency asymptotics, e.g., migration and diffraction tomography (Berkhout, Devaney), numerical solution of boundary value problems (Oliger, Wahlbin), and modeling issues (Driessel). The reader should be aware, however, that the distinctions among these categories are quite blurred. The theory of inverse problems in wave propagation is in a state of ferment induced by the pressure of applications interest and the rapidly expanding computational possibilities, and its intellectual organization will doubtless change substantially over the next years.

The progress reports of the three ONR-sponsored research groups (Iowa State University, Colorado School of Mines, Cornell University) address many of the issues outlined above.

The Conference was sponsored by the Office of Naval Research. The Organizing Committee was comprised of Yih-Hsing Pao (Chairman), Charles Holland, Fadil Santosa, and William Symes. The Conference Coordinator was Janice B. Conrad, whose importance in making this meeting possible cannot be overstated.

There were ten invited lectures, ten solicited papers, and three technical progress reports of the ONR-sponsored research groups. These proceedings contain the papers presented by the lecturers, the reports, and extended abstracts of the solicited papers.

We would like to express our gratitude to Professors Roger Newton, Cathleen Morawetz, Victor

Barcilon, Lars Wahlbin, and Emil Wolf for participating as chairpersons, and we acknowledge their invaluable contribution to the conference. Lynne Lehman and Janice Conrad expertly put many of the manuscripts and abstracts into camera-ready form, and we express our grateful appreciation for their work.

WILLIAM W. SYMES
FADIL SANTOSA
September 1984

A SURVEY OF THE VOCAL TRACT INVERSE PROBLEM: THEORY, COMPUTATIONS AND EXPERIMENTS

MAN MOHAN SONDHI*

Abstract. In this talk I will discuss methods of estimating the shape of a human vocal tract while it is moving as during normal speech. After discussing several approaches to this problem, I will show that methods based on transient acoustical measurements at the lips offer many advantages over the others. I will then outline the theoretical basis for such methods and finally describe an experimental set up and associated numerical procedures which enable one to measure and display the shape in real time.

Introduction. As an introduction to the inverse problem for the vocal tract, let me begin by sketching briefly the manner in which we produce speech sounds. Figure 1 shows a sketch of the side view of a human vocal tract. The various speech sounds are produced by adjusting both the shape of the vocal tract and the manner in which it is acoustically excited. The shape is controlled by moving the jaw, tongue and lips to positions appropriate to the desired speech sound. The acoustic excitation can be produced in a variety of ways. For the vowel sounds such as ah, ee, etc., the excitation is primarily a sequence of quasi-periodic pulses at one end of the tract. These pulses are produced by the vocal cords by successive interruptions of the steady air flow provided by the lungs. For fricative sounds (e.g., s, sh, th, f, etc.) the excitation is noise-like and is produced by turbulent air flow at some point along the tract. For plosives (e.g., p, t, k) the excitation is provided by the sudden release of pressure built up behind a closure in the tract. For nasal sounds (e.g., m, n, ng) the excitation is similar to the one used for producing vowels; however, the acoustic properties of the tract are modified by coupling the nasal cavity through an opening at the velum. During fluent speech the shape of the vocal tract changes continuously and the various sources of excitation are used in a coordinated manner.

Because of its fundamental role in speech production, the time-varying shape of the vocal tract during normal speech has been of considerable interest in speech research. Early attempts at estimating this shape consisted of taking x-ray movies of the side view of the vocal tract of a subject reading some test material. I will not discuss these methods here except to mention that they have several drawbacks: they require considerable exposure to x-rays; they require considerable amount of effort in identifying the outlines of the tongue, palate, etc., on each frame of the movie; and they provide no information on the shape in the planes parallel to the x-ray beam.

Another approach is to estimate the shape of the vocal tract from measurements of its acoustical properties. This approach (in particular when the measurements are in the time domain) leads to the inverse problem that I will discuss in some detail.

For the purposes of studying acoustic wave propagation, the tract is assumed to be straightened out as shown in Fig. 1b, with the lips at $x=0$ and the glottis at $x=L$. Also, in spite of the fact that the shape changes continuously during speech, the tract may be assumed to be

* AT&T Bell Laboratories, Murray Hill, New Jersey 07974.

stationary. This is because the motion is very slow on the time scale of acoustic phenomena of interest. (In fact, for the time domain measurements to be described later, it is sufficient to assume that the motion of the tract is negligible in any 1 msec. time interval. This assumption is quite accurate.) Thus in the simplest approximation the vocal tract is assumed to be a stationary variable area tube with hard walls, filled with a perfect gas. (Later I will introduce losses too.) Further, the wave propagation in the tube is assumed to be planar. In view of this assumption, which is accurate up to about 4 kHz, the actual shape of the cross-section at any point along the tube is irrelevant. All that matters is the cross-sectional area at that point. Let the area function $A(x)$ be the cross-sectional area at a distance x from the lips. Then the pressure $P(x,t)$ and the volume velocity $U(x,t)$ satisfy the differential equations

$$\frac{\partial}{\partial x} P(x,t) = -\frac{1}{A(x)} \frac{\partial}{\partial t} U(x,t) \tag{1a}$$

and

$$\frac{\partial}{\partial x} U(x,t) = -A(x) \frac{\partial}{\partial t} P(x,t). \tag{1b}$$

In these equations the units of mass, length and time have been chosen such that the velocity of sound, the density of air and $A(0)$ are all equal to unity.

Broadly speaking, one may define the inverse problem for the vocal tract as the estimation of $A(x)$ from acoustical measurements at the lips. Now the easiest measurement to make is, of course, a measurement of the speech wave itself, i.e., the pressure picked up by a microphone in front of the lips. Hence, paraphrasing the title of Kac's famous article, one might ask: "Can we hear the shape of the vocal tract?" The answer to this question is "no" for several reasons. First of all even in the simplest case of vowels, the speech signal is the convolution of the excitation at the glottis with the transfer function of the tract from glottis to lips. Since only the transfer function carries information about the vocal tract, the effect of the excitation must first be removed by deconvolution. Without further assumptions this is impossible because the excitation is unknown. What is worse is that even if the transfer function is known (based on some assumed or estimated excitation) it does not give $A(x)$ uniquely.

It is therefore evident that some other external measurements must be made if the unique area function is to be reconstructed. The first attempts at such measurements were made by Schroeder and Mermelstein in the 1960's [1-3]. To describe their proposal it is necessary to write equations (1a) and (1b) in the frequency domain by taking their Fourier transforms. Thus

$$\frac{\partial}{\partial x} P(x,\omega) = -\frac{j\omega}{A(x)} U(x,\omega) \tag{2a}$$

and

$$\frac{\partial}{\partial x} U(x,\omega) = -j\omega A(x) P(x,\omega), \tag{2b}$$

or, upon eliminating U between equations (2a) and (2b)

$$\frac{\partial}{\partial x} A(x) \frac{\partial}{\partial x} P(x,\omega) + \omega^2 A(x) P(x) = 0. \tag{2c}$$

Equation (2c) is recognized to be a Sturm-Liouville equation.

Let $\hat{P}(x,\omega)$, $\hat{U}(x,\omega)$ be a solution to eqns (2a,2b) satisfying a homogeneous boundary condition at $x=L$ corresponding to the properties of the termination at the glottis. Then the input impedance at the lips is defined as $H(\omega) = \hat{P}(0,\omega)/\hat{U}(0,\omega)$. It can be shown that the *poles* of the input impedance give the infinite sequence of eigenvalues of (2c) corresponding to the above boundary condition at the glottis and the condition $U=0$ at the lips. The *zeros* on the other hand form an interlaced sequence of eigenvalues corresponding to the same condition at the glottis and the condition $P=0$ at the lips. The unique relationship between such sets of eigenvalues and the function $A(x)$ is well-known to this audience. It was first proved by Borg [4], and ramifications of this connection have since been studied by many investigators, notably by Gelfand and Levitan [5]. [Incidentally, the transfer function of the tract from glottis to lips provides only one set of eigenvalues -- the zeros. That is why the area function is not determined uniquely by the transfer function.]

The articles by Schroeder and Mermelstein mentioned above presented a practical method of estimating $A(x)$ from estimates of the poles and zeros of the input impedance. One obvious problem that they faced is the fact that only a finite number of poles and zeros can be measured. On the assumption that the area function of the tract differs only slightly from that of a uniform tube, they proved a one-to-one relation between the $k-th$ pole (zero) and the $k-th$ even (odd) coefficient of the Fourier cosine series for $\ln A(x)$ on the interval $(0,L)$. Thus a "bandlimited" version of $A(x)$ is reconstructed. The larger the number of known poles and zeros, the finer the detail in the reconstruction. For large deviations from uniform (which is the rule for the present problem) the simple relationship breaks down. However, Mermelstein [3] showed that starting with a uniform tube, the Fourier coefficients can be iteratively modified so as to match the eigenvalues of the resulting tube with the measured poles and zeros. Of course, the convergence of the iterative scheme and the uniqueness of the limiting configuration are well nigh impossible to prove for such schemes.

In [6] Gopinath and I put these ideas on a more solid footing. We used a partial fraction expansion of the input impedance i.e., wrote it in terms of poles and *residues* instead of poles and zeros. This, of course, is equivalent to specifying the spectral function for equation (2c), so we were able to adapt a method originally suggested by Krein [7] to determine $A(x)$. We showed that the unknown higher frequency poles and residues may be set equal to their asymptotic values (which are known if the length L is known) without significantly affecting the reconstruction. We were also able to extend the analysis to cover the case of area functions with certain types of jump discontinuities provided the locations and magnitudes of the jumps are known.

The frequency domain methods summarized above overcome two major inherent difficulties in methods based on the speech wave. They eliminate all uncertainty regarding the excitation source and they eliminate the nonuniqueness problem alluded to earlier. On synthetic data or on known cavities they yield remarkably good results. However for the vocal tract problem they have serious shortcomings. The main drawback of these methods is that one must know the length of the vocal tract and the boundary condition at the glottis. Neither of these is known with any degree of reliability and each strongly affects the recovered area function. Another difficulty is that frequency domain measurements must be made over fairly long time intervals (10-20 msec) for reliable estimates of the eigenfrequencies. Over such long time intervals the stationarity assumption implied in equations (1a, 1b) and (2a, 2b, 2c) is not accurate.

All these problems are eliminated if one uses inversion methods based on transient, time-domain measurements. As shown in several publications, ([8,9] by myself and Gopinath, [10,11] by Resnick) such methods do not require knowledge of the tract length L or the boundary condition at the glottis. Also, a single measurement for about one msec suffices to reconstruct a vocal tract of average length (about 17 cm). Sequences of such measurements in rapid succession can therefore be made to track the changes in shape of a slowly varying vocal tract. In practice the transient measurements are made with the vocal tract coupled by means of a flexible coupler to one end of an impedance tube which is excited at the other end by acoustic pulses (see later sections for details).

During the last ten or twelve years some experiments have been reported which utilize these methods. Most of these ([8,10,11] mentioned above, [12] by me, [13] by Descout et al, and

[14] by Tousignant et al) provide little more than a qualitative validation of the theory. The experiments by Lefevre et al [15] are more ambitious and catalog the measured shapes for all French vowels. Recently [16] in collaboration with J.R. Resnick I have done some experimental work on this problem which has led to the following new developments: (1) We can now make a measurement, calculate $A(x)$ and display it in about 50 msec. This speed allows us to display the dynamic variations of the tract in real time. (2) We have considerably improved the accuracy of the measurements over earlier attempts; as partial evidence of this we have been able to synthesize quite intelligible speech from the measured areas. In the rest of the talk I will outline the relevant theory and then describe the experiment and the numerical algorithms that we had to implement to realize the above-mentioned advances. Anyone interested in greater detail may consult reference [16]. Those of you interested in a more formal mathematical description of transient methods may consult a paper by Burridge [17].

The Lossless Case. The transient responses that are of interest here are as follows:

[a] Input impedance. From the fact that equations (1a, 1b) are linear and time-invariant it follows that the pressure at $x=0$ is related to the volume velocity at the same point by a relation of the form

$$P(0,t) = \int_0^t H(t-\tau)U(0,\tau)d\tau. \tag{3}$$

The function $H(t)$ which is $P(0,t)$ when $U(0,t) = \delta(t)$, is called the input impedance of the unknown cavity*.

It can be shown [8] that H must have the form $H(t) = \delta(t)+h(t)$, so that eq (3) has the form

$$P(0,t) = U(0,t) + \int_0^t h(t-\tau)U(0,\tau)d\tau. \tag{4}$$

[b] Reflectance. A convolutional relationship exists also between the pressures in the right- and left- going waves at $x=0$. Thus if $P_R(0,t)$ and $P_L(0,t)$ are, respectively the pressures in these waves at $x=0$, then

$$P_L(0,t) = \int_0^t R(t-\tau)P_R(0,\tau)d\tau. \tag{5}$$

The function $R(t)$, which is $P_L(0,t)$ when $P_R(0,t) = \delta(t)$, is called the reflectance.

[c] Step reflectance. The function $S(t)$ defined by

$$S(t) = \int_0^t R(\tau)d\tau \tag{6}$$

is called the step reflectance for obvious reasons. It is just $P_L(0,t)$ when $P_R(0,t)$ is a unit step.

It is worth noting that a nonlinear convolutional relationship exists between the functions $h(t)$ and $R(t)$ defined above. This relationship can be formally derived as follows: Suppose that the right going pressure wave is $P_R(x,t) = \delta(t-x)$, in the region $x \leqslant 0$. Then if $R(t)$ is the reflectance of the unknown cavity, it is clear that $P_L(x,t) = R(t+x)$. Remembering that $A(x)=1$, when $x \leqslant 0$, it follows from equations (1,2) that

$$P(x,t) = \delta(t-x) + R(t+x)$$

$$U(x,t) = \delta(t-x) - R(t+x).$$

Setting $x=0$ in these equations and substituting into eq (4) gives the relation

* Everything to the right of the origin in Fig. 2 will be referred to as the unknown cavity, even though the origin is chosen one or two cm before the end of the impedance tube.

$$2R(t) = h(t) - \int_0^t h(\tau)R(t-\tau)d\tau. \tag{7}$$

The area function $A(x)$ can be derived from each of the functions $h(t)$, $R(t)$ and $S(t)$. The derivation from $R(t)$ is discussed in reference [10]. However we will not use it in this paper, because it turns out that $S(t)$ is a much better function to use than $R(t)$.

The relationship between $A(x)$ and $h(t)$ is derived in reference [8], by means of a simple physically motivated argument. It is shown there that if $f(x,t)$ is the solution of the integral equation

$$f(x,t) + \frac{1}{2}\int_{-x}^{x} h(|t-\tau|)f(x,\tau)d\tau = 1 \quad |t| \leqslant x \tag{8}$$

then the volume function (i.e., the integral of the area function) is given by

$$V(x) = \int_0^x A(y)dy = \int_0^x f(x,t)dt. \tag{9}$$

Alternatively,

$$A(x) = f^2(x,x). \tag{10}$$

The derivation of $A(x)$ in terms of $S(t)$ is given in reference [11] and may be summarized as follows: With the tube initially quiescent, let $U(x,t)$ and $P(x,t)$ be the solutions of the nonlinear initial value problem

$$\frac{\partial}{\partial x}U(x,t) + \frac{\partial}{\partial t}U(x,t) + \frac{U(x,x)}{P(x,x)}\left(\frac{\partial}{\partial x}P(x,t) + \frac{\partial}{\partial t}P(x,t)\right) = 0 \tag{11a}$$

$$\frac{\partial}{\partial x}U(x,t) - \frac{\partial}{\partial t}U(x,t) - \frac{U(x,x)}{P(x,x)}\left(\frac{\partial}{\partial x}P(x,t) - \frac{\partial}{\partial t}P(x,t)\right) = 0. \tag{11b}$$

in the region

$$x < t < 2L-x \; ; 0 < x < L, \tag{12}$$

with initial data

$$P(0,t) = 1 + S(t) \tag{13a}$$

$$U(0,t) = 1 - S(t), \tag{13b}$$

and, of course, the causality condition, $U(x,t) = P(x,t) = 0$ for $t < x$. The area function is then obtained from

$$A(x) = \frac{U(x,x)}{P(x,x)}. \tag{14}$$

It is interesting to note that eqs (11a,11b) differ from usual partial differential equations in that they contain the values of the unknowns U and P at both (x,t) and (x,x) !

The Lossy Case. If the walls of the vocal tract are not rigid, and/or the air is assumed to be viscous, then additional terms have to be added to the right hand sides of equations (1a,1b). Thus

$$\frac{\partial}{\partial x}P(x,t) = -\frac{1}{A(x)}\frac{\partial}{\partial t}U(x,t) - \tilde{p}(x,t) \tag{15a}$$

and

$$\frac{\partial}{\partial x}U(x,t) = -A(x)\frac{\partial}{\partial t}P(x,t) - \tilde{u}(x,t) \tag{15b}$$

where $\tilde{p}(x,t)$ is the pressure drop per unit length required to overcome the viscous drag, and $\tilde{u}(x,t)$ is the volume velocity shunted by the wall motion per unit length. It is possible to model [18,19] $\tilde{p}$ and $\tilde{u}$ by linear transformations of U and P respectively. Thus

$$\tilde{p}(x,t) = \int_0^t U(t-\tau)z(x,\tau)d\tau \tag{16a}$$

and

$$\tilde{u}(x,t) = \int_0^t P(x,t-\tau)y(x,\tau)d\tau. \tag{16b}$$

The input impedance, reflectance and step reflectance are defined exactly as in the lossless case; only P and U are now solutions of equations (15a,15b) instead of equations (2a,2b). As for the inverse problem, it can be shown [8] that the input impedance (hence, also, the reflectance) for the lossy tube is indistinguishable from that of *some* lossless tube. It is therefore possible to estimate $A(x)$ in the lossy case only if $z(x,t)$ and $y(x,t)$ are known. When estimates of these functions are available (see e.g., [19] pp 34-35, and [18]) the theory for the lossless case can be modified to compensate for these perturbations.

In reference [9] the special case is considered when the functions have the separable form

$$z(x,t) = \frac{1}{A(x)}\alpha(t) \tag{17a}$$

and

$$y(x,t) = A(x)\beta(t). \tag{17b}$$

(Estimates of this form have been derived in [18]). The result derived in [2] may be summarized as follows: Let $H(t)$ and $H_0(t)$ be the input impedance of the lossy tube and that of the *same* tube with the losses removed. Let $\tilde{H}(s)$, $\tilde{H}_0(s)$, $\tilde{\alpha}(s)$, $\tilde{\beta}(s)$ be the Laplace transforms of $H(t)$, $H_0(t)$, $\alpha(t)$, $\beta(t)$ respectively. Then

$$\tilde{H}(s) = \tilde{H}_0(\sqrt{(s+\tilde{\alpha})(s+\tilde{\beta})})\sqrt{\frac{s+\tilde{\alpha}}{s+\tilde{\beta}}} \tag{18}$$

In the appendix of reference [16] we showed in some detail how equation (18) can be used to give $H_0(t)$ in terms of the measured response $H(t)$. With $H_0(t)$ known, $A(x)$ is computed from it as in the lossless case. The appendix of [16] also provides some details of an application of equation (18) to the case of viscous loss and wall vibration.

The derivations of $A(x)$ from $R(t)$ and $S(t)$ are capable of handling much more general types of loss distributions. The recovery of the area function of a lossy tube from $R(t)$ is derived in [10]. An extension and modification of the ideas presented there gives the derivation in terms of $S(t)$ [20], which we now summarize. Suppose that equations (16a,16b) have the following form

$$\tilde{p}(x,t) = \gamma(x,A)U(x,t) \tag{19a}$$

and

$$m(x,A)\frac{\partial \tilde{u}}{\partial t} + b(x,A)\tilde{u} + K(x,A)\int_0^t \tilde{u}(x,\tau)d\tau = P(x,t) \tag{19b}$$

Then $A(x)$ can be computed in a manner quite analogous to the lossless case even for such general loss distributions. Thus with the tube initially quiescent, let $U(x,t)$ and $P(x,t)$ be solutions of the set of equations

$$\frac{\partial U}{\partial x} + \frac{U(x,x)}{P(x,x)}\frac{\partial P}{\partial t} + \tilde{u} = 0 \tag{20a}$$

$$\frac{\partial P}{\partial x} + \frac{P(x,x)}{U(x,x)} \frac{\partial U}{\partial t} + \tilde{p} = 0 \tag{20b}$$

$$\tilde{p} = \gamma^*(x) U \tag{20c}$$

$$m^*(x) \frac{\partial \tilde{u}}{\partial t} + b^*(x)\tilde{u} + K^*(x) \int_0^t \tilde{u}(x,\tau) d\tau = P \tag{20d}$$

$$\tilde{u}(x,x) = 0, \quad \frac{\partial}{\partial t}\tilde{u}(x,x) = 0 \tag{20e}$$

with initial data given in equations (13a,13b), and the causality condition $U(x,t)=P(x,t)=\tilde{u}(x,t)=0$ for $t<x$. Then the area function is given by eq (14). In equations (20c) and (20d)

$$m^*(x) = m(x, \frac{U(x,x)}{P(x,x)}), \quad etc. \tag{20f}$$

We have developed inversion algorithms based on both the impedance and the step reflectance, and have incorporated various types of losses too. We have also written an algorithm for the direct problem (which arises when we want to compute a transient response of a known cavity, and, more importantly, when we want to synthesize speech from the vocal tract area functions). These algorithms are all described in detail in Ref [16]. I will outline just one or two of them here. Before proceeding to the algorithms, however, let me briefly describe the experiment and the measurements.

The Experiment. A schematic diagram of the experimental set-up is shown in Fig. 2. The measurements are all made on-line under the control of an Eclipse 250 computer. Besides the computer itself the essential components of the experimental set-up are [a] the impedance tube, [b] the transducer, [c] the microphone , anti-aliasing filter and Analog/Digital converter and [d] the acoustic coupler.

[a] The measurements are made by means of an impedance tube. The exact dimensions of this tube are not critical to the success of the experiments. The one we are using is made of brass, is 40 inches long and has an inner diameter of 1 inch. This inner diameter gives a cross-sectional area of about 5 cm^2, which is an average value of the area of the lip opening.

[b] After experimenting with several types of transducers we found an electrostatic transducer to be the most satisfactory for our purposes. Such a transducer is constructed essentially like a foil electret microphone. The only difference is that an *uncharged* foil is used. (A 1 mil mylar foil was found satisfactory.) If a step voltage (of any polarity) is applied between the back plate and the metallized outer surface of the foil, the foil is attracted toward the back plate. This produces a pressure pulse of rarefaction which travels down the impedance tube. The shape of this pulse as measured at the output of the filter in Fig. 3, is shown in Fig. 2.

In order to make repeated measurements the train of rectangular voltage pulses sketched in Fig. 4 is applied to the transducer. Since the trailing edge of each voltage pulse also produces an acoustic pulse (of condensation), it is important that the duration of the voltage pulse be larger than the time interval for which each measurement is to be made. (As mentioned in the Introduction, this time interval is about 1.5 msec.)

[c] The microphone used is a standard Bruel and Kjaer 1/4 inch condenser microphone with a pre-amplifier. The output of the pre-amplifier is passed through a bandpass Rockland filter with an upper cut-off frequency of 8 kHz and a lower cut-off frequency of 3 Hz. This serves as an anti-aliasing filter. The lower cut-off frequency was found to be necessary in order to eliminate drifts due to breathing and other low frequency interference.

The sampling rate of the Analog/Digital converter is set at 50,000/sec, which is considerably higher than the Nyquist rate for the 8 kHz pulse bandwidth. This high sampling rate allows reconstruction of the area function every .35 cm without interpolating the discretized signals.

The microphone is located about 2 ft. from the right-hand end of the impedance tube. This allows the higher modes generated upon reflection from the vocal tract (or other unknown cavity) to be attenuated before reaching the microphone.

[d] We experimented with several different types of couplers. Our latest and most successful model is molded to fit snugly over the lips. It is made of Dow Corning Silastic type J RTV.

The Measurements. As shown above, the area function can be computed from either the input impedance or the reflectance at the lips. Ideally the input impedance should be measured by producing a δ-function of volume velocity at the lips, and the reflectance by producing an incident pressure wave which is a δ-function. In practice each of these is derived by appropriate deconvolution from measurements of the incident pressure wave and of the corresponding reflected wave.

The incident wave is just the rarefaction pressure wave produced by the transducer. It is highly repeatable and need be measured only once every few days. After this calibration measurement of the incident wave has been made, a single measurement of the reflected wave suffices to compute the area function of the vocal tract (or any other cavity) coupled to the end of the impedance tube.

Figure 5 shows a sketch of the first few milliseconds of the microphone signal from the start of the step function. The portion of the waveform actually used for reconstructing the area function is the portion between points A and B shown as a solid heavy line. The arrows indicate the direction of travel of the waves which produce various portions of the microphone signal.

Each measurement of the reflected wave can be converted to an area function. For a time-varying vocal tract this represents a snapshot of the area function nominally at the time at which the incident pulse arrives at the lips. The maximum rate at which these snapshots may be taken depends on the mode of the experiment.

In the display mode of the experiment the area is computed and displayed before the next pulse is applied to the transducer. As mentioned in the Introduction the time required for the computations is about 50-60 msec, so that about 18 frames per second can be displayed.

In the recording mode the measurements are merely stored for later processing. In this mode the maximum rate is governed almost exclusively by the time required for the acoustic energy introduced by the incident pulse to decay to an insignificant level. With the experimental set-up as shown in Fig. 2, this time is about 50 msec. By placing a wedge (of, say, urethane foam) in the tube this decay time can be reduced to about 20 msec. Unfortunately this also has the effect of reducing the bandwidth of the acoustic pulse arriving at the vocal tract. This in turn adversely affects the accuracy and resolution of the reconstructed area functions. For the present experiment we decided to use the slow pulse rate in order to maintain the maximum possible accuracy.

The Algorithms. In this section I will outline a few selected numerical procedures that we have developed for this experiment. For details and other algorithms see ref [16].

Deconvolution. A critical step in the estimation of area functions is the estimation of the transient response itself, from the measurements of the incident and the reflected waveforms. The problem we face is this: We want to infer some unknown transient response $z(t)$, given an input signal $x(t)$ and an output signal $y(t)$, where

$$y(t) = \int_0^t x(t-\tau) z(\tau) d\tau, \quad 0 \leqslant t \leqslant T. \tag{21}$$

The difficulty in finding $z(t)$ arises from the fact that this problem is "ill-posed"; that is, small errors in measuring $y(t)$ and $x(t)$ can produce massive errors in the values of $z(t)$. To see this, suppose we sample our signals at time $t=0, 1, 2, \ldots$ and approximate the integral as a sum. We can then write

$$\mathbf{y} = \mathbf{X}\mathbf{z} \tag{22}$$

where $\mathbf{y}$ and $\mathbf{z}$ are vectors of sample values and $\mathbf{X}$ is a triangular matrix with components

$$\begin{aligned} X_{ij} &= x(i-j) \quad i \geqslant j \\ &= 0 \qquad\quad i < j. \end{aligned} \tag{23}$$

Assuming $x(0) \neq 0$, the matrix $\mathbf{X}$ is nonsingular, hence the exact solution of this problem is simply

$$\mathbf{z} = \mathbf{X}^{-1}\mathbf{y}. \tag{24}$$

However, $\mathbf{X}^{-1}$ is easily seen to be extremely sensitive to $x(0)$, so that in practice the above solution tends to oscillate wildly and/or blow up.

To avoid this problem we have adopted a regularization procedure based on the singular value decomposition [21]. Let $\mathbf{X}$ have the singular value decomposition

$$\mathbf{X} = \mathbf{U}\boldsymbol{\Lambda}\mathbf{V}' \tag{25}$$

where $\mathbf{U}$ and $\mathbf{V}$ are orthonormal matrices with columns $\mathbf{u}_j$ and $\mathbf{v}_j$ respectively, $\boldsymbol{\Lambda}$ is a diagonal matrix with entries λ_j and $'$ denotes matrix transpose. Then it can be easily seen that $\mathbf{z}$ can be written as

$$\mathbf{z} = \sum_{j=1}^{N} \frac{\mathbf{y}'\mathbf{u}_j}{\lambda_j}\mathbf{v}_j. \tag{26}$$

where N is the order of the matrix. Then the regularized estimate of z is just

$$\hat{\mathbf{z}} = \sum_{j \epsilon J} \frac{\mathbf{y}'\mathbf{u}_j}{\lambda_j}\mathbf{v}_j, \tag{27}$$

where J is the set of indices for which λ_j is greater than some value which is preset on the basis of the estimated accuracy of the measurements. We also tried Tikhonov's regularization procedure [22] which estimates $\mathbf{z}$ by

$$\tilde{\mathbf{z}} = (\mathbf{X}'\mathbf{X} + \alpha I)^{-1}\mathbf{X}'\mathbf{y} \tag{28}$$

where α is a preset constant. From the singular value decomposition (25) of $\mathbf{X}$ it is straightforward to show that

$$\tilde{\mathbf{z}} = \sum_{j} \frac{\lambda_j \, \mathbf{y}'\mathbf{u}_j}{\lambda_j^2 + \alpha}\mathbf{v}_j. \tag{29}$$

Equation (29) shows that Tikhonov's method also attenuates the expansion coefficients of the eigenvectors corresponding to small eigenvalues; it fades them out gradually rather than stopping abruptly at some value.

In practice we found that for the signals we are interested in, $\hat{\mathbf{z}}$ is indistinguishable from $\tilde{\mathbf{z}}$ provided the threshold value for eq (27) is selected appropriately. Since eqs (28) or (29) require more computation than eq (27), we selected the latter.

In order to estimate reflectance by this method the signal $x(t)$ of eq (21) is identified with the incident signal and the signal $y(t)$ with the reflected signal. Note that the singular value decomposition need be performed only once; the reflectance corresponding to any reflected signal is then computed by taking the dot products of the reflected signal with the precomputed eigenvectors as required in eq (27).

In principle, the other transient responses can also be estimated this way by appropriate identification of the signals x and y. In practice the procedure is inconvenient for estimating the impedance directly. This is because to estimate impedance, x in eq (21) must be identified with the velocity, which is the difference between the incident and reflected waves. Thus the singular value decomposition of eq (27) would have to be performed for each measurement of reflected signal. This is clearly inefficient. Instead, therefore, we estimate the reflectance as described above, and then compute the impedance from the nonlinear convolution of eq (7).

The Direct Problem and Speech Synthesis. We need to solve the direct problem for two purposes. One is to compute the transient responses of cavities with known shapes. The other is to synthesize speech from area functions estimated from our measurements. The numerical procedures required for these two problems are quite similar. Details of an algorithm for the direct problem may be found in Ref [16]. However, since it has little bearing on the inverse problem, I will not discuss it here.

The Inverse Problem. Of the various algorithms which we have developed, I will confine myself here to the one based on eq (8). This is the algorithm we use for the real time display; of course, the same algorithm suffices for the case where the losses have the special assumed distribution given by eqs (17a-17b). (One need only correct the impulse according to eq (18) before using the inversion algorithm.)

We need an algorithm for solving eq (8) for $f(x,t)$. As a first step the discretized impulse response is obtained from the estimated step response by discretizing eqs (6) and (7). Simple first differences are used. Let T be the time step, let $S_n = S(nT)$, and let r_n and h_n be the areas under the functions $h(t)$ and $r(t)$ respectively for the n-th time step. Then, $S_0 = h_0 = 0$ and

$$r_n = S_n - S_{n-1} \tag{30a}$$

$$\tfrac{1}{2}h_1 = r_1 \tag{30b}$$

$$\tfrac{1}{2}h_n = r_n + \sum_{1}^{n-1} \tfrac{1}{2}h_k r_{n-k}. \tag{30c}$$

Next we must discretize eq (8). For the algorithm given here it is convenient to define $\hat{f}(x,t) = f(x,t-x)$ with t now ranging from 0 to $2x$. Thus eq (8) may be written as

$$\hat{f}(x,t) + \frac{1}{2}\int_0^{2x} h(|t-\tau|)\hat{f}(x,\tau)d\tau = 1. \tag{31}$$

Now let $x = \dfrac{NT}{2}$, and divide the integral into N sub-intervals of length T. Then the discretized version of eq (8)' is

$$\mathbf{H}\hat{\mathbf{f}} = \mathbf{1}. \tag{32}$$

Here $\mathbf{H}$ is an $N \times N$ Toeplitz matrix, $\mathbf{1}$ is an N-vector with every component equal to 1 and $\hat{\mathbf{f}}$ is an N-vector with components $f_i = \hat{f}(\frac{NT}{2}, iT), i = 1,...N$. The elements of $\mathbf{H}$ are

$$H_{ii} = 1 \qquad i=1, 2, \cdots N \tag{33a}$$

$$H_{ij} = \tfrac{1}{2}h_{|i-j|} \qquad i,j=1, 2, \cdots N;\ i \neq j, \tag{33b}$$

with h_n defined by eqs (30a-30c). Now eq (9) and the fact that f is an even function of t gives the discrete approximation

$$V(\frac{NT}{2}) = \frac{1}{2}\int_0^{NT} \hat{f}(\frac{NT}{2},t)dt = \frac{T}{2}\mathbf{1}'\hat{\mathbf{f}}. \tag{34}$$

Thus the volume function is obtained. Solving the matrix equation (32) for $N=1, 2, \cdots M$ gives $V(\frac{NT}{2})$ for the same range of values of N. Define $V_N = \frac{2}{T}V(\frac{NT}{2})$. Then the area function is obtained as the first difference of the sequence V_N.

Ordinarily the sequence of matrix inversions would be quite time consuming. However, since $\mathbf{H}$ is Toeplitz, and the right hand side of eq (32) is the vector $\mathbf{1}$, the sequence of inversions can be performed very efficiently. To see how we can capitalize on these properties let us index $\mathbf{H}$ with N. Then from equations (32) and (34) we get

$$V_N = \mathbf{1}'\mathbf{H}_N^{-1}\mathbf{1}. \tag{35}$$

But this means that V_N is the sum of all the elements of the matrix $\mathbf{H}_N^{-1}$. Since H is a toeplitz matrix, eq (35) can be shown (see below) to yield

$$A(\frac{NT}{2}) = [V_{N+1} - V_N] = \frac{1}{d_0}[\sum_0^N d_i]^2 \tag{36}$$

where $d_0, \ldots, d_N$ are the elements of the first row of $\mathbf{H}_{N+1}^{-1}$.

Now there is a well known technique due to Trench [23] (based on earlier work of Levinson) which gives a recursion for the first row of $\mathbf{H}_{N+1}^{-1}$ in terms of the first rows of $\mathbf{H}_{N+1}$ and $\mathbf{H}_N^{-1}$. This algorithm provides a very efficient way to compute the areas. When programmed on the array processor of the S250 computer it has the speed of computation necessary for the display mode of the experiment described in Sections 1 and 3.

Eq (36) can be derived as follows by using a fairly straightforward bordering technique. Let us partition the matrix $\mathbf{H}_{N+1}$ into the matrix $\mathbf{H}_N$ and its border. Let us partition $\mathbf{H}_{N+1}^{-1}$ into similar blocks as shown in eq (37).

$$\begin{bmatrix} 1 & \mathbf{h}' \\ \mathbf{h} & \mathbf{H}_N \end{bmatrix} \begin{bmatrix} d_0 & \mathbf{d}' \\ \mathbf{d} & \mathbf{D} \end{bmatrix} = \mathbf{I}. \tag{37}$$

Here $\mathbf{h}$ is the vector with components $h_1, h_2, \ldots h_N$, $\mathbf{d}$ is the vector with components $d_1, d_2, \cdots d_N$ and $\mathbf{D}$ is an $N \times N$ matrix. From eq (37) it follows that

$$d_0 + \mathbf{h}'\mathbf{d} = 1 \tag{38a}$$

$$\mathbf{d}' + \mathbf{h}'\mathbf{D} = 0 \tag{38b}$$

$$d_0\mathbf{h} + \mathbf{H}_N\,\mathbf{d} = 0 \tag{38c}$$

$$\mathbf{h}\mathbf{d}' + \mathbf{H}_N\,\mathbf{D} = \mathbf{I} \tag{38d}$$

Let $\mathbf{H}_N^{-1} = \mathbf{C}$. Then from eq (38c) we get $\mathbf{d} = -d_0\mathbf{C}\mathbf{h}$. Substituting this in eq (38a) gives

$$d_0 = \frac{1}{1 - \mathbf{h}'\mathbf{C}\mathbf{h}}. \tag{39}$$

From eq (38d)

$$\mathbf{D} = \mathbf{C} + \frac{1}{d_0}\mathbf{d}\mathbf{d}'. \tag{40}$$

Premultiplying this equation by $\mathbf{1}'$ and postmultiplying by $\mathbf{1}$ gives

$$\mathbf{1}'\mathbf{D}\mathbf{1} = \mathbf{1}'\mathbf{C}\mathbf{1} + \frac{1}{d_0}(\mathbf{1}'\mathbf{d})^2. \tag{41}$$

The first term on the right is just V_N from eq (35). Moreover the left hand side is simply V_{N+1} minus the sum of the terms on the border of $\mathbf{H}_N$ in eq (37). Thus, rearranging terms

$$[V_{N+1} - V_N] = d_0 + 2\,\mathbf{1}'\mathbf{d} + \frac{1}{d_0}(\mathbf{1}'\mathbf{d})^2. \tag{42}$$

This last equation can be written as

$$A\left(\frac{NT}{2}\right) = \frac{1}{d_0}(d_0 + \mathbf{1}'\mathbf{d})^2. \tag{43}$$

Thus the area function is expressed in terms of the elements of the first row of $\mathbf{H}_{N+!}^{-1}$ as asserted.

It is interesting to note that the expression within parentheses in eq (43) is just the sum of the terms of the first row of $\mathbf{H}_{N+1}^{-1}$. Hence from the equation corresponding to eq (32) for the index $N+1$ this can be interpreted as $\hat{f}\left(\frac{(N+1)T}{2},(N+1)T\right)$. From eq (10), this last quantity is just $\sqrt{A\left(\frac{(N+1)T}{2}\right)}$. Thus it is also possible to write eq (43) as

$$A\left(\frac{(N+1)T}{2}\right) = d_0 A\left(\frac{NT}{2}\right). \tag{43'}$$

The values obtained from eqs (43) and (43)' become closer to each other as T becomes smaller. However, the values obtained from eq (43) are always somewhat closer to the correct values when tested on known shapes. Therefore, we have chosen to use eq (43) even though it requires slightly more computation.

In our actual application N was about 65. Also, the impulse response was corrected for viscous loss. For this purpose $\tilde{\beta}$ and $\tilde{\alpha}$ in eq (18) were chosen appropriately for viscous loss and the correction computed according to the formulas of Appendix 1. The details are given in Ref [5]. (Corrections for wall losses were found to be significant only for the speech synthesis problem, not for the inverse problem.)

Experimental Results. It is not possible in a lecture or a written description to capture the essence of the display mode of our experiment. The value of the instantaneous feedback and the ability to visualize the shape of one's vocal tract can be appreciated only by participating in the experiment. The dynamic variations in the shape of the vocal tract too cannot be presented. I can only show you a few results from fixed cavities and from selected frames of movies of vocal tract motion which we have made with our technique.

Figs (6a,6b) show the reconstruction of the shape of a rubber mold with a cavity typical of the "oh" shape. Here the reconstruction is from the step response measured by the experimental procedure described above. Fig. 6a shows the reconstruction by the (lossless) algorithm based on eqs (11) and (13). Fig. 6b is the reconstruction from the algorithm based on eqs (20a-20d). The loss values were not known for the rubber walls, so they were chosen equal to typical values for a vocal tract.

In Fig. 7 we show a selection of frames from a movie of the shape of a vocal tract as it varies for the utterance "ah - ee - ah". Two of the frames show the "ah" and "ee" shapes during steady portions and the third is an intermediate transitory position. While the shape for "ah" is reasonable, the shape for "ee" does not appear to be so. Beyond the constriction, towards the glottis, the area function stays small rather than increasing towards the glottis as expected. Although there can be considerable variation from person to person, this large a variation appears to be an artifact of our experiment.

There are several possible explanations for this apparent artifact. These include (i) bandlimiting of the reflectance, (ii) nonplanar wave motion, and (iii) viscosity. We investigated each of these effects by simulation. The results of these simulations are presented in [5]. The main conclusion is that with the present experimental setup, bandlimitation of the reflectance is the major culprit. Thus for improving the present experiment, one of the first requirements is a transducer which puts out a better sound pulse. The pulse must have a much faster rise time, and should approximate a step function over a time interval of about a millisecond.

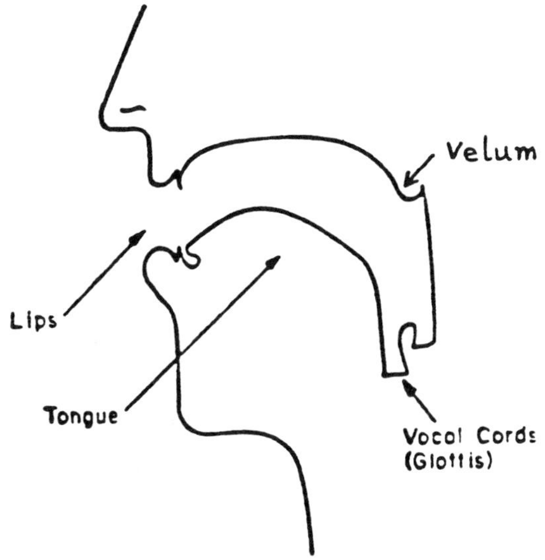

Fig. 1a Sketch of the side view of a vocal tract.

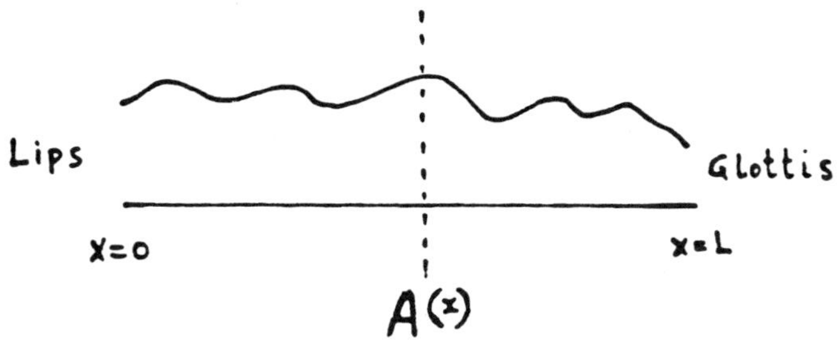

Fig. 1b The vocal tract straightened out with one boundary along the x-axis.

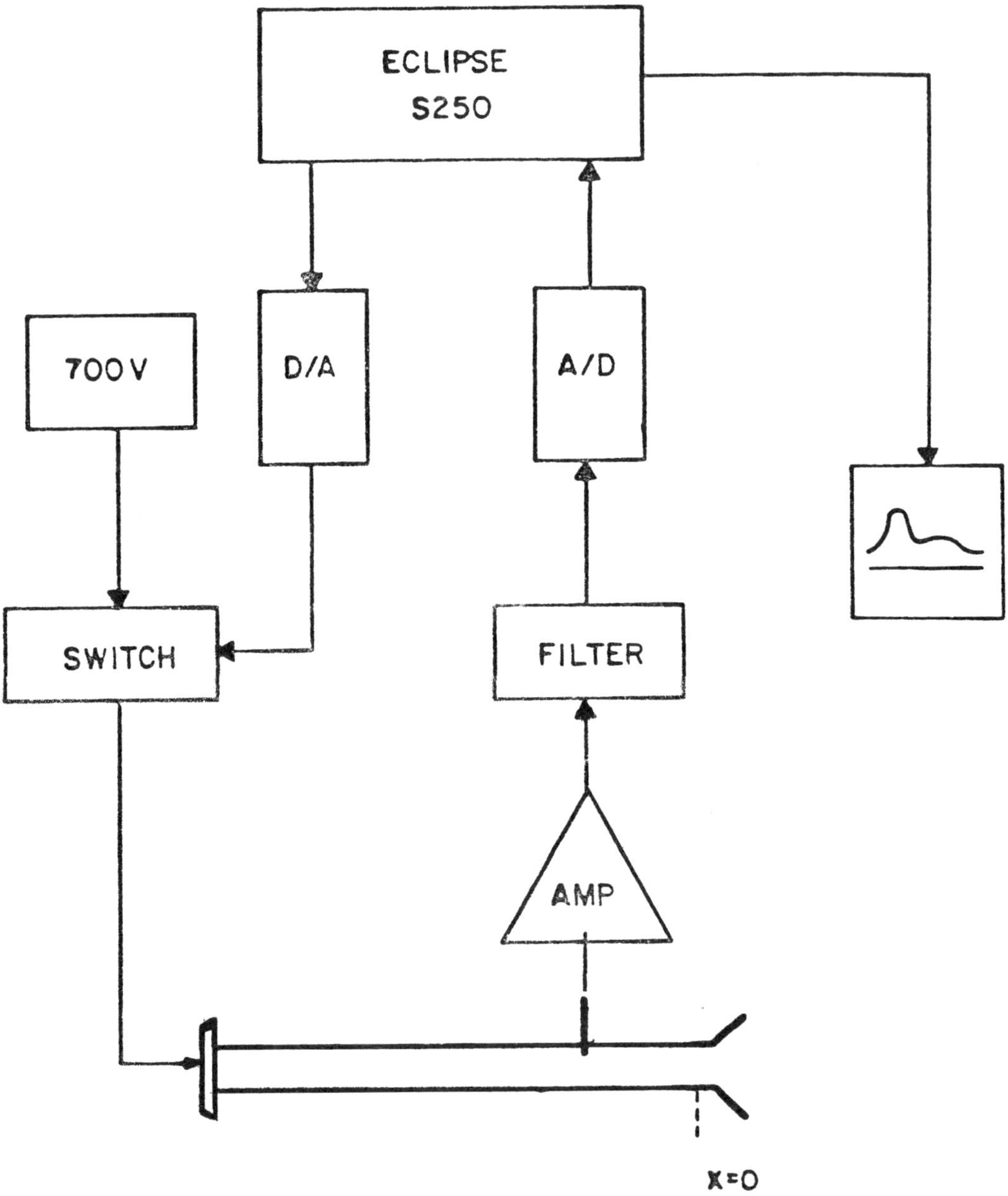

Fig. 2 A schematic diagram of the experiment.

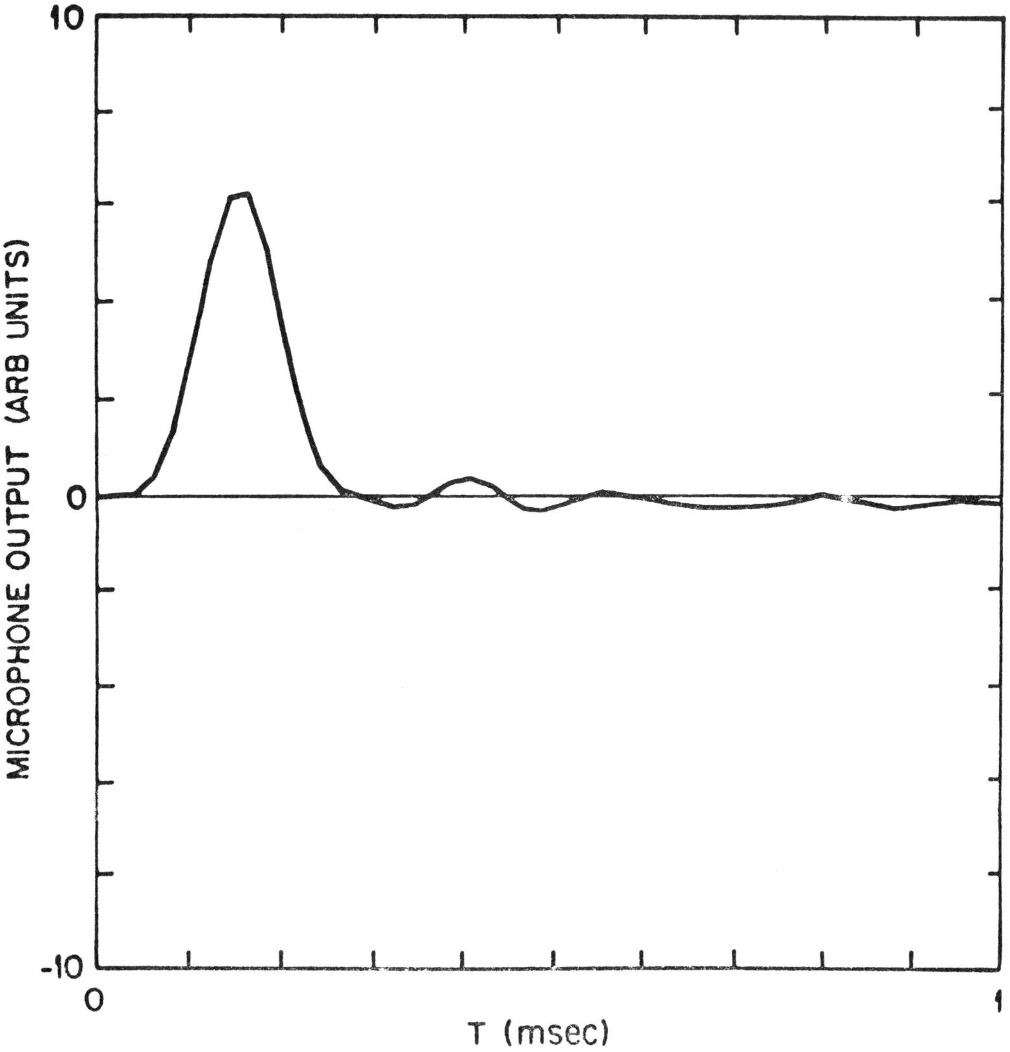

Fig. 3 The shape of the pressure pulse produced by the electrostatic transducer. The ordinate is in volts at the output of the microphone amplifier.

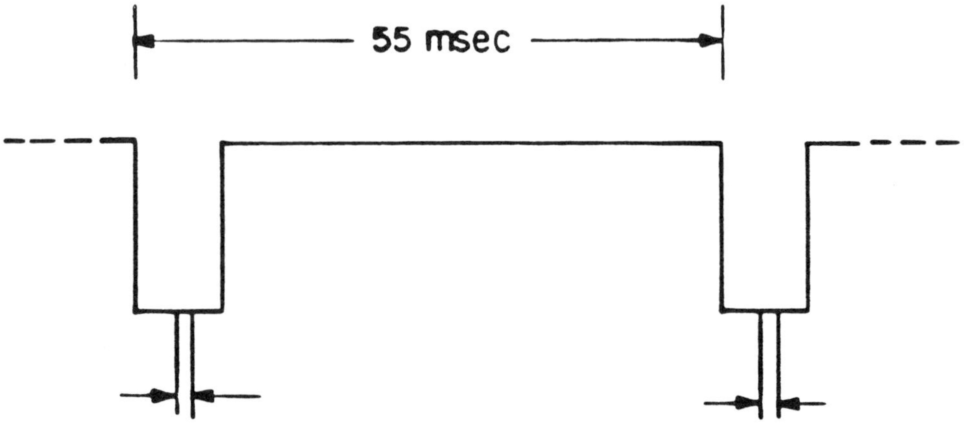

Fig. 4 A diagram of the pulse train applied to the transducer. (The actual pulses, of course, have finite rise and decay times.) The negative pulses are of about 700 volts, and of duration 8 msec. In each period the time interval between arrows (about 1.5 msec) is the interval of interest for the measurement of reflectance.

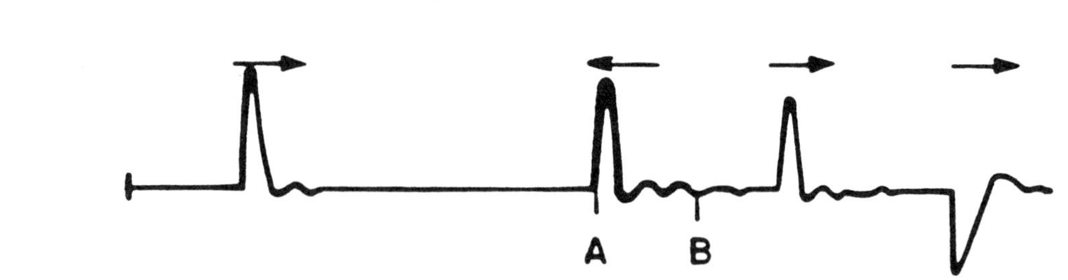

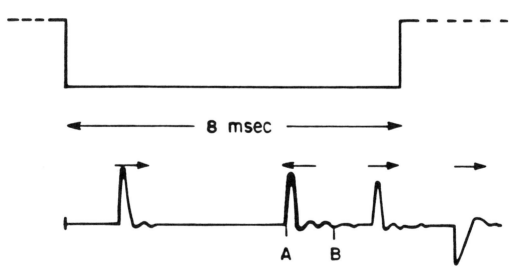

Fig. 5 Microphone output for a few msec from the start of a voltage pulse to the transducer. The small arrows indicate the direction of travel of the waves producing different portions of the signal.

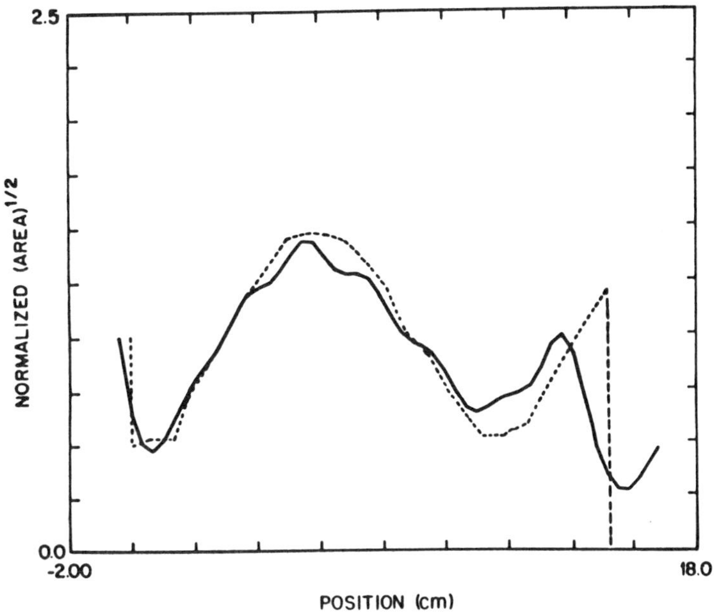

Figure 6a

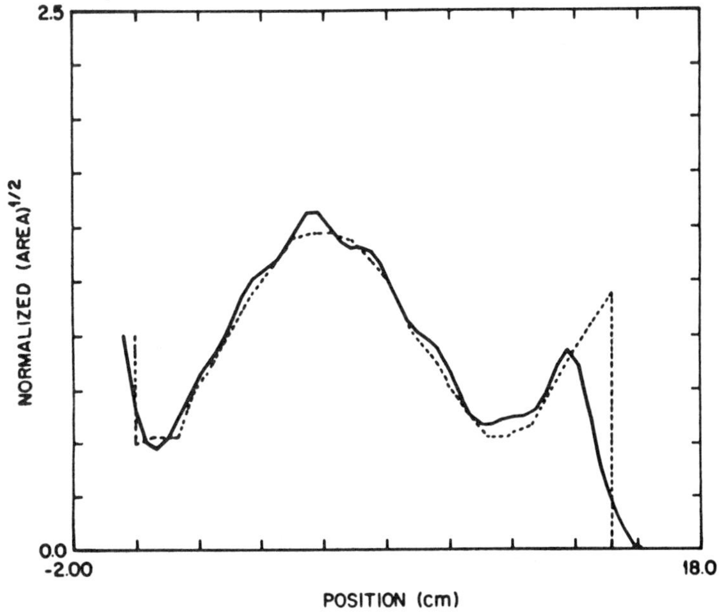

Figure 6b

Fig. 6 Reconstruction of the shape of a rubber mold from measurements. The shape of the rubber mold (shown as a dotted curve) corresponds to an "oh" sound. The step reflectance for the mold was measured by the experimental technique described in Section 3). The solid curve of Fig. (a) was reconstructed from the lossless algorithm based on eqs (11-13); the solid curve of Fig. (b) was obtained from the algorithm based on eqs (20a-20d), with the losses chosen to be those typical of a vocal tract.

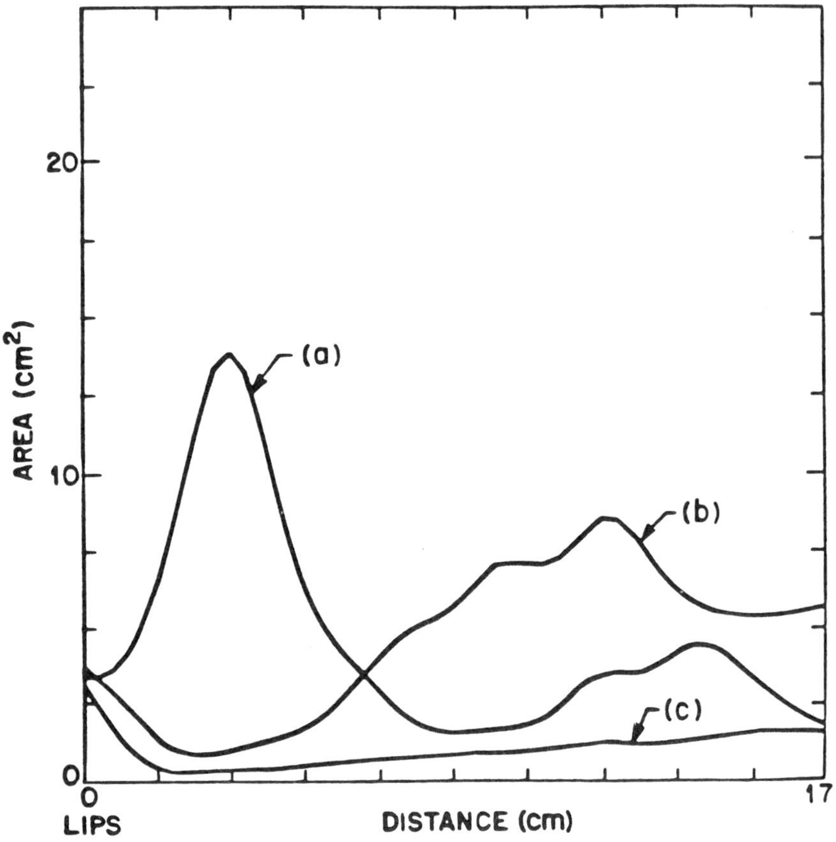

Fig. 7 Sample frames from a movie of the utterance "ah-ee-ah" measured on one subject. Curve (a) is an "ah" shape, curve (b) a transition shape and curve (c) an "ee" shape.

REFERENCES

[1] P. MERMELSTEIN and M.R. SCHROEDER, "Determination of Smoothed Cross-Sectional Area Functions of the Vocal Tract from Formant Frequencies," paper A-24, Proceedings of the Fifth Int'l Congress on Acoustics, 1965, Liege,Belgium., D.E. Commins, editor (Imprimierie Georges Thone, Liege,1965) Vol 1a.

[2] M.R. SCHROEDER, "Determination of the Geometry of the Human Vocal Tract," J. Acoust, Soc. of Am., **41,** 1002-1010 (1967).

[3] P. MERMELSTEIN, "Determination of the Vocal Tract from Measured Formant Frequencies," J. Acoust. Soc. of Am., **41,** 1283-1294 (1967)

[4] G. BORG, Acta Mathematica **78,** 1-96 (1946).

[5] I.M. GELFAND and B.M. LEVITAN, "On Determination of a Differential Equation from its Spectral Function," Izv. Akad. Nauk, SSSR (Seriya Mathematicheskaya) **15,** 309-360 (1951). [English translation: Amer. Math. Soc. Translations, Series 2 **1,** 253-304 (1955).

[6] B. GOPINATH and M.M. SONDHI, "Determination of the Shape of the Human Vocal Tract from Acoustical Measurements," Bell Sys Tech Jour **49,** 1195-1214 (1970).

[7] M.G, KREIN, "Solution of the Inverse Sturm-Liouville Problem," Dokl. Akad. Nauk SSSR (N.S.) **76,** 21-24 (1951).

[8] M.M. SONDHI and B. GOPINATH, "Determination of Vocal Tract Shape from Impulse Response at the Lips," Jour. Acoust. Soc. of Am. **49,** 1867-1873 (1971).

[9] M.M. SONDHI and B. GOPINATH, "Determination of the Shape of a Lossy Vocal Tract," paper 23C10, Seventh Int'l Congr. on Acoustics, Budapest, Hungary (1971).

[10] J.R. RESNICK, "Acoustical Inverse Scattering as a Means for Determining the Area Function of a Lossy Vocal Tract: Theoretical and Experimental Model Studies," Doctoral Dissertation, The Johns Hopkins University (1979).

[11] J.R. RESNICK, "Use of Step Response in Inverse Scattering," J. Acoust. Soc. Am., **68,** s85 (1980).

[12] M.M. SONDHI, "Experimental Determination of the Area Function of a Lossy Dynamically Varying Vocal Tract," Jour. Acoust. Soc. of Am., **53,** 294 (1973). (Abstract of Talk).

[13] R. DESCOUT, B. TOUSIGNANT and M. LECOURS, "Vocal Tract Area Function Measurements: Two Time-Domain Methods," Int'l Conf. on Acoust. Speech and Sig. Process., 1976. Pages 76-78 of the Proceedings.

[14] B. TOUSIGNANT, J.-P. LEFEVRE and M. LECOURS, "Speech Synthesis from Vocal Tract Area Function Acoustical Measurements," paper 24.12, Int'l Conf. Acoust. Speech and Sig. Process. (1979).

[15] J.-P. LEFEVRE, B. TOUSIGNANT and M. LECOURS, "Etude des configurations vocaliques des voyelles francaises a partir de mesures acoustiques," Acustica, **52,** 227-231 (1983).

[16] M.M. SONDHI and J.R. RESNICK, "The Inverse Problem for the Vocal Tract: Numerical Methods, Acoustical Experiments, and Speech Synthesis" Jour. Acoust. Soc. AM., vol 73, 1983, pp 985-1002.

[17] R. BURRIDGE, "The Gelfand-Levitan, the Marchenko, and the Gopinath-Sondhi integral equations of inverse scattering theory, regarded in the context of inverse impulse-response problems," Wave Motion, **2,** 305-323 (1980).

[18] M.M. SONDHI, "Model for Wave Propagation in a Lossy Vocal Tract," Jour. Acoust. Soc. Am., vol 55, 1974, pp 1071-1075.

[19] J.L. FLANAGAN, *Speech Analysis Synthesis and Perception,* Springer-Verlag, New York (1972), Chapter 3.

[20] J.R. RESNICK, to be published

[21] A. BEN-ISRAEL and T.N.E. GREVILLE, *Generalized Inverses: Theory and Applications,* John Wiley and Sons, New York (1974).

[22] A.N. TIKHONOV and V.Y. ARSENIN, *Solutions of Ill-Posed Problems,* V.H. Winstons and Sons, Washington, D.C. (1977).

[23] S. ZOHAR, "Toeplitz Matrix Inversion: the Algorithm of W.F. Trench," J.A.C.M. **16,** 592-601 (1967).

CONVERGENCE OF DISCRETE INVERSION SOLUTIONS

KENNETH P. BUBE*

Abstract. We study the convergence of numerical methods for a one-dimensional inverse problem in reflection seismology. An impulsive plane wave in pressure is applied at the surface of a piecewise smooth stratified elastic half space, and the resulting particle velocity is measured at the surface. The characteristic impedance of the medium is to be recovered as a function of travel time.

We consider two methods for the numerical solution of this inverse problem: a discrete version of the Gopinath-Sondhi integral equation and a downward continuation method. Both methods are based on using second order difference schemes to reconstruct a solution of the elastic wave equations, and both are exact for the popular Goupillaud layered media. A formally second order accurate equation based on the theory of geometrical acoustics is used to recover the impediance. If the impulse response is measured by sampling the step response or ramp response appropriately, then these two methods are second order convergent for piecewise smooth media with discontinuities occurring only at integer multiples of the mesh width.

We conclude with an example showing that if the discontinuities of the medium occur at noninteger multiples of the mesh width, the convergence need not be uniform, and convergence in the L^p norms for $1 \leq p < \infty$ need not be any better than order $1/p$ in the mesh width.

1. Introduction. Since we cannot expect to solve many inverse problems in geophysics and other applications

* Department of Mathematics, University of California, Los Angeles, 405 Hilgard Avenue, Los Angeles, CA 90024. This work was supported in part by the National Science Foundation under Grant MCS-82-00788.

exactly by analytic methods, numerical methods are used in practice to obtain discrete approximations to the solution of a continuum inverse problem from discrete data sets. With real data and real physical geometries, a host of approximations are made in the processing sequence, and there are many difficulties (e.g., band-limited data) to be overcome. At the foundation of these procedures is an assumption of convergence, not often stated because of the focus on these other difficulties: the solutions of the discrete methods would, at least under ideal conditions, converge in some sense to the solution of the continuum inverse problem as some parameter of the discretization (e.g., mesh-width) tends to zero.

In this paper, we address this convergence question for the one-dimensional inverse problem of reflection seismology. The correspondence between discrete and continuum inverse problems has been addressed for this and a number of related problems in the literature [1, 2, 3, 4, 9, 12, 18 and others]. We assume here a one-dimensional nonabsorbing medium, probed by plane waves at normal incidence, reflection data (source and measurements at the surface), and free-surface boundary conditions, with full band-width data sampled in the time domain. It is not our purpose here to prove that convergence is obtained in practice with all the practical difficulties taken into account, but to show that this underlying assumption of convergence under ideal conditions is indeed valid for this inverse problem.

Consider now the one-dimensional inverse problem of reflection seismology. Assume the earth is a horizontally homogeneous elastic medium in which plane P waves travel vertically. An impulse in the normal component of traction is applied uniformly on the ground surface and the impulse response (the resulting particle velocity at the surface) is measured as a function of time. The characteristic impedance of the medium is to be recovered as a function of travel time.

If the impulse response $h(t)$ is known as a function of time t for $0 \leq t \leq 2X$, then the impedance profile $\zeta(x)$ can be determined as a function of travel time x for $0 \leq x \leq X$. Gopinath and Sondhi [10, 14] developed an integral equation method to solve this problem. Symes [17] has shown it is possible to solve a nonstandard boundary value problem continuing surface values downward to solve this inverse problem. In §2, we review some results on the continuum inverse problem.

In practice we measure only a discrete sample h_k $(0 \leq k \leq n)$ of some time function related to $h(t)$. (We use the phrase "sampling the impulse response" loosely to mean measuring any such discrete sample.) From this discrete sample, we compute an approximation ζ_j $(0 \leq j \leq n)$ to some values of the impedance profile. The hope is that the computed discrete impedance profile ζ_j $(0 \leq j \leq n)$ converges to the true impedance profile as the sampling interval goes to zero.

With appropriate choices of (i) sampling the impulse response, (ii) discrete inversion, and (iii) interpretation of the discrete impedance profile as approximate values of the true impedance profile, the error in the computed discrete impedance profile converges uniformly to zero like the square of the sampling interval. This <u>second order convergence</u> holds under the assumption that the true impedance profile $\zeta(x)$ is <u>piecewise smooth</u>, provided that all discontinuities of $\zeta(x)$ occur at travel times x which are interger multiples of the mesh width.

The order of convergence depends critically on the choice of <u>sampling</u>, <u>discrete inversion</u>, and <u>interpretation</u>. In §3, we choose a discrete system by making a second order difference approximation to the differential equations using the method of characteristics on a uniform grid with mesh width $\Delta t = \Delta x = \Delta$. The choice of interpretation is then natural to guarantee the formal second order accuracy of the difference approximation: ζ_j approximates $\zeta(x)$ at $x=(j+\frac{1}{2})\Delta$, the midpoint between the grid points $x = j\Delta$ and $x=(j+1)\Delta$.

Once formulated, the discrete system can stand on its own without further reference to the continuum problem. Given a discrete impedance profile ζ_j $(0 \leq j \leq n)$, the discrete impulse response h_k (at the grid points $t = 2k\Delta$, $0 \leq k \leq n$) is determined by the solution of a discrete initial-boundary value problem. There is a well determined one-to-one correspondence between positive discrete impedance profiles ζ_j $(0 \leq j \leq n)$ and discrete impulse response sequences h_k $(0 \leq k \leq n)$ characterized by a positive definite property.

To compute an approximation ζ_j $(0 \leq j \leq n)$ to the values of the true impedance profile from the discrete sample h_k $(0 \leq k \leq n)$ of the impulse response, we treat h_k $(0 \leq k \leq n)$ as a discrete impulse response sequence in the discrete system and solve the inverse problem for the discrete system to obtain ζ_j $(0 \leq j \leq n)$. This is our choice of <u>discrete inversion</u>. Observe that the second order accuracy of the difference approximation does not automatically guarantee the second order convergence of the computed discrete impedance profile.

Several algorithms are known for solving the discrete inverse problem. The two methods we study here are a discrete version of the Gopinath-Sondhi integral equation and a downward continuation algorithm. Both methods reconstruct a solution of the discrete system using difference schemes en route to solving the discrete inverse problem; see Bube and Burridge [4,5] for details. Gerver [9], Claerbout [7], Symes [17], Santosa and Schwetlick [13], and Sondhi and Resnick [15] study related methods.

We study two ways of <u>sampling</u>. The first, which we call <u>sampling the step response</u>, is to measure

$$(1.1)\qquad h_k = \int_{-\Delta}^{\Delta} h(2k\Delta + \tau)\, d\tau \qquad (0 \le k \le n)\ ,$$

with sampling interval 2Δ; it is equivalent to sampling the step response (the time integral of the step response) at odd multiples $t = (2k + 1)\Delta$ $(0 \le k \le n)$ of Δ, and the taking the first time difference; see §4. If an impulse $\delta(t)$ in pressure were applied at the surface, the step response would then be the particle displacement at the surface. This sampling is chosen to make the equation for the recovery of ζ_j in the downward continuation algorithm (applied to the step response) formally second order accurate [4]. In practice, it is not possible to set off an impulse $\delta(t)$ or a step function in pressure; we shall not address here the important questions of band-limited data and deconvolution. Berryman and Greene [1] have pointed out previously that the step response should be sampled if convergence is desired. Our specific choice leads to second order convergence.

We show in §4 how to sample the velocity response for a large class of nonimpulsive pressure sources to obtain <u>formal</u> second order accuracy in the downward continuation algortihm. The second way of sampling studied here, which we call <u>sampling the ramp response</u>, is to measure

$$(1.2)\qquad h_k = \int_{-2\Delta}^{2\Delta} \left[1 - \frac{|\tau|}{2\Delta}\right] h(2k\Delta + \tau)\, d\tau \qquad (0 \le k \le n)$$

with sampling interval 2Δ; it is equivalent to sampling the ramp response (the second time integral of the impulse response) at even multiples $t = (2k + 2)\Delta$ $(0 \le k \le n)$ of Δ, and then taking $\frac{1}{2\Delta}$ times the second time difference; see §4.

The difference approximation made in choosing the discrete system is closely related to so-called Goupillaud-layered media [11]. These are media with homogeneous layers of equal travel-time thickness d. If d is an integer multiple of the mesh width Δ, then the difference approximation is exact. Moreover, since the impulse response of a Goupillaud-layered medium is a sum of delta functions centered at even multiples of d, both ways of sampling chosen above measure the amplitudes of these delta functions and the solution of the discrete inverse problem will be exact. Notice that the choice of interpretation assigns ζ_j to the midpoint of the layer between $x = j\Delta$ and $x = (j + 1)\Delta$, so we avoid attempting to approximate $\zeta(x)$ right at a discontinuity.

One popular way to discretize the inverse problem is to assume that the sampled impulse response actually comes from a Goupillaud-layered medium with $d = \Delta$. This assumption is not necessary to use the discrete system, but it is a common way of visualizing the discrete system. For the two ways of sampling chosen above, our results show that this assumption yields second order convergence for piecewise smooth media if all discontinuities of $\zeta(x)$ occur at integer multiples of Δ and we relax our interpretation of ζ_j from giving $\zeta(x)$ for the whole interval $j\Delta < x < (j + 1)\Delta$ to just giving $\zeta(x)$ at $x = (j + \frac{1}{2})\Delta$.

The convergence proofs are based on comparing the discrete and continuum versions of the Gopinath-Sondhi integral equation. There are two main steps. In §5, we present results showing that the sampled impulse response has the positive definite property necessary to solve the discrete inverse problem. If the step response is sampled, we assume that Δ is sufficiently small and all discontinuities of $\zeta(x)$ occur at integer multiples of Δ. If the ramp response is sampled, Δ can be arbitrary and the locations of the discontinuities of $\zeta(x)$ can be arbitrary and we still get the necessary positive definiteness. Then in §6 we present estimates of the difference between the computed discrete impedance and the corresponding values of the true impedance; here for both ways of sampling we need to assume that all discontinuities of $\zeta(x)$ occur at integer multiples of Δ. It follows directly that the discrete Gopinath-Sondhi integral equation method is second order convergent for both ways of sampling. It then follows immediately that the versions of the downward continuation algorithm based on the discrete impulse response, the discrete step response, or the discrete ramp response are also second order convergent; for each way of sampling, all these methods compute the same discrete impedance profile. In §6 we also consider what order of errors in the measurements can be allowed without disturbing the order of convergence.

Finally in §7, we present an example which shows that if the discontinuities of $\zeta(x)$ occur at noninteger multiples of Δ, the convergence need not be uniform in x. We do get convergence for this example in the L^p norms for $1 \leq p < \infty$, but the error is only of order $1/p$ in Δ as Δ tends to zero.

2. The Continuum Inverse Problem. Normally incident P waves in a horizontally homogeneous elastic half space $z \geq 0$ propagate according to the hyperbolic system

$$(2.1) \qquad \rho w_t + p_z = 0 , \qquad p_t + \rho c^2 w_z = 0 .$$

Time t and depth z are the independent variables, with the surface of the medium at $z = 0$. The dependent variables are the z-component of particle velocity w and the zz-component of stress p which we call pressure (positive in compression). The coefficients, assumed to be functions of z alone, are the density ρ and the P-wave speed c.

Changing the depth variable to one-way travel time

$$x = \int_0^z \frac{dz'}{c(z')} ,$$

system (2.1) becomes the system

$$(2.2) \qquad \zeta(x) w_t + p_x = 0 , \qquad p_t + \zeta(x) w_x = 0 ,$$

in the region $x \geq 0$, $-\infty < t < \infty$, where

$$(2.3) \qquad \zeta(x) = \rho(x) c(x)$$

is the <u>impedance</u>. We assume that $\zeta(x)$ is positive, bounded away from 0 and $+\infty$, and piecewise smooth, i.e., $\zeta(x)$ has sufficiently many continuous derivatives except at a discrete set of points $x = x_1, x_2, \ldots$ (with $0 < x_1 <$

$x_2 < \ldots$ and $x_\nu \to \infty$ as $\nu \to \infty$), where $\zeta(x)$ and its derivatives may have jump discontinuities but still have left- and right-hand limits. System (2.2) is the basic set of differential equations for the continuum problem. This system is hyperbolic with t as the time-like variable.

Given the impedance profile $\zeta(x)$, the continuum <u>forward problem</u> is to determine the velocity response at the surface to a pressure pulse set off at the surface. This is a standard initial-boundary value problem. Let $w_1(x,t)$, $p_1(x,t)$ be the solution of system (2.2) satisfying homogeneous initial conditions

$$(2.4) \quad w_1(x,t) = p_1(x,t) = 0 \qquad (x \geq 0 , \quad t < 0) ,$$

and the impulsive pressure boundary conditions

$$(2.5) \quad p_1(0,t) = \delta(t) .$$

The velocity response at the surface,

$$(2.6) \quad w_1(0,t) \equiv h(t)$$

is called the <u>impulse response</u>. It can be shown by geometrical acoustics that $h(t)$ can be written in the form

$$(2.7) \quad h(t) = \sum_{\mu \geq 0} \alpha_\mu \, \delta(t - \tau_\mu) + \bar{h}(t) ,$$

where $0 = \tau_0 < \tau_1 < \tau_2 < \ldots,$ $\tau_\mu \to \infty$ as $\mu \to \infty$,

$$(2.8) \quad \alpha_0 = \frac{1}{\zeta(0)} ,$$

and $\bar{h}(t)$ is zero for $t < 0$ and piecewise smooth. Also, $h(t)$ for $0 \leq t \leq 2X$ depends only on $\zeta(x)$ for $0 \leq x \leq X$ by the finite speed of propagation for solutions of system (2.2).

The continuum <u>inverse problem</u> is to determine the impedance profile $\zeta(x)$ from the impulse response $h(t)$. We restrict our attention to a finite interval: given $h(t)$ for $0 \leq t \leq 2X$, determine $\zeta(x)$ for $0 \leq x \leq X$.

There are three concepts common to several methods of solving this inverse problem. The system (2.2) of differential equations is also hyperbolic with x as the time-like variable, so it allows propagation in the positive x direction. Advancing in the positive x direction, it is possible to construct a solution of system (2.2) with a singularity propagating along the characteristic line $t = x$ from the impulse response. Then geometrical acoustics can be used to recover the impedance profile from this constructed solution. Two methods are of interest here. The Gopinath-Sondhi integral equation [10, 14] constructs a noncausal solution of system (2.2) which is constant outside the triangular region $|t| \leq x$. The downward continuation method (see Symes [17]) constructs a causal solution of system (2.2) with support in the triangular region $0 \leq x \leq t$. (A causal solution is zero for $t < 0$ by definition, and hence is zero for $t < x$.)

We now describe the Gopinath-Sondhi integral equation. Let $\tilde{w}_2(x,t)$, $\tilde{p}_2(x,t)$ be the solution of system (2.2) satisfying the Cauchy data on the t-axis

$$(2.9)\qquad \tilde{w}_2(0,t) = -\mathrm{sgn}(t)\ , \quad \tilde{p}_2(0,t) = 0\ ,$$

where sgn(t) is 1 for $t > 0$ and -1 for $t < 0$. By the finite speed of propagation,

$$(2.10)\qquad \tilde{p}_2(x,t) = 0 \quad (|t| > x)\ .$$

The theory of geometrical acoustics implies that $\tilde{p}_2(x,t)$ has jump discontinuities across the characteristic lines $t = \pm x$. Also, $\tilde{p}_2(x,t)$ is smooth in $|t| < x$ except for jump discontinuities in its derivatives across the lines $x = x_\nu$ and jump discontinuities across a discrete set of half-lines of slope ± 1 pointing in the positive x direction. This set of half-lines can be described recursively as follows:

(i) the half lines $t = \pm x$, $x \geq 0$ are in the set;

(ii) wherever a half-line in the set crosses a line $x = x_\nu$, a new half-line with opposite slope originates at that point.

This set of half-lines and the lines $x = x_\nu$ separate the quarter plane $|t| \leq x$ into polygonal regions; for each of these open polygonal regions, $\tilde{p}_2(x,t)$ can be extended to be smooth in its closure. A Green's function argument [6] (or other arguments [14]) lead to the Gopinath-Sondhi integral equation

$$(2.11)\qquad \frac{1}{2}\int_{-x}^{x}\big(h(t-\tau) + h(\tau-t)\big)\,\tilde{p}_2(x,\tau)\,d\tau = 1 \quad (|t| < x)\ .$$

Consider a fixed x with $0 < x \leq X$. Equation (2.11) is an integral equation for $p_2(x,t)$ as a function of t for $-x < t < x$. Let $L^2(-x,x)$ denote the Hilbert space of real-valued square integrable functions defined on the interval $(-x,x)$ with the inner product

$$(2.12) \qquad (\phi,\psi)_x = \int_{-x}^{x} \phi(t)\psi(t)dt \ .$$

Define operators H_x and R_x from $L^2(-x,x)$ into itself by

$$(2.13) \qquad (H_x\psi)(t) = \int_{-x}^{x} h(t-\tau)\psi(\tau)d\tau \ ,$$

$$(2.14) \qquad (R_x\psi)(t) = \frac{1}{2}\int_{-x}^{x} \big(h(t-\tau) + h(\tau-t)\big)\ \psi(\tau)d\tau \ .$$

Then

$$(2.15) \qquad R_x = \frac{1}{2}\,(H_x + H_x^T) \ ,$$

where H_x^T is the operator transpose of H_x. It can be shown [14, 17] that R_x is a positive definite operator on $L^2(-x,x)$, whose inverse has norm bounded in terms of the total variation of $\log\zeta$ from 0 to x. Thus equation (2.11) can be solved to obtain $p_2(x,t)$ $(-x < t < x)$. Standard arguments imply

$$(2.16) \qquad c_0(\psi,\psi)_x \leq (\psi,R_x\psi)_x \leq c_1(\psi,\psi)_x$$

for all ψ in $L^2(-x,x)$, and hence since R_x is symmetric

$$(2.17) \qquad ||R_x|| \leq c_1 \ , \quad ||R_x^{-1}|| \leq \frac{1}{c_0} \ ,$$

where

$$(2.18) \qquad c_1 = ||R_X|| \ , \quad c_0 = ||R_X^{-1}||^{-1} > 0 \ .$$

Geometrical acoustics [5] (or other arguments [14]) imply

(2.19)

$$\zeta(x) = \frac{1}{2}\frac{d}{dx}\int_{-x}^{x}\tilde{p}_2(x,t)dt \quad (0 \le x \le X\ ;\ x \ne x_\nu\ ,\ 1 \le \nu \le N)\ .$$

(The simpler formula in [14] expressing $\zeta(x)$ in terms of $p_2(x,t)^2$ is no longer valid when $\zeta(x)$ has discontinuities.) The inverse problem of recovering $\zeta(x)$ $(0 \le x \le X)$ from $h(t)$ $(0 \le t \le 2X)$ is therefore well-posed. In fact, any $h(t)$ $(0 \le t \le 2X)$ of the form (2.7) (with the stated assumptions on $\tilde{h}(t)$) for which R_X is positive definite yields a piecewise smooth positive $\zeta(x)$ $(0 \le x \le X)$ when equation (2.11) is solved and then equation (2.19) is applied; moreover the given $h(t)$ is the impulse response for this $\zeta(x)$ [14].

We now describe the downward continuation method. Let $w(x,t)$, $p(x,t)$ be a solution of system (2.2). If $w(0,t)$ and $p(0,t)$ are known for $0 \le t \le 2X$ (and $\zeta(x)$ is known for $0 \le x \le X$), then the solution is determined in the triangle $0 \le x \le t \le 2X - x$. If the solution is known in this triangle and the solution is known to be <u>causal</u> with a singularity propagating on the line $t = x$, then $\zeta(x)$ $(0 \le x \le X)$ can be recovered using geometrical acoustics. These two ideas can be used together to recover simultaneously the solution in this triangle and $\zeta(x)$ for $0 \le x \le X$, if we know $w(0,t)$ and $p(0,t)$ for $0 \le t \le 2X$ for such a causal solution.

For example, let $\tilde{w}_1(x,t)$, $\tilde{p}_1(x,t)$ be the solution of system (2.2) satisfying homogeneous initial conditions

(2.20) $\tilde{w}_1(x,t) = \tilde{p}_1(x,t) = 0 \quad (x \ge 0\ ,\ t < 0)$,

and the step pressure boundary conditions

(2.21) $\tilde{p}_1(0,t) = H(t)$,

where the step function $H(t)$ is 1 for $t > 0$ and 0 for $t < 0$. This solution is the time integral of $w_1(x,t)$, $p_1(x,t)$. The <u>step response</u> is

(2.22) $\tilde{w}_1(0,t) = \int_{-\infty}^{t} h(\tau)d\tau \equiv g(t)$.

A jump discontinuity propagates along the line $t = x$, and geometrical acoustics [4] implies

$$(2.23)\quad \zeta(x) = \tilde{p}_1(x,x+)/\tilde{w}_1(x,x+) \quad (x \neq x_\nu\ ,\ 1 \leq \nu)\ .$$

If the step response $g(t)$ is measured for $0 \leq t \leq 2X$, then $\tilde{w}_1(0,t)$ and $\tilde{p}_1(0,t)$ are known for $0 \leq t \leq 2X$. The method advances the solution downward in the triangle $0 \leq x \leq t \leq 2X - x$, using equation (2.23) to recover $\zeta(x)$ $(0 \leq x \leq X)$ as the solution is advanced downward.

We are also interested in the time integral of $\tilde{w}_1(x,t)$, $\tilde{p}_1(x,t)$. Let $\hat{w}_1(x,t)$, $\hat{p}_1(x,t)$ be the solution of system (2.2) satisfying homogeneous initial conditions

$$(2.24)\quad \hat{w}_1(x,t) = \hat{p}_1(x,t) = 0 \quad (x \geq 0\ ,\ t < 0)$$

and the ramp pressure boundary conditions

$$(2.25)\quad \hat{p}_1(0,t) = tH(t)\ .$$

The <u>ramp response</u> is

$$(2.26)\quad \hat{w}_1(0,t) = \int_{-\infty}^{t} g(\tau)d\tau = \int_{-\infty}^{t}(t-\tau)h(\tau)d\tau \equiv b(t)\ .$$

This solution is continuous with a jump discontinuity in the t derivative propagating along the line $t = x$. Geometrical acoustics implies

$$(2.27)\quad \zeta(x) = \lim_{t\to x+} \hat{p}_1(x,t)/\hat{w}_1(x,t) \quad (x \neq x_\nu\ ,\ 1 \leq \nu)\ .$$

3. The Discrete Inverse Problem. In this section we describe the discrete system and inverse problem; see Bube and Burridge [4] for further details.

Consider a regular grid in the half plane $x \geq 0$, $-\infty < t < \infty$ with mesh width $\Delta t = \Delta x = \Delta$ and grid points at $x = j\Delta$ for integers $j \geq 0$ and $t = i\Delta$ for all integers i. Consider grid functions w_j^i, p_j^i (corresponding to $w(x,t)$, $p(x,t)$ at $x = j\Delta$, $t = i\Delta$) and a discrete impedance profile ζ_j (corresponding to $\zeta(x)$ at $x = (j + \frac{1}{2})\Delta$ for integers $j \geq 0$). To obtain the basic equations for the discrete system we use the method of characteristics for system (2.2). Setting $R = (j\Delta, i\Delta)$, $S = ((j+1)\Delta, (i+1)\Delta)$, and $T = ((j+1)\Delta,(i-1)\Delta)$, system (2.2) implies

$$(3.1)\quad \int_R^S \zeta\, dw + dp = 0 = \int_R^T \zeta\, dw - dp ,$$

where the paths of integration are the line segments $\overline{RS}$ and $\overline{RT}$. Approximating the integrals of $\zeta\, dw$ by the midpoint rule leads to

$$(3.2)\quad \zeta_j w_j^i + p_j^i = \zeta_j w_{j+1}^{i+1} + p_{j+1}^{i+1} ,$$

$$(3.3)\quad \zeta_j w_j^i - p_j^i = \zeta_j w_{j+1}^{i-1} - p_{j+1}^{i-1} .$$

These are the basic equations for the discrete system, valid for $j \geq 0$ and $-\infty < i < \infty$. There are two separate grids which are not connected by equations (3.2-3): the even grid ($i+j$ even), and the odd grid ($i+j$ odd). Given a positive discrete impedance profile ζ_j for $j \geq 0$, we call a solution of the discrete system a pair of grid functions w_j^i, p_j^i, both defined on either the even grid or the odd grid, which satisfy equations (3.2-3) on that grid.

Equations (3.2-3) allow propagation in the positive t direction given initial conditions and one boundary condition. They also allow propagation in the positive x direction according to the second order accurate difference scheme

$$(3.4)\quad w_j^i = \tfrac{1}{2}\left((w_{j-1}^{i+1} + w_{j-1}^{i-1}) - (p_{j-1}^{i+1} - p_{j-1}^{i-1})/\zeta_j\right) ,$$

$$(3.5)\quad p_j^i = \tfrac{1}{2}\left((p_{j-1}^{i+1} + p_{j-1}^{i-1}) - \zeta_j (w_{j-1}^{i+1} - w_{j-1}^{i-1})\right) .$$

The <u>discrete forward problem</u> is to determine the discrete velocity rsponse at the surface ($j = 0$) to a discrete pressure pulse set off at the surface. This is a discrete initial-boundary value problem. Let $(w_1)_j^i$, $(p_1)_j^i$ be the solution of system (3.2-3) on the even grid satisfying homogeneous initial conditions

$$(3.6)\quad (w_1)_j^i = (p_1)_j^i = 0 \qquad (i < 0 ,\ i+j \text{ even}) ,$$

and the impulsive pressure boundary conditions

$$(3.7)\quad (p_1)_0^{2k} = \delta_{k0} ,$$

where δ_{k0} is 1 if $k = 0$ and 0 otherwise. The discrete velocity response at the surface

$$(3.8)\quad (w_1)_0^{2k} \equiv h_k$$

is called the <u>discrete impulse response</u>. Clearly h_k is zero for $k < 0$, and by equation (3.3) at $i = j = 0$,

$$(3.9)\quad h_0 = 1/\zeta_0 .$$

Also h_k for $0 \le k \le n$ depends only on ζ_j for $0 \le j \le n$.

The <u>discrete inverse problem</u> is to determine the discrete impedance profile ζ_j $(0 \le j \le n)$ from the discrete impulse response h_k $(0 \le k \le n)$. As in the continuum problem, we restrict our attention to a finite interval. The same three concepts common to methods of solution of the continuum inverse prolem apply also to the discrete inverse problem. The two methods of interest here are the discrete counterparts of the Gopinath-Sondhi integral equation and the downward continuation method.

The discrete Gopinath-Sondhi integral equation is based on the solution $(\tilde{w}_2)_j^i$, $(\tilde{p}_2)_j^i$ of system (3.2, 3.3) on the odd grid satisfying the discrete Cauchy data

$$(3.10)\quad (\tilde{w}_2)_0^i = -\mathrm{sgn}(i) , \quad (\tilde{p}_2)_0^i = 0 , \quad (i \text{ odd}) .$$

By system (3.4, 3.5),

$$(3.11)\quad (\tilde{p}_2)_j^i = 0 \quad (|i| > j , \quad i+j \text{ odd}) .$$

Let H_j and R_j be the $(j+1) \times (j+1)$ matrices

$$(3.12)\quad H_j = \begin{pmatrix} h_0 & 0 & \cdots & 0 \\ h_1 & h_0 & \ddots & \vdots \\ \vdots & \ddots & \ddots & 0 \\ h_j & \cdots & h_1 & h_0 \end{pmatrix} , \qquad R_j = \begin{pmatrix} r_0 & r_1 & \cdots & r_j \\ r_1 & r_0 & \ddots & \vdots \\ \vdots & \ddots & \ddots & r_1 \\ r_j & \cdots & r_1 & r_0 \end{pmatrix} ,$$

where

(3.13) $r_k = \frac{1}{2} (h_k + h_{-k})$.

Note that

(3.14) $R_j = \frac{1}{2} (H_j + H_j^T)$.

A discrete Green's function argument leads to the discrete Gopinath-Sondhi integral equation

(3.15) $R_j \phi^j = 1^j \quad (0 \le j \le n)$,

where ϕ and 1^j are the (j+1) - vectors

(3.16)

$$\phi^j = [\phi_0^j, \phi_1^j, \ldots, \phi_j^j]^T \equiv [(\tilde{p}_2)_{j+1}^j, (\tilde{p}_2)_{j+1}^{j-2}, \ldots, (\tilde{p}_2)_{j+1}^{-j}]^T ,$$

(3.17) $1^j \equiv [1,1,\ldots,1]^T$.

It can be shown that whenever the ζ_j $(0 \le j \le n)$ are positive, R_j is positive definite for $0 \le j \le n$. Thus equation (3.15) can be solved for ϕ^j for $0 \le j \le n$. Fast algorithms which also recover ζ_j $(0 \le j \le n)$ can be derived (based on difference schemes) to solve all $n + 1$ linear systems recursively. The discrete counterpart of equation (2.19) is

(3.18) $\zeta_j = (1^j)^T \phi^j - (1^{j-1})^T \phi^{j-1} \quad (0 \le j \le n)$,

where the zero-length vectors 1^{-1} and ϕ^{-1} are zero by definition. We can thus recover ζ_j $(0 \le j \le n)$ from h_k $(0 \le k \le n)$.

Any discrete sequence h_k $(0 \le k \le n)$ for which R_n defined by (3.12, 3.13) is positive definite yields a positive discrete impedance profile ζ_j $(0 \le j \le n)$ when equation (3.15) is solved and then equation (3.18) is applied; moreover the given h_k $(0 \le k \le n)$ is the discrete impulse response for this ζ_j $(0 \le j \le n)$.

Consider now the downward continuation algorithm in the discrete system. Let w_j^i , p_j^i be any <u>causal</u> solution of system (3.2-3) on the even grid (so both are zero for $i < j$) , and suppose $p_0^0 \ne 0$. Then w_j^j , p_j^j are both nonzero for $0 \le j \le n$, and (3.3) with $i = j$ implies

$$(3.19)\qquad \zeta_j = p_j^j/w_j^j .$$

Suppose w_0^{2k} and p_0^{2k} are known for $0 \le k \le n$ for such a solution. For $j = 0,1,2,\ldots,n$ do the following: (i) obtain ζ_j from (3.19), and then (ii) advance w and p from j to $j+1$ (within the triangle $0 \le j \le i \le 2n - j$) using system (3.4-5).

In particular, if h_k $(0 \le k \le n)$ is known, then $(w_1)_0^{2k} = h_k$ and $(p_1)_0^{2k} = \delta_{k0}$ are known for $0 \le k \le n$, and we can apply this algorithm to $(w_1)_j^i$, $(p_1)_j^i$ to recover ζ_j $(0 \le j \le n)$.

We also consider the downward continuation algortihm applied to the discrete step response. Let $(\tilde{w}_1)_j^i$, $(\tilde{p}_1)_j^i$ be the solution of system (3.2-3) on the odd grid satisfying homogeneous initial conditions

$$(3.20)\qquad (\tilde{w}_1)_j^i = (\tilde{p}_1)_j^i = 0 \quad (i < 0,\ i+j \text{ odd}) ,$$

and the step pressure boundary conditions

$$(3.21)\qquad (\tilde{p}_1)_0^{2k+1} = \begin{cases} 1 & (k \ge 0) , \\ 0 & (k < 0) . \end{cases}$$

This is the time sum of the solution $(w_1)_j^i$, $(p_1)_j^i$. The <u>discrete step response</u> is

$$(3.22)\qquad (\tilde{w}_1)_0^{2k+1} = \sum_{\ell \le k} h_\ell \equiv g_k .$$

Both $(\tilde{w}_1)_j^{j+1}$, $(\tilde{p}_1)_j^{j+1}$ are nonzero for $0 \le j \le n$, and (3.3) with $i = j+1$ implies

$$(3.23)\qquad \zeta_j = (\tilde{p}_1)_j^{j+1}/(\tilde{w}_1)_j^{j+1} .$$

Suppose g_k $(0 \le k \le n)$ is known. Then $(\tilde{w}_1)_0^{2k+1} = g_k$ and $(\tilde{p}_1)_0^{2k+1} = 1$ are known for $0 \le k \le n$. For $j = 0,1,2,\ldots,n$ do the following: (i) obtain ζ_j from (3.23), and then (ii) advance $\tilde{w}_1$ and $\tilde{p}_1$ from j to $j+1$ (within the triangle $j+1 \le i \le 2n+1-j$) using system (3.4, 3.5).

In addition, we apply the downward continuation algorithm to the discrete ramp response. Let $(\hat{w}_1)_j^i$, $(\hat{p}_1)_j^i$ be the solution of system (3.2, 3.3) on the even grid satisfying homogeneous initial conditions

$$(3.24) \quad (\hat{w}_1)_j^i = (\hat{p}_1)_j^i = 0 \quad (i < 0 \ , \quad i+j \text{ even}) \ ,$$

and the ramp pressure boundary conditions

$$(3.25) \quad (\hat{p}_1)_0^{2k} = \begin{cases} k & (k \geq 0) \ , \\ 0 & (k < 0) \ . \end{cases}$$

This solution is the time sum of the solution $(\tilde{w}_1)_j^i$, $(\tilde{p}_1)_j^i$. The <u>discrete ramp response</u> is

$$(3.26) \quad (\hat{w}_1)_0^{2k+2} = \sum_{\ell \leq k} g_\ell = \sum_{\ell \leq k} (k + 1 - \ell) h_\ell \equiv b_k \ .$$

Both $(\hat{w}_1)_j^i$, $(\hat{p}_1)_j^i$ are zero for $i \leq j$, and $(\hat{w}_1)_j^{j+2}$, $(\hat{p}_1)_j^{j+2}$ are both nonzero for $0 \leq j \leq n$. Equation (3.3) with $i = j+2$ implies

$$(3.27) \quad \zeta_j = (\hat{p}_1)_j^{j+2} / (\hat{w}_1)_j^{j+2} \ .$$

Suppose b_k $(0 \leq k \leq n)$ is known. Then $(\hat{w}_1)_0^{2k+2} = b_k$ and $(\hat{p}_1)_0^{j+2} = k + 1$ are known for $0 \leq k \leq n$. For $j = 0,1,2,\ldots,n$ do the following: (i) obtain ζ_j from (3.27), and then (ii) advance $\hat{w}_1$ and $\hat{p}_1$ from j to $j+1$ (within the triangle $j+2 \leq i \leq 2n+2-j$) using system (3.4, 3.5).

Note from (3.22) and (3.26) that

$$(3.28) \quad h_k = g_k - g_{k-1} \ ,$$

$$(3.29) \quad g_k = b_k - b_{k-1} \ ,$$

$$(3.30) \quad h_k = b_k - 2\, b_{k-1} + b_{k-2} \ .$$

Since h_k, g_k, and b_k are all zero for $k < 0$, knowing b_k $(0 \leq k \leq n)$ is equivalent to knowing g_k $(0 \leq k \leq n)$, which in turn is equivalent to knowing h_k $(0 \leq k \leq n)$.

Observe also that with the normalizations chosen in this section,

$$(3.31) \quad (w_1)_j^i \simeq 2\Delta w_1(j\Delta, i\Delta) ,$$

$$(3.32) \quad (\tilde{w}_1)_j^i \simeq \tilde{w}_1(j\Delta, i\Delta) ,$$

$$(3.33) \quad (\hat{w}_1)_j^i \simeq \frac{1}{2\Delta} \hat{w}_1(j\Delta, i\Delta) ,$$

(where $\simeq$ means "formally approximates"), and similarly for p_1 , $\tilde{p}_1$, and $\hat{p}_1$. In particular,

(3.34)

$$h_k \simeq 2\Delta h(2k\Delta) , \quad g_k \simeq g((2k+1)\Delta) , \quad b_k \simeq \frac{1}{2\Delta} b((2k+2)\Delta) .$$

4. Choice of Sampling. If equations (3.23) and (2.23) are compared, it appears that (3.23) is only first order accurate when we view the discrete system as an approximation to the continuum system. We show in this section that actually (3.23) and (3.27) are formally second order accurate, and this observation guides our choice of sampling.

Let $w(x,t)$, $p(x,t)$ be a causal solution of system (2.2) generated by the pressure boundary conditions

$$(4.1) \quad p(0,t) = f_0(t) \equiv t^{\ell} f(t) H(t) ,$$

where $\ell \geq 0$ is an integer, $f(t)$ is smooth with $f(0) \neq 0$, and $H(t)$ is the unit step function defined after (2.21). Then $f_0(t)$ has $\ell - 1$ continuous derivatives, and the ℓ'th derivative of $f_0(t)$ has a jump discontinuity at $t = 0$. Define

(4.2)

$$f_k(t) = \int_{-\infty}^{t} f_{k-1}(\tau) d\tau = \int_{-\infty}^{t} \frac{1}{(k-1)!} (t-\tau)^{k-1} f_0(\tau) d\tau \quad (k \geq 1) .$$

Then $f_k(t) = 0$ for $t < 0$, and for t near zero,

$$(4.3) \quad f_k(t) = O(t^{k+\ell}) .$$

Hence for $t > 0$ near zero,

$$(4.4) \quad f_k(t)/f_0(t) = 0(t^k) .$$

The method of geometrical acoustics yields an asymptotic expansion in smoothness near the line $t = x$ (except near the lines $x = x_\nu$, $1 \le \nu$)

$$(4.5) \quad \begin{pmatrix} w(x,t) \\ \\ p(x,t) \end{pmatrix} \sim \sum_{k \ge 0} f_k(t-x) \begin{pmatrix} w^k(x) \\ \\ p^k(x) \end{pmatrix} ;$$

see [4] for further details. The functions $w^k(x)$, $p^k(x)$ are smooth except for possible jump discontinuities at the points $x = x_\nu$, $1 \le \nu$. It can be shown [4] that $p^0(x)$ and $w^0(x)$ are bounded away from 0 for $0 \le x \le X$, and

$$(4.6) \quad p^0(x)/w^0(x) = \zeta(x) \qquad (x \ne x_\nu , \quad \nu \ge 1) ,$$

$$(4.7) \quad \frac{w^0(x)p^1(x) - p^0(x)w^1(x)}{(w^0(x))^2} = \frac{1}{2} \zeta'(x) \qquad (x \ne x_\nu , \quad \nu \ge 1) .$$

Suppose we sample $w(0,t)$ and $p(0,t) = f_0(t)$ with sampling interval 2Δ and time shift σ where

$$(4.8) \quad 0 \le \sigma = 0(\Delta) ,$$

i.e., we measure $w(0,t)$, $p(0,t)$ at $t = \sigma + 2k\Delta$ $(0 \le k \le n)$. Consider applying the downward continuation algorithm on the shifted grid $x = j\Delta$, $t = \sigma + i\Delta$ (i,j integers with $i+j$ even), treating w , p as zero for $i < j$. The difference scheme (3.4-5) on this new grid is still second order accurate. We want to choose σ so that equation (3.19) for recovering ζ_j on this new grid is at least formally second order accurate, i.e., so that

$$(4.9) \quad \frac{p(x,x+\sigma)}{w(x,x+\sigma)} = \zeta(x + \frac{1}{2}\Delta) + 0(\Delta^2)$$

away from $x = x_\nu$, $\nu \ge 1$. Clearly if $\ell > 0$, $w(x,x) = p(x,x) = 0$, so any such σ must be positive.

By (4.4-5), (4.8), for sufficiently small Δ ,

$$(4.10)\qquad \frac{1}{f_0(\sigma)}\left\{\begin{bmatrix} w(x,x+\sigma) \\ p(x,x+\sigma)\end{bmatrix} - f_0(\sigma)\begin{bmatrix} w^0(x) \\ p^0(x)\end{bmatrix} - f_1(\sigma)\begin{bmatrix} w^1(x) \\ p^1(x)\end{bmatrix}\right\} = O(\Delta^2)$$

Defining

$$(4.11)\qquad \delta = f_1(\sigma)/f_0(\sigma)\ ,$$

which is $O(\Delta)$ by (4.4, 4.8), we obtain

$$\frac{p(x,x+\sigma)}{w(x,x+\sigma)} = \frac{p^0(x)+\delta p^1(x)}{w^0(x)+\delta w^1(x)} + O(\Delta^2) = \frac{p^0}{w^0}\ \frac{w^0p^1-p^0w^1}{(w^0)^2}\ \delta + O(\Delta^2)$$

$$= \zeta(x) + \frac{\delta}{2}\zeta'(x) + O(\Delta^2) = \zeta(x+\tfrac{1}{2}\delta) + O(\Delta^2)\ .$$

So to get (4.9), we choose σ so that $\delta = \Delta$.

For example, suppose $f_0(t) = t^\ell H(t)$ where $\ell \geq 0$ is an integer. Then $f_1(t) = t^{\ell+1}H(t)/(\ell+1)$, so $f_1(\sigma) = \sigma/(\ell+1)$ for $\sigma > 0$. We choose

$$(4.12)\qquad \sigma = (\ell + 1)\Delta\ .$$

In particular, for $\ell = 0$, this leads us to sample the step response by setting

$$(4.13)\qquad g_k = g((2k+1)\Delta) \qquad (0 \leq k \leq n)\ .$$

Then when we apply the downward continuation algorithm to the step response as described in §3, both the difference scheme (3.4-5) and the recovery of the coefficient (3.23) are formally second order accurate. In the following sections, when we say <u>the step response is sampled</u>, we mean that g_k $(0 \leq k \leq n)$ is obtained by sampling $g(t)$ as in (4.13); then h_k $(0 \leq k \leq n)$ is obtained by differencing g_k as in (3.28). This choice of sampling $h(t)$ is equivalent to equation (1.1).

For $\ell = 1$, we are led to sample the ramp response by setting

(4.14) $b_k = \frac{1}{2\Delta} b((2k+2)\Delta) \qquad (0 \le k \le n)$.

Then when we apply the downward continuation algorithm to the ramp response as described in §3, both the difference scheme (3.4-5) and the recovery of the coefficient (3.27) are formally second order accurate. In the following sections, when we say the ramp response is sampled, we mean that b_k $(0 \le k \le n)$ is obtained by sampling $b(t)$ as in (4.14); then h_k $(0 \le k \le n)$ is obtained by differencing b_k as in (3.30). This choice of sampling is equivalent to equation (1.2).

5. Solvability of the Discrete Problem Using Sampled Data. In this section, we shall see that the discrete impulse response, obtained either from sampling the step response using (4.13) or the ramp response using (4.14), has the positive definite property necessary to solve the discrete inverse problem. If the step response is sampled, this is true for sufficiently small Δ whenever the discontinuities of $\zeta(x)$ occur at integer multiples of Δ (Corollary 1). If the ramp response is sampled, we this is true for all Δ and for arbitrary locations of the discontinuities of $\zeta(x)$ (Theorem 2). See Bube [2,3] for proofs.

For $\underset{\sim}{y} = [y_0, y_1, \ldots, y_n]^T \in \mathbb{R}^{n+1}$, define its norm

$$||\underset{\sim}{y}||_2 = (|y_0|^2 + |y_1|^2 + \ldots + |y_n|^2)^{1/2} .$$

For real symmetric $(n+1) \times (n+1)$ matrices A and B, we write $A \le B$ to mean $\underset{\sim}{y}^T A \underset{\sim}{y} \le \underset{\sim}{y}^T B \underset{\sim}{y}$ for all $\underset{\sim}{y} \in \mathbb{R}^{n+1}$. Define the matrix 2-norm

$$||A||_2 = \max_{\underset{\sim}{y} \ne 0} ||A\underset{\sim}{y}||_2 / ||\underset{\sim}{y}||_2 .$$

In what follows, all of the variables in the discrete system depend on Δ ; we do not include this dependence explicitly in our notation. We assume that the upper bound n on k and j in the discrete system satisfies

(5.1) $n = [X/\Delta] - 1$,

where $[X/\Delta]$ is the greatest integer less than or equal to X/Δ. Then all the grid points used for the discrete equations (3.15-16) are in the region $|t| < x \leq X$. Also,

$$(5.2) \quad n = O(\Delta^{-1}) .$$

Theorem 1 Suppose the step response is sampled, and all discontinuities of $\zeta(x)$ for $0 < x \leq X$ occur at integer multiples of some positive number d. Then there exists a constant c_2 such that

$$(5.3) \quad (c_0 - c_2\Delta)I \leq R_j \leq (c_1 + c_2\Delta)I$$

whenever $(j+1)\Delta \leq X$ and Δ divides d.

Corollary 1 Suppose the step response is sampled, and all discontinuities of $\zeta(x)$ for $0 \leq x \leq X$ occur at integer multiples of some positive number d. Then there are positive constants Δ_0, c_0^*, and c_1^* so that whenever $0 < \Delta \leq \Delta_0$, Δ divides d, and $(j+1)\Delta \leq X$,

$$(5.4) \quad c_0^* I \leq R_j \leq c_1^* I ,$$

and hence

$$(5.5) \quad ||R_j||_2 \leq c_1^* , \quad ||R_j^{-1}||_2 \leq 1/c_0^* ;$$

thus the discrete inverse problem can be solved, yielding a positive discrete impedance profile.

Theorem 2 Suppose the ramp response is sampled. Then

$$(5.6) \quad c_0 I \leq R_j \leq c_1 I$$

whenever $(j+1)\Delta \leq X$, and hence

$$(5.7) \quad ||R_j||_2 \leq c_1 , \quad ||R_j^{-1}||_2 \leq 1/c_0 ;$$

thus the discrete inverse problem can be solved, yielding a positive discrete impedance profile.

6. Convergence. In this section we present the convergence theorems (Theorems 3 and 4). See Bube [2,3] for proofs. We assume throughout this section that all discontinuities of $\zeta(x)$ for $0 \leq x \leq X$ occur at integer multiples of some fixed positive number d, and that Δ divides d. We assume that n and X are related by (5.1).

Theorem 3 (Convergence when the step response is sampled.) Suppose $\zeta(x)$ is positive and piecewise smooth for $0 \leq x \leq X$

and that all its discontinuities for $0 < x \leq X$ occur at integer multiples of d. Suppose Δ divides d, and

$$(1.1) \quad h_k = \int_{-\Delta}^{\Delta} h(2k\Delta + \tau)d\tau$$

is measured for $0 \leq k \leq X/\Delta - 1$, where $h(t)$ is the impulse response for $\zeta(x)$. If $0 < \Delta \leq \Delta_0$ (where Δ_0 is given by Corollary 1), then the discrete inverse problem can be solved, yielding a positive discrete impedance profile ζ_j $(0 \leq j \leq X/\Delta - 1)$ which satisfies

$$(6.1) \quad \zeta_j = \zeta((j + \tfrac{1}{2})\Delta) + O(\Delta^2) ,$$

uniformly in such j and Δ.

<u>Theorem 4</u> (Convergence when the ramp response is sampled.) Suppose $\zeta(x)$ is positive and piecewise smooth for $0 \leq x \leq X$ and that all its discontinuities for $0 < x \leq X$ occur at integer multiples of d. Suppose Δ divides d, and

$$(1.2) \quad h_k = \int_{-2\Delta}^{2\Delta} \left(1 - \frac{|\tau|}{2}\right) h(2k\Delta + \tau)d\tau$$

is measured for $0 \leq k \leq X/\Delta - 1$, where $h(t)$ is the impulse response for $\zeta(x)$. Then the discrete inverse problem can be solved, yielding a positive discrete impedance profile ζ_j $(0 \leq j \leq X/\Delta - 1)$ which satisfies

$$(6.2) \quad \zeta_j = \zeta((j + \tfrac{1}{2})\Delta) + O(\Delta^2) ,$$

uniformly in such j and Δ.

<u>Corollary 2</u> If h_k is measured using equation (1.1) [equivalently equations (4.13) and (3.28)--sampling the step response] or using equation (1.2) [equivalently equation (4.14) and (3.30)--sampling the ramp response], and the discrete Gopinath-Sondhi integral equation method outlined in equations (3.15) and (3.18) is used to solve the discrete inverse problem, then the method is second order convergent (in the sense of equation (6.1)).

Corollary 3 If h_k is measured as in Corollary 2 and the downward continuation algorithm (applied to the discrete impulse response) described after equation (3.19) is used to solve the discrete inverse problem, then the method is second order convergent (in the sense of equation (6.1)).

Corollary 4 If the step response $g(t)$ is sampled as in equation (4.13) and the downward continuation algorithm (applied to the step response) described after equation (3.23) is used to solve the discrete inverse problem, then the method is second order convergent (in the sense of equation (6.1)).

Corollary 5 If the ramp response $b(t)$ is sampled as in equation (4.14) and the downward continuation algorithm (applied to the ramp response) described after equation (3.27) is used to solve the discrete inverse problem, then the method is second order convergent (in the sense of equation (6.1)).

If the sampled step or ramp response has errors, both the positive definiteness results of §5 and the convergence results of this section remain valid if the errors are small enough.

Theorem 5 Theorems 3 and 4 and their Corollaries remain valid if (1.1) or (1.2) are not satisfied exactly, but with errors e_k in h_k $(0 \leq k \leq n)$, provided that

$$(6.3) \quad e \equiv \sum_{k=0}^{n} |e_k| = O(\Delta^2) \ ,$$

and $e < c_0^*$ when the step response is sampled, and $e < c_0$ when the ramp response is sampled.

In particular, if

$$(6.4) \quad g_k = g((2k+1)\Delta) + O(\Delta^3)$$

or

$$(6.5) \quad b_k = \frac{1}{2\Delta} (b((2k+2)\Delta) + O(\Delta^4)) \ ,$$

the results remain valid. Because h_k is obtained from g_k and b_k by differencing, errors with lower powers of Δ but some degree of smoothness are also allowable. For example, if

$$(6.6) \quad g_k = g((2k+1)\Delta) + \Delta^2 f((2k+1)\Delta) + O(\Delta^3)$$

where $f(t)$ is piecewise Lipschitz continuous, the results also remain valid.

7. Discontinuities at Noninteger Multiples of the Mesh Width. In this section we present an example which shows that if the discontinuities of the impedance $\zeta(x)$ are not restricted to being integer multiples of the mesh width Δ, then convergence need not be uniform in x as Δ tends to zero.

Let

$$(7.1) \quad \zeta(x) = \begin{cases} 1 & \text{for } 0 \le x < 1 , \\ 2 & \text{for } x > 1 , \end{cases}$$

and $X = 2$. Then in the interval $0 \le t \le 2X = 4$,

$$(7.2) \quad h(t) = \delta(t) - \frac{2}{3}\,\delta(t-2) + \frac{2}{9}\,\delta(t-4) ,$$

$$(7.3) \quad g(t) = H(t) - \frac{2}{3}\,H(t-2) + \frac{2}{9}\,H(t-4) ,$$

$$(7.4) \quad b(t) = tH(t) - \frac{2}{3}\,(t-2)H(t-2) + \frac{2}{9}\,(t-4)H(t-4) .$$

Suppose $\Delta = (m + \frac{1}{2})^{-1}$ for some positive integer m. Then by (5.1), $n = 2m$. If we sample the step response using equation (4.13), $g(t)$ has a jump discontinuity at $t = 2 = (2m+1)\Delta$; suppose we take g_m to be $-1/3$, the average of the right and left hand limits of $g(t)$ at $t = 2$. Using (3.28), we obtain

$$(7.5) \quad \begin{aligned} h_0 &= 1 , \\ h_k &= 0 && \text{for } 1 \le k \le m-1 , \\ h_m &= h_{m+1} = -\frac{1}{3} , \\ h_k &= 0 && \text{for } m+2 \le k \le 2m = n . \end{aligned}$$

If we sample the ramp response using (4.14) and then apply (3.30), we get the same h_k's. The solution of the discrete inverse problem is

$$
\begin{aligned}
&\zeta_k = 1 && \text{for } 0 \le k \le m-1 , \\
&\zeta_m = 1.4 , \\
&\zeta_{m+1} = 287/145 \simeq 1.9793 , \\
&\zeta_{m+2} = 9799/4901 \simeq 1.9994 ,
\end{aligned}
\tag{7.6}
$$

and for $3 \le j \le m$, ζ_{m+j} approaches 2 geometrically as j increases. Note that ζ_m approximates $\zeta(x)$ right at the discontinuity $x = 1 = (m + \frac{1}{2})\Delta$.

Suppose we now increase m, and thereby decrease $\Delta = (m + \frac{1}{2})^{-1}$. For $j \ge 0$, ζ_{m+j} is independent of m (i.e., if we increase m, ζ_m for the new Δ is still 1.4, ζ_{m+1} is still 287/145, etc.). If we let $m \to \infty$, then $\Delta \to 0$, but for each fixed m and $\Delta = (m + \frac{1}{2})^{-1}$,

$$
\max_{\substack{0 \le k \le 2m \\ k \ne m}} \left| \zeta_k - \zeta\left(\left(k + \frac{1}{2}\right)\Delta\right) \right| = 3/145 \simeq .0207 ,
\tag{7.7}
$$

where we have ignored the value of ζ_m right at the discontinuity. The difficulty is, however, localized near the discontinuity. For each fixed $\varepsilon > 0$, if we take the max in (7.7) only over all k's for which $1 + \varepsilon \le (k + \frac{1}{2})\Delta < 2$, then the max does tend to zero as $m \to \infty$ and $\Delta = (m + \frac{1}{2})^{-1} \to 0$. Thus we do get convergence in weaker norms. For example, if $1 \le p < \infty$, the appropriate discrete L^p norm

$$
\left[\Delta \sum_{\substack{0 \le k \le 2m \\ k \ne m}} \left| \zeta_k - \zeta\left(\left(k + \frac{1}{2}\right)\Delta\right) \right|^p \right]^{1/p}
\tag{7.8}
$$

of the error tends to zero, but (7.8) is only $O(\Delta^{1/p})$. If we extend X beyond 2, more complicated problems of a similar nature occur near values of x corresponding to values of t where multiples appear in the impulse response.

These difficulties do not appear just because we are sampling g(t) right at a discontinuity. Suppose $\Delta = (m + \beta)^{-1}$ for some fixed β with $0 < \beta < 1$. If $\beta \neq \frac{1}{2}$ and we sample the step response, we recover the discontinuity correctly, but troubles will appear later from the multiples. If we sample the ramp response, we get

$$(7.9) \quad h_m = \frac{2}{3}(1 - \beta), \quad h_{m+1} = -\frac{2}{3}\beta$$

and thus

$$(7.10) \quad \zeta_m = \frac{4-\beta}{2+\beta}, \quad \zeta_{m+1} = \frac{32+12\beta-9\beta^2+\beta^3}{16+6\beta-3\beta^2-\beta^3}.$$

Simple calculus shows that we get the most error in ζ_{m+1} when $\beta \simeq .664$.

Observe also that when $\beta = \frac{1}{2}$ (so ζ_m approximates $\zeta(x)$ right at the discontinuity), $\zeta_m = 1.4$, not the average 1.5 or 1 and 2, which we might expect from the literature [1]. In general, if

$$(7.11) \quad \zeta(x) = \begin{cases} \zeta_- & \text{for } 0 \le x < 1, \\ \zeta_+ & \text{for } x > 1, \end{cases}$$

and $\beta = \frac{1}{2}$, then

$$(7.12) \quad \frac{\zeta_m - \zeta_-}{\zeta_+ - \zeta_-} = \frac{2}{3 + \dfrac{\zeta_+}{\zeta_-}},$$

which ranges from near zero if $\zeta_- \ll \zeta_+$ to near $\frac{2}{3}$ if $\zeta_- \gg \zeta_+$.

These examples illustrates the more complicated behavior of the convergence of these discrete methods when discontinuties of $\zeta(x)$ occur at noninteger multiples of Δ.

References

[1] J.G. BERRYMAN and R.R. GREENE, Discrete inverse methods for elastic waves, Geophys., 45 (1980), pp. 213-233.

[2] K.P. BUBE, Convergence of difference methods for one-dimensional inverse problems, to appear in IEEE Trans. Geosci. Rem. Sens. (1984).

[3] K.P. BUBE, Convergence of numerical inversion methods for discontinuous impedance profiles, submitted to SIAM J. Numer. Anal. (1984).

[4] K.P. BUBE and R. BURRIDGE, The one-dimensional inverse problem of reflection seismology, SIAM Rev., 25 (1983), pp. 497-559.

[5] K.P. BUBE and R. BURRIDGE, Difference methods for solving one-dimensional inverse problems, in IFAC Identification and System Parameter Estimation 1982, ed. G.A. Bekey and G.N. Savidis, Pergamon Press, Oxford, 1983, pp. 957-961.

[6] R. BURRIDGE, The Gelfand-Levitan, the Marchenko, and the Gopinath-Sondhi integral equations of inverse scattering theory, regarded in the context of inverse impulse response problems, Wave Motion, 2 (1980), pp. 305-323.

[7] J.F. CLAERBOUT, Fundamentals of Geophysical Data Processing, McGraw-Hill, New York, 1976.

[8] R. COURANT and D. HILBERT, Methods of Mathematical Physics, Vol. II, John Wiley, New York, 1962.

[9] M. GERVER, Inverse problems for the one-dimensional wave equation, Geophys. J. Roy. Astr. Soc., 21 (1970), pp. 337-357.

[10] B. GOPINATH and M.M. SONDHI, Inversion of the telegraph equation and the synthesis of nonuniform lines, Proc. IEEE, 59 (1971), pp. 383-392.

[11] P. GOUPILLAUD, An approach to inverse filtering of near-surface layer effects from seismic records, Geophys., 26 (1961), pp. 754-760.

[12] R.J. KRUEGER, Inverse problems for non-absorbing media with discontinuous material properties, J. Math. Phys., 23 (1982), pp. 396-404.

[13] F. SANTOSA and H. SCHWETLICK, The inversion of acoustical impedance profile by methods of characteristics, Wave Motion, 4 (1982), pp. 99-110.

[14] M.M. SONDHI and B. GOPINATH. Determination of vocal-tract shape from impulse response at the lips, J. Acoust. Soc. Amer., 49 (1971), pp. 1867-1873.

[15] M.M. SONDHI and J.R. RESNICK, The inverse problem for vocal tract: numerical methods, acoustical experiments, and speech synthesis, J. Acoust. Soc. Amer., 73 (1983), pp. 985-1002.

[16] W.W. SYMES, Stable solution of the inverse reflection problem for a smoothly stratified elastic medium, SIAM J. Math. Anal., 12 (1981), pp. 421-453.

[17] W.W. SYMES, Impedance profile inversion via the first transport equation, J. Math. Anal. Appl., 94 (1983), pp. 435-453.

[18] J.A. WARE and K. AKI, Continuous and discrete inverse-scattering problems in a stratified elastic medium, I. Plane waves at normal incidence, J. Acoust. Soc. Amer., 45 (1969), pp. 911-921.

INVERSION OF BAND LIMITED REFLECTION SEISMOGRAMS

D. W. OLDENBURG*

Abstract. This paper examines the problem of recovering the acoustic impedance from band-limited normal incidence reflection seismograms. The convolutional model for the seismogram is adopted at the outset and it is therefore required that initial processing has removed multiples and recovered true amplitudes as well as possible. Recognizing the inherent non-uniqueness in the inversion, we proceed by constructing an impedance model which satisfies the processed seismogram, has a minimum of structural variation, honors any point velocity constraints that are provided, and incorporates information from stacking velocities. The constructed impedance is consistent with all available geological and geophysical information and therefore constitutes our best estimate of the true earth impedance.

1. Introduction. In this paper I would like to describe a method by which reflection seismograms may be inverted to produce an estimate of the acoustic impedance as a function of depth. At the outset, I want to mention that the work in this paper is the result of collaboration with a number of individuals, particularly S. Levy, K. Stinson and T. Scheuer. Much of the material presented here is extracted from papers written with these colleagues (Oldenburg, Scheuer, and Levy (1983); Oldenburg, Levy, and Stinson (1984)).

Before I outline an attack on the inversion problem I would like to briefly describe some aspects of the data. I think it is valuable to keep in mind how the data are generated and processed, for an understanding of this part of the problem can influence the method by which the inversion is carried out.

A typical seismic experiment consists of setting off an explosion (or using a Vibroseis source) and recording seismic energy reflected from the subsurface with an array of geophones. Sources and receivers are moved systematically so that multi-fold coverage of subsurface is obtained. The first portion of the traces are usually muted to remove the effects of surface waves. The data are amplitude corrected to

*University of British Columbia, Vancouver, B.C., CANADA V6T 1W5

counteract (approximately) the effects of spherical divergence and attenuation. Preliminary static corrections are applied to remove the effects of near surface geology, and then, in order to enhance the signal to noise ratio, Normal Moveout Corrections (NMO) are applied to common midpoint gathers. The data are stacked and subjected to predictive deconvolution in an effort to remove the effects of multiples (this may also be applied before stack). The output may then undergo additional processing (e.g. noise filtering, velocity filtering, residual statics correction, deconvolution, and migration). The final output is a CDP stacked section in which each trace is assumed to be a zero phase bandpassed version of the true reflectivity function.

The numerous steps and assumptions involved in the standard processing of seismic data means that the data can only be regarded as an approximation to the bandpassed reflectivity function. Any procedure which attempts to invert these data must necessarily make allowances for data inaccuracies.

2. Mathematical Problem. Let $\rho(z)$ and $v(z)$ be the density and velocity of a one-dimensional isotropic earth. Let $u(z,t)$ be the vertical displacement where z is the vertical spatial coordinate measured positive down and t is the time. The initial differential equation is

$$\partial_{tt}u = \frac{1}{\rho}\,\partial_z(\rho v^2 \partial_z u) \tag{1}$$

It is supposed that an initial plane wave pressure pulse is input into the system and a reflection response (displacement velocity or its Fourier transform) is measured at the earth's surface. The goal of the inverse problem is to use this measured response to recover the impedance $\xi = \rho v$ as a function of depth or travel time.

There have been many different methods proposed to solve this problem. It is not the intent of this paper to summarize those methods nor to list all of the excellent research carried out in this regard. Nevertheless, some fruitful approaches to the problem include: (1) transforming the wave equation into the Schrodinger equation and solving for the scattering potential using the methods of Gel'fand and Levitan or Marchenko (e.g. Weidelt (1972), Newton (1980), Symes (1981), Santosa (1982), Santosa and Schwetlick (1982), and references therein); (2) transforming the wave equation to a Riccati equation and solving for the potential (Brekhovskikh (1960), Gjevik et al. (1976), Nilsen and Gjevik (1978)); (3) use of time domain methods (Kay (1960), Weston (1972), Sondhi and Gopinath (1971)); (4) use of discretized methods with layers of equal travel time (Claerbout (1968), Ware and Aki (1969), Hubral (1978), Berryman and Greene (1981)); (5) using only first order differential equations of motion and a vector formulation of the problem to give a vector Marchenko equation (Howard, 1983). There exist other scattering formulations and some of these are listed in the excellent review article by Newton (1981). In addition, there are different formulations of the inverse problem (e.g. Bamberger et al. (1982), Sabatier (1978), Barcilon (1974)).

There are, evidently, many ways of approaching the inverse problem. Unfortunately, all of these methods suffer from a severe drawback when applied to field data. To illustrate, it suffices to consider an approximate solution to the acoustic impedance inversion. To first order, we may write

$$\xi(t) = \xi(0)\exp[2 \int_0^t r(u)du] \tag{2}$$

This expression can be arrived at the beginning with discrete reflection coefficients in a layered earth (Waters, 1978) or by considering the first term in a solution based upon solving the Marchenko equation by successive approximations (Howard, 1983).

The ideal seismogram is a zero phase bandpassed version of the reflectivity function and may be written as

$$\langle r(t)\rangle = r(t) \otimes a(t) \tag{3}$$

where a(t) is a zero phase averaging function whose spectrum is unity within the information band of the seismic experiment. Let

$$\eta(t) = \ln\left(\frac{\xi(t)}{\xi(0)}\right) \tag{4}$$

Substituting the processed seismogram (equation (3)) into (2) we obtain an approximate solution

$$\tilde{\eta}(t) = 2 \int_0^t \langle r(u)\rangle du \tag{5}$$

It is however straight forward to show that

$$\tilde{\eta}(t) \simeq \eta(t) \otimes a(t) \tag{6}$$

and thus the logarithm of the normalized acoustic impedance estimated from the data will also be bandlimited. It follows that the recovered impedance, obtained by exponentiating (6) will also be missing low frequencies. The deleterious effect of missing frequency information is illustrated in Fig. 1.

The true acoustic impedance and reflectivity are shown in Figs. (1a) and (1b). Conventional seismic processing will recover a seismogram shown in Fig. (1e). Inserting this result into equation (5) and exponentiating produces the approximate acoustic impedance shown in Fig. (1g).

Other problems that arise when attempting to invert the seismogram using an inverse scattering approach include the difficulty of incorporating data inaccuracies and an inability to include addition information about the impedance. We shall therefore attempt to solve the inverse problem in a different manner.

3. Construction. In order to develop an alternate solution I shall begin by assuming that the processed seismogram is a reasonable approximation to a bandlimited reflectivity function. The discretized form of equation (3) may be written as

$$\langle r(t_j)\rangle = \langle r_j\rangle = \int r(t)a(t_j-t)dt = \int r(t)g_j(t)dt \quad j=1,N \tag{7}$$

where $g_j(t)$ is $a(t_j-t)$, that is a time displaced version of the averaging function $a(t)$. Equation (7) is a Fredholm equation of the first kind and there exist many approaches by which it can be solved. In this paper we shall be interested only in model construction, that is, given the (inaccurate) data $\langle r_j\rangle$, and the averaging function $a(t)$, we want to compute a model $r(t)$ which reproduces those data to the degree that is consistent with their errors. The primary obstacle for this approach is the nonuniqueness of the solution. There are infinitely many acceptable models and they will not necessarily be similar. This is most easily seen by writing the Fourier transform of the seismogram as $R_{<>}(f) = R(f)A(f)$. Clearly, $R(f)$ can take on any value at those frequencies where $A(f)$ is zero, and hence greatly different time functions $r(t)$ can give rise to the same observed seismogram.

A reduction in nonuniqueness, and the attainment of a geophysically realistic model, may be achieved by taking the following steps:

(1) Find an acoustic impedance which has the minimum amount of variation.

(2) Specify the functional form of $r(t)$. Here we shall specify that the constructed $r(t)$ is that corresponding to a layered earth.

(3) Input geologic information in the form of point velocity (or impedance) constraints.

(4) Input geophysical information that is obtained from RMS velocities.

(5) Solve (1) as a single linear inverse problem after specifying (2) and including information from (3) and (4) as available. In this solution it is important to make allowance for errors in the seismogram as well as uncertainties in the point velocity constraints and in the RMS velocity information.

I would now like to consider in turn each of the steps (1)-(4)

1. A Minimum Structure Model. The acoustic impedance we are

attempting to find is a function in a Hilbert space and it therefore has infinitely many degrees of freedom. Because only a finite number of constraints upon this function are available, the approach taken in model construction is to minimize a functional norm of the model subject to the data constraints. The type of model constructed will depend strongly upon the norm used and hence selection of the appropriate norm for each problem is crucial. We first ask the question of what type of functions $\xi(t)$ are we trying to find?

Earth impedance functions generally increase with depth but they also exhibit high frequency variations which are of geologic importance. The manner of change in the impedance may be gradual (for instance the smooth increase resulting from compaction of homogeneous material) or it may proceed in discrete jumps resulting from the juxtaposition of layers having different composition. We realize at the outset that impedances having arbitrarily large variations can be constructed. Consider for instance the seismic data response due to a "thin bed". Such a structure produces two closely spaced reflectivity coefficients of opposite polarity and equal amplitude. The convolution of the averaging function with such a "dipole" yields an arbitrarily small output if the bed is sufficiently thin. The dipole is thus an annihilator for the problem and any number of thin beds can be added to the acoustic impedance without significantly affecting the fit to the data.

It may be a matter of personal philosophy but I think it is generally good practice to find that model which exhibits the least amount of structural variation and still satisfies the data constraints. The earth impedance $\xi(t)$ would have at least as much structural variation as that on the constructed impedance $\xi_c(t)$, but features observed on $\xi_c(t)$ have a good probability of also being on $\xi(t)$. If it is structural variation (for instance, particular low impedance zones) which are of importance, then this philosophy yields a conservative approach; $\xi_c(t)$ will not include structural variation that is not demanded by the data.

A minimum structure impedance can be found by minimizing the appropriate norm of the model. For instance, minimizing

$$\phi_1 = \int \left|\frac{d\xi(t)}{dt}\right| dt \tag{8}$$

$$\phi_2 = \int \left|\frac{d\eta(t)}{dt}\right| dt \tag{9}$$

subject to the data constraints would suffice. We could also use the ℓ_2 norm but we shall use the ℓ_1 norm here because it will be convenient to solve our problem using linear programming techniques.

Also, we shall use only the norm given in equation (9) because it leads to particularly simple results. Using the definition of $\eta(t)$

we obtain

$$d\eta(t)/dt = 2\ r(t) \tag{10}$$

Therefore minimizing ϕ_2 is equivalent to minimizing

$$\phi = \int |r(t)|\, dt \tag{11}$$

2. <u>The Functional Form of r(t)</u>. A common assumption for geophysicists is that the earth is adequately represented by a sequence of homogeneous layers. With that hypothesis, the reflectivity function must have the mathematical form

$$r(t) = \sum_j r_j\ \delta(t-\tau_j) \tag{12}$$

In (12) r_j is the reflection coefficient at the bottom of the j'th layer and r_j is the two-way travel time to that layer.

The importance of this equation is three-fold. First, it reduces the nonuniqueness because there exist many acceptable functions r(t) which cannot be written in this form. Second, equation (12) is a sum of isolated delta functions and therefore it will have energy at frequencies outside the information band. Third, this mathematical form of r(t) conforms, at least sometimes, to physical reality. This means that whenever the earth really is a layered sequence, and we have found an r(t) in the form of (12) which reproduces the data, there is some hope that the true earth reflectivity function has been discovered. The difficulty with writing (12) is that τ_j are not known a priori and direct substitution of (12) into (11) would yield a nonlinear problem to be solved for r_j and τ_j. Making the problem nonlinear is highly undesirable. We overcome this by permitting a reflection coefficient at each possible time and therefore write

$$r(t) = \sum_{n=0}^{N-1} r_n \delta(t-n\Delta) \tag{13}$$

where N is the number of data and Δ is the digitization interval. Substitution of (13) into equation (11) yields

$$\phi = \sum_{n=0}^{N-1} |r_n| \tag{14}$$

as an objective function to be minimized. When carried out, this minimization yields a reflectivity function having the fewest number of nonzero reflection coefficients, or equivalently, an earth impedance having the fewest number of distinct layers.

The minimization of equation (14) is precisely the problem dealt with in the excellent paper by Levy and Fullagar (1981). Taking the digital Fourier transform of equation (13) and separating into real and imaginary parts we have

$$\operatorname{Re}\{R_j\} = \sum_n r_n \cos(2\pi jn/N)$$

and (15)

$$\operatorname{Im}\{R_j\} = \sum_n r_n \sin(2\pi jn/N)$$

These equations provide a linear relationship between the Fourier transform of the reflectivity function and the reflection coefficients. It follows that if the real and/or imaginary parts of R_j are known for some frequencies, then a linear programming (LP) algorithm can be used to minimize (14) subject to the constraint equations (15). Inserting the resultant $\{r_n\}$ into the acoustic impedance formula

$$\xi_{k+1} = \xi_k \left(\frac{1+r_k}{1-r_k} \right)$$

generates an impedance with minimum structure.

In the formulation presented here, it is assumed that A(f), the Fourier transform of the averaging function, has unit amplitude within the frequency band (f_L, f_H) and is also zero phase. Consequently, estimates of R_j are known for frequencies within the information band. Nevertheless, these will not be known precisely and so the constraint equations must be altered accordingly. The constraints supplied to the linear programming are modified in the usual manner to accommodate these errors.

3. Point Impedance Constraints. Thus far, we have attempted to reduce nonuniqueness in the model construction by requiring the reflectivity function to have a mathematical form consistent with a layered earth model and by confining our attention to impedance functions displaying a minimum of structure. But even with these restrictions, it is unlikely that the nonuniqueness has been completely conquered. Our confidence in the solution would be greatly enhanced if additional constraints could be included. Such constraints, which might be derivable from a nearby acoustic log, are readily incorporated into the linear programming construction.

We first substitute equation (13) into (2) to obtain

$$\frac{\xi(t)}{\xi(0)} = \exp\left[2 \sum_n r_n H(t-n\Delta)\right] \tag{16}$$

where H(t) is the Heaviside step function. Taking the logarithm we have

$$\eta(t) = \ln\left[\xi(t)/\xi(0)\right] = 2 \sum_n r_n H(t-n\Delta) \tag{17}$$

Equation (17) is the desired result. If $\eta(t)$ is known for any record time t, then the linear constraint to be added to the data constraints is

$$\eta(t) - \sigma_1(t) \leq 2 \sum_n r_n H(t-n\Delta) \leq \eta(t) + \sigma_2(t)$$

where $\sigma_1(t)$ and $\sigma_2(t)$ are error bounds for $\eta(t)$. These are readily derivable if we wish the constructed impedance at time t to lie within the bounds $\xi_1(t) \leq \xi(t) \leq \xi_2(t)$ and if bounded estimates of $\xi(0)$ are provided.

The importance of impedance constraints cannot be overemphasized. Ultimately, we wish to incorporate as many geologic constraints as possible in order to increase the probability that a good representation of the true earth has been constructed.

4. Constraints for RMS (Stacking) Velocities. A precursory step to stacking data traces is to perform velocity analysis and estimate stacking velocities for the NMO correction. When the earth consists of a set of uniform horizontal plane layers, the stacking velocities can be equated to RMS velocities. The basic equation relating the RMS velocity V(t) to the interval velocity v(t) is

$$V^2(t) = \frac{1}{t} \int_0^t v^2(u)du \tag{18}$$

When the RMS velocities are known only at discrete times t_j j=1,N, the appropriate equations are

$$V_j^2 = V^2(t_j) = \frac{1}{t_j} \int_0^{t_j} v^2(t)dt \qquad j=1,N \tag{19}$$

It is seen that the RMS velocity at time t_j contains information about the true velocity structure from the surface to a time t_j. The goal of this section is to manipulate this information into a form that can be incorporated as a constraint into the linear programming formulation of the acoustic impedance problem.

Let t_m be a maximum time of interest for the problem. The data equations (19) may be written as

$$e_j = \int_0^{t_m} v^2(t)G_j(t)dt \tag{20}$$

The only unique information obtainable from these data are the unique Backus-Gilbert averages of the form

$$\langle v^2(t_0)\rangle = \int_0^{t_m} v^2(t)A(t,t_0)dt \tag{21}$$

where $A(t,t_0)$ is the averaging function centered on t_0 having the form

$$A(t,t_0) = \Sigma\, a_j(t_0)G_j(t) \tag{22}$$

The variance of $\langle v^2(t_0)\rangle$ is

$$\varepsilon^2(t_0) = \sum_j a_j^2\sigma_j^2 \tag{23}$$

where σ_j is the standard deviation of each datum e_j. The quadratic form to be minimized to obtain the optimal $\{a_j\}$ is

$$\phi(t_0) = \cos\theta \int_0^{t_m} (A(t,t_0)-\delta(t-t_0))^2 dt + \sin\theta\; \Sigma a_j^2\sigma_j^2 + \lambda(1-\int A(t,t_0)dt) \tag{24}$$

In equation (24) $\theta (0 \leq \theta \leq \pi/2)$ is the tradeoff parameter used to sacrifice resolution in $\langle v^2(t_0)\rangle$ for gain in statistical accuracy, and λ is a Lagrange multiplier for the unimodular constraint. Estimates of $\langle v^2(t_0)\rangle$ and its standard error ε are available for any value of θ. In practical applications where the data are inaccurate, the value of θ should always be greater than zero, that is some resolution should always be sacrificed.

The LP reflectivity construction algorithm can incorporate constraints in the form

$$\sum_{i=1}^{n} \beta_{ik} r_i = \lambda_k \pm \delta\gamma_k \tag{25}$$

where r_i are the unknown reflectivity coefficients in a time window (t_1,t_2), β_{ik} are a set of constants for the k'th constraint, γ_k is the constraint value, and $\delta\gamma_k$ is an estimated error.

Using the linearized form of the acoustic impedance (equation (2)) and neglecting the effects of density, we can rewrite equation (21) as

$$\langle v^2(t_0)\rangle = \int_0^{t_m} v^2(0)\; e^{2\eta(t)} A(t,t_0) \pm \delta_{\langle\rangle} \tag{26}$$

where $\delta_{\langle\rangle}$ is the error in $\langle v^2(t_0)\rangle$ arising from observational uncertainties. By expanding the exponential, using the unimodularity of the averaging function, and completing some algebra, we may write

$$\int_{t_1}^{t_2} r(u)\beta(u,t_0)du = -\tfrac{1}{2}\{I_1+I_2+I_3\} + \tfrac{1}{4}\{\langle v^2(t_0)\rangle/v^2(0) - 1 - \sum_{k=2}^{\infty} 2^k L_k\} + \delta_{\langle\rangle}/4v^2(0) \tag{27}$$

where

$$L_k = 1/k! \int_0^{t_m} \eta^k(t)\; A(t,t_0)dt$$

and

$$I_1 = \int_0^{t_1} \eta(t)A(t,t_0)dt, \quad I_2 = \int_{t_2}^{t_m} \eta(t)A(t,t_0)dt \text{ and } I_3 = \eta(t_1)\int_{t_1}^{t_2} A(t,t_0)dt$$

A discretized version of the integral on the left hand side of equation (27) is of the form required for the constraint. It remains only to evaluate the right hand side. A basic difficulty arises at this point because such an evaluation demands a knowledge of $\eta(t)$. Fortunately, only integrals of $\eta(t)$ are required and therefore good estimates to the values of the integrals can be obtained if an approximate interval velocity could be found. We therefore turn our attention for the moment to the construction of a velocity structure which emulates the broad scale features of the earth velocity.

Let us first integrate equation (18) by parts to get

$$V_j^2 - v^2(0) = \int_0^{t_m} \frac{1}{t_j} (t_j-t)\ H(t_j-t)m(t)dt \tag{28}$$

where H(t) is the Heaviside step function and

$$m(t) = dv^2(t)/dt \tag{29}$$

Another datum equation can be found by integrating (29) to get

$$v^2(t_k) - v^2(0) = \int_0^{t_m} m(t)H(t_k-t)dt \tag{30}$$

Equation (30) is important if some a priori knowledge about the velocity at time t_k is available. It will permit us to construct interval velocities subject to point constraints.

Our intermediate goal is to use these equations to construct an interval velocity which has a minimum amount of structural detail. This can be done by minimizing

$$\phi = \int_0^{t_m} m^2(t)\ dt \tag{31}$$

subject to the data constraints in equation (28) and (30). The constructed model $v_c(t)$ is then found from equation (29).

4. Examples. The results of applying the linear programming construction to the data in Fig. 1 are shown in Fig. 2. Only the 10-35 Hz band was used and the recovered acoustic impedance, shown in Fig. 2d, is nearly identical to the true $\xi(t)$ shown in Fig. 1a. Since no extra constraints were added, this example shows that the minimum structure formulation and the LP solution can produce correct results.

A more complex case is shown in Fig. 3. The earth reflectivity shows many reflectors. The LP construction (again using only the 10-35 Hz band) is shown in panel (d). The computed acoustic impedance in panel (f) is seen to be a good representation to the true impedance shown in panel (c).

There will be cases where the earth velocity changes slowly with depth and therefore additional constraints will be required before good results can be achieved from the inversion. Such an earth model is shown in Fig. 4. Also shown there are the RMS values for that velocity computed from equation (19), and the reflection seismogram, band-limited to 10-35 Hz (density will be assumed to be constant). Direct application of the LP construction produces a velocity profile shown in Fig. 5a. A flattest interval velocity constructed by using the 17 RMS values in Fig. 4 is shown in Fig. 5c. Backus-Gilbert averages using that velocity structure were incorporated into the inversion and the final results are shown in Fig. 5b. The comparison with the true velocity is extremely good.

The usefulness of the RMS data degrades with the increased errors on the RMS velocities. Shown in Fig. 6 are the results of incorporating stacking velocities contaminated with substantial error. However, the inclusion of six point velocity constraints (Fig. 6d) or even one point velocity constraint (Fig. 6f) produces a very good representation of the true velocity structure.

5. Conclusion. The primary goal of this paper has been to show how reflection seismograms could be inverted in a stable and efficient manner. The mathematical representation for a layered earth, point velocity constraints (with errors) and RMS velocities, can all be incorporated directly into a linear programming solution which finds an acoustic impedance having a minimum amount of structure. Importantly, the solution is linear and therefore computationally efficient. Inversion of a seismogram containing 500 points can be accomplished in about 50 seconds on a Vax 11/780.

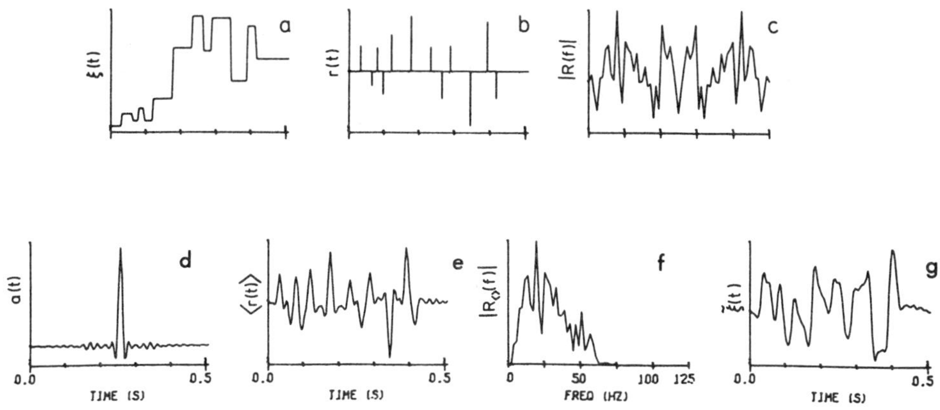

Figure 1 Shown in successive panels are: (a) the acoustic impedance; (b) reflection coefficients; (c) amplitude spectrum of the reflection coefficients; (d) averaging function from a high-resolution deconvolution; (e) final processed seismogram; (f) amplitude spectrum of (e); (g) estimated acoustic impedance.

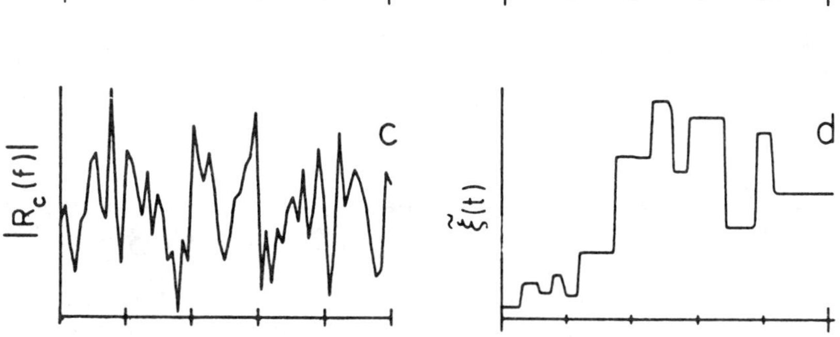

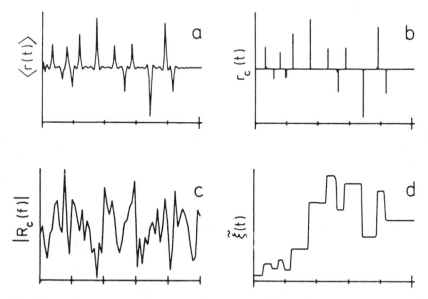

Figure 2 Shown in successive panels are: (a) input seismogram; (b) LP reflectivity recovered by using only the 10-35 Hz band; (c) amplitude spectrum of (b); (d) acoustic impedance recovered from the reflectivities in (b).

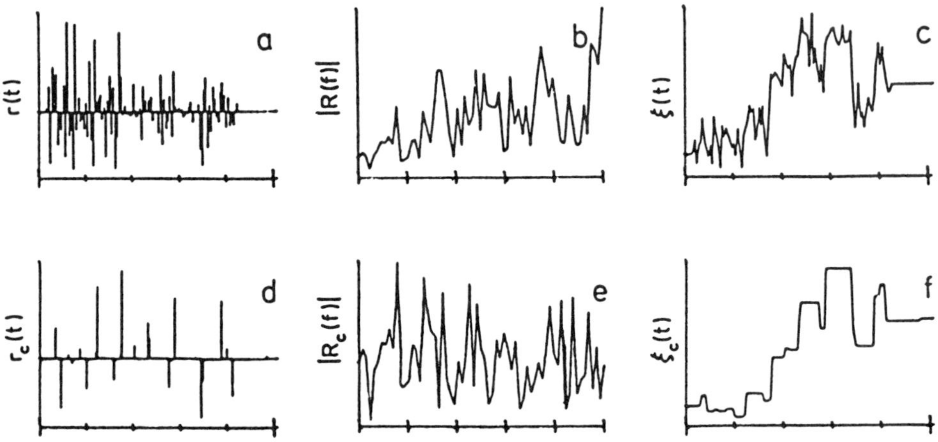

Figure 3 Plotted successively in each row are the reflectivity function, its spectrum, and the acoustic impedance. The true values of these functions are in panels (a)-(c). Panels (d)-(f) contain the results from the LP algorithms applied to inaccurate data using only the 10-35 Hz frequency band.

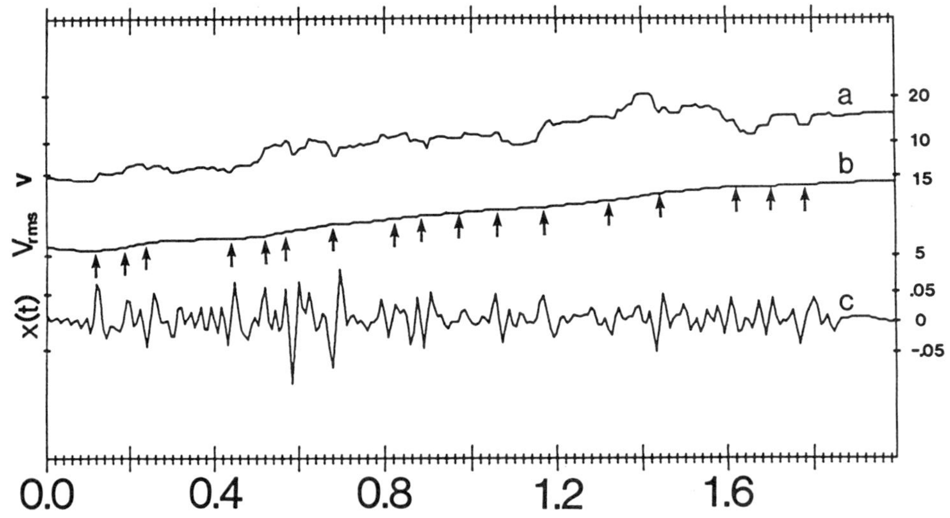

Figure 4 The true velocity structure is shown in (a). The RMS velocities corresponding to the velocity in (a) are shown in (b). The arrows indicate those RMS velocities that will be used in the inversion. For both (a) and (b) the scales on the right are in kft/sec. The reflection seismogram, band-limited 10-35 Hz is shown in (c).

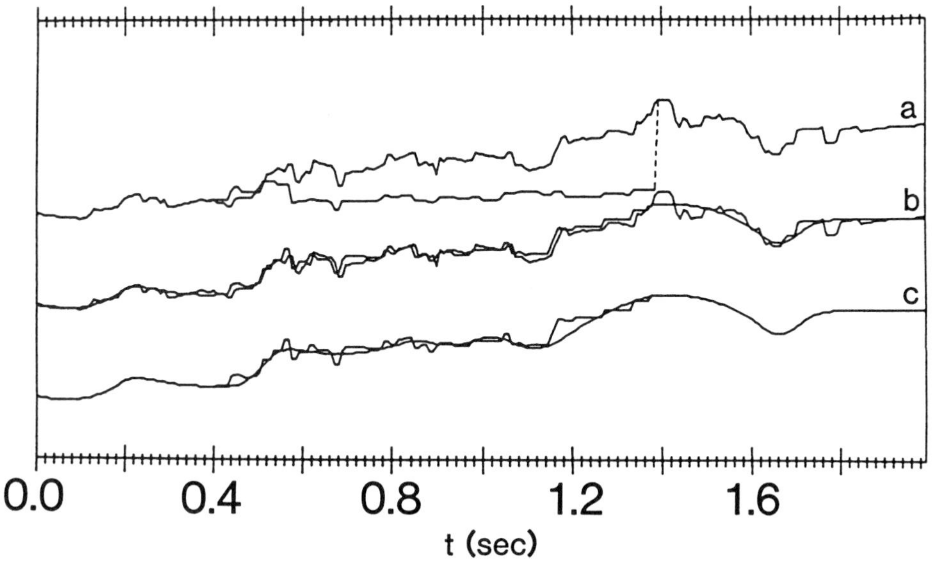

Figure 5 The true velocity structure, and the recovered interval velocity when no RMS constraints were used, are shown in (a). The inversion was carried out only on the time window (0.4-1.4) seconds. Shown in (b) is the recovered interval velocity when RMS constraints were incorporated into the inversion. The recovered velocity in the time window (0.4-1.4) seconds is inserted into the flattest model obtained by inverting the RMS data alone. The entire profile (0.0-2.0) seconds is then superposed upon the true velocity structure. The correspondence between the flattest model and the LP construction is shown in (c) where these two functions have been superposed.

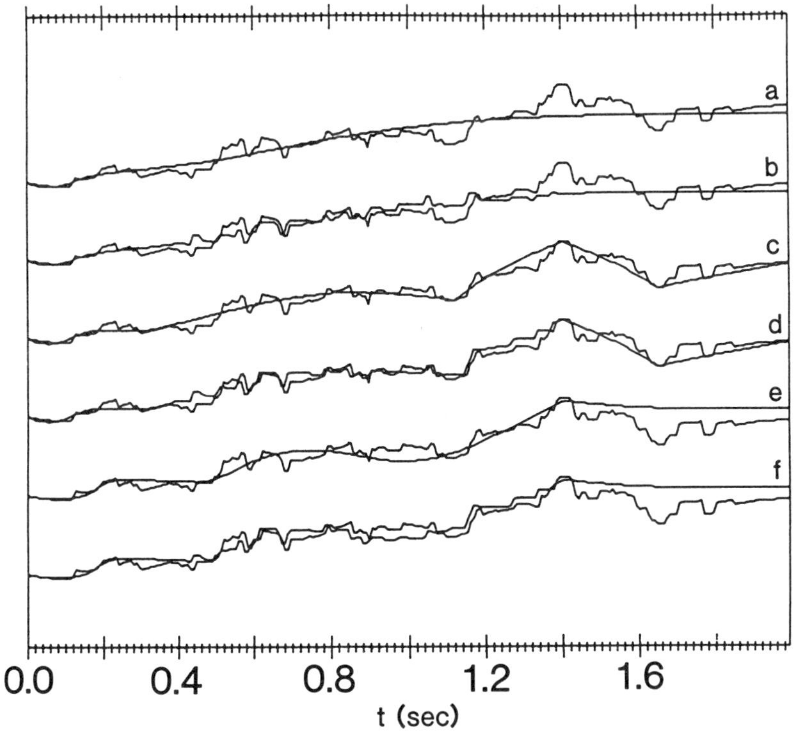

Figure 6 The interval velocity obtained by inverting inaccurate RMS velocities are shown in (a) and the acoustic impedance constructed by using the seismic data and this interval velocity are shown in (b). (c)-(d) are the same as (a)-(b) except that six point velocity constraints are included in the inversion of the RMS velocities. (e)-(f) are the same as (a)-(b) except that only one point velocity constraint is included. The true velocity has been superposed on all curves (a)-(f).

REFERENCES

[1] A. BAMBERGER, G. CHAVENT, CH HEMON and P. LAILLY, Inversion of normal incidence seismograms, Geophysics, 47(1982), pp. 757-770.

[2] J.G. BERRYMAN and R.R. GREENE, Discrete inverse methods for elastic waves in layered media, Geophysics, 45(1980), pp. 213-233.

[3] V. BARCILON, Iterative solution to the inverse Sturn-Liouville problem, J. Math. Phys., 15(1974), pp. 429-436.

[4] J.F. CLAERBOUT, Synthesis of a layered medium from its transmission response, Geophysics, 33(1968), pp. 264-269.

[5] B. GJEVIK, A. NILSEN and J. HOYEN, An attempt at the inversion of reflection data, Geophys. Prosp., 24(1976), pp. 492-505.

[6] M.S. HOWARD, Inverse scattering for a layered acoustic medium using the first-order equations of motion, Geophysics, 48(1983), pp. 163-170.

[7] P. HUBRAL, On getting reflection coefficients from waves, Geophys. Prosp., 26(1978), pp. 627-630.

[8] I. KAY, The inverse scattering problem when the reflection coefficient is a rational function, Comm. Pure Appl. Math., 13(1960), pp. 371-393.

[9] S. LEVY and P.K. FULLAGAR, Reconstruction of a sparse spike train from a portion of its spectrum and application to high resolution deconvolution, Geophysics, 46(1981), pp. 1235-1243.

[10] R.G. NEWTON, Inverse scattering I: one-dimension, J. Math. Phys., 21(1980), pp. 493-505.

[11] R.G. NEWTON, Inversion of reflection data for layered media: a review of exact methods, Geophys. J.R. Astr. Soc., 65(1981), pp. 191-215.

[12] A. NILSEN and B. GJEVIK, Inversion of reflection data, Geophys. Prosp., 26(1978), pp. 421-432.

[13] D.W. OLDENBURG, S. LEVY, K. STINSON, RMS velocities and recovery of the acoustic impedance, Geophysics, 1984, in press.

[14] D.W. OLDENBURG, T. SCHEUER and S. LEVY, Recovery of the acoustic impedance from reflection seismograms, Geophysics, 48(1983), pp. 1318-1337.

[15] P.C. SABATIER, Spectral and scattering inverse problems, J. Math. Phys., 19(1978), pp. 2410-2425.

[16] F. SANTOSA, Numerical scheme for the inversion of acoustical impedance profile based on Gel'fand-Levitan method, Geophys. J.R. Astr. Soc., 70(1982), pp. 229-243.

[17] F. SANTOSA and H. SCHWETLICK, The inversion of acoustical impedance profile by methods of characteristics, Wave Motion, 4(1982), pp. 99-110.

[18] M.M. SONDHI and B. GOPINATH, Determination of vocal-tract shape from impulse response at the lips, J. Acoust. Soc. Am., 49(1971), pp. 1867-1873.

[19] W. SYMES, Stable solution of the inverse reflection problem for a smoothly stratified elastic medium, SIAM J. Math. Anal., 12(1981), pp. 421-453.

[20] J.A. WARE and K. AKI, Continuous and discrete inverse scattering problems in a stratified elastic medium I. Plane waves at normal incidence, J. Acoust. Soc. Am., 45(1969), pp. 911-921.

[21] K.W. WATERS, Reflection seismology: A tool for energy resources exploration, John Wiley & Sons, 1978.

[22] P. WEIDELT, The inverse problem of geomagnetic induction, Z, Geophys. 38(1972), pp. 257-289.

[23] V.H. WESTON, On the inverse problem for a hyperbolic dispersive partial differential equation, J. Math. Phys., 13(1972), pp. 1952-1955.

SOME RECENT RESULTS IN INVERSE SCATTERING THEORY

J. P. CORONES*, R. J. KRUEGER* AND V. H. WESTON**

Abstract. Two approaches to inverse scattering problems are briefly outlined in this paper. The first is based on a wave splitting or input-output approach to scattering problems. In this context, scattering operators which map incident to reflected fields are shown to provide a general means of reconstructing material properties. The second approach addresses the problem of using sparse data in the reconstruction of the acoustic properties of a scatterer. Detailed analyses regarding these approaches are provided in the references.

1. Introduction. This paper presents a summary of some recent results in inverse scattering theory. Included here is a discussion of a general approach to the computation and use of reflection kernels to reconstruct material properties of media interrogated by elastic or electromagnetic waves. This approach is applicable to a wide range of problems involving such wave phenomena, including problems for both lossless and lossy media, and more generally, dispersive media. Various examples and aspects of the method will be given. In addition, a new approach to the inverse scattering problem for acoustic waves in more than one space dimension is summarized. This approach incorporates an essential feature of practical inversion theory, namely, inversion using sparse data.

The work summarized here was done as part of a larger project supported by the Office of Naval Research. Although it is not possible to discuss the full breadth of results obtained in this project thus far, a complete list of references ([1]-[19]) of work completed to date is included.

*Applied Mathematical Sciences, Ames Laboratory, Iowa State University, Ames, IA 50011. Research was supported in part by the Applied Mathematical Sciences subprogram of the Office of Energy Research, U.S. Department of Energy, under Contract No. W-7405-ENG-82, and in part by the Office of Naval Research, Contract N0014-83-K-0038.

**Department of Mathematics, Purdue University, West Lafayette, IN 47907. Research was supported by the Office of Naval Research, Contract N0014-83-K-0038.

2. Inverse Scattering Using Reflection Kernels. This section deals with an exact inverse scattering technique. Subsection (a) describes the conceptual framework of the approach. Briefly, this involves introducing an operator which relates the input to a scatterer (e.g., incident field) to the output. In subsection (b) the question of converting the PDE which describes the scattering experiment to one in terms of input-output pairs is addressed. This can be thought of as a change of basis for the problem, and it results in a particular representation for the scattering operator. In turn, this representation leads to a reconstruction algorithm for the problem of interest. An example involving inverse scattering for a lossy medium is given. Subsection (c) provides a more detailed example of the above approach by considering the case of electromagnetic propagation in a dispersive (i.e., frequency dependent) medium. Finally, subsection (d) looks at the sensitivity of the reconstruction algorithm to changes in data. An example is provided which suggests that it is feasible to quantify profile degradation due to bandlimited data.

Details regarding the discussion in subsections (a) and (b) can be found in [4]-[8]. Further analysis of the problem in (c) is provided in [2], and subsection (d) is discussed more fully in [9].

a. Input-Output Pairs and the Scattering Matrix. Scattering theories are often phrased in terms of the S-matrix (scattering matrix). Typically, in time independent settings incident fields are prescribed at large (infinite) distances from the target and radiation conditions are prescribed at spatial infinity. Roughly speaking, the S-matrix yields the scattered field at infinity as a functional of the incident fields. In time dependent settings the S-matrix connects the fields at $t = -\infty$ with these at $t = +\infty$. In a general sense, these S-matrix formulations provide an input-output description of the system. The physically independent fields (incident fields at infinity) are the inputs and the physically dependent fields (scattered fields) are the outputs.

To use this formulation in practice it is necessary for the target to be in the far field of the source and the receiver to be in the far field of the scattered wave generated by the target. It is further necessary for both source and receiver to be in a homogeneous host.

In many situations these restrictions are not satisfied. For example, in acoustic non-destructive evaluation experiments there is no a priori reason to believe that target flaws are far from the interrogated surface. Alternatively, the material under investigation may itself be strongly scattering (e.g., laminated composites). In these cases, it is still desirable to deal with an object that connects the physically independent variables (waves incident upon the material) to the physically dependent variables (waves scattered by the material). The operator that yields outgoing waves given the incoming waves is again termed the S-matrix.

A conceptually simple, though very interesting class of models is provided by a slab of material assumed to have parallel sides of infinite extent in the y and z directions with left edge at $x=a$ and right edge at $x=b$, $a<b$. Assume this slab is illuminated from either side by acoustic or electromagnetic waves and that the medium in which the slab is situated is homogeneous for $x<a$ and $x>b$.

Suppose that the incident field at the left edge of the slab is denoted by $I(a)$ (I for input) and that $I(b)$ denotes the input field at the right edge of the slab. Both $I(a)$ and $I(b)$ are in general partially transmitted and partially reflected by the medium. If $O(a)$ and $O(b)$ denote the fields exiting the slab at a and b, respectively, then the S-matrix is the operator that connects the input and output fields. Thus,

$$\begin{bmatrix} O(a) \\ O(b) \end{bmatrix} = \tilde{S}(a,b) \begin{bmatrix} I(a) \\ I(b) \end{bmatrix} = \begin{bmatrix} \tilde{R}^{+}(a,b) & \tilde{T}^{-}(a,b) \\ \tilde{T}^{+}(a,b) & \tilde{R}^{-}(a,b) \end{bmatrix} \begin{bmatrix} I(a) \\ I(b) \end{bmatrix}$$

Note that the S-matrix iself depends only on the medium and not on the particular input or output fields.

If the input and output fields satisfy

$$\frac{d}{dx} \begin{bmatrix} I(x) \\ O(x) \end{bmatrix} = \begin{bmatrix} \tilde{\alpha} & \tilde{\beta} \\ \tilde{\gamma} & \tilde{\delta} \end{bmatrix} \begin{bmatrix} I(x) \\ O(x) \end{bmatrix}$$

then it is possible to show that studied as a function of its first argument, the S-matrix satisfies

$$\frac{\partial \tilde{S}(x,x')}{\partial x} = \begin{bmatrix} \tilde{R}^{+}(x,x') & \tilde{I} \\ \tilde{T}^{+}(x,x') & \tilde{0} \end{bmatrix} \begin{bmatrix} -\tilde{\alpha} & -\tilde{\beta} \\ \tilde{\gamma} & \tilde{\delta} \end{bmatrix} \begin{bmatrix} \tilde{I} & \tilde{0} \\ \tilde{R}^{+}(x,x') & \tilde{T}^{-}(x,x') \end{bmatrix}$$

Of particular interest is the equation for $\tilde{R}^{+}$,

$$(1) \qquad \frac{\partial \tilde{R}^{+}(x,x')}{\partial x} = \tilde{\gamma} + \tilde{\delta}\tilde{R}^{+} - \tilde{R}^{+}\tilde{\alpha} - \tilde{R}^{+}\tilde{\beta}\tilde{R}^{+}$$

Notice that the equation for $\tilde{R}^{+}$ does not involve $\tilde{R}^{-}$ or $\tilde{T}^{\pm}$.

b. Change of Basis and the Resulting Reconstruction Algorithm. To utilize Eq. (1) in a computational context, it is necessary to give a representation of the $\tilde{R}^{+}$ operator. This must be done in each concrete example, specified by a wave equation. The definition of the $\tilde{R}^{+}$ operator itself of course depends on how input and output fields are defined again starting with the wave equation of interest. This corresponds to a change of basis from field quantities to input, output quantities. Although there is no unique way to do the latter, a variety of mathematically well-defined and physically intuitive results have been obtained.

The general guideline (in one dimension) is that when media are homogeneous, waves traveling to the right and left should decouple. Thus beginning with the wave equation for propagation in a lossy medium,

(2) $$u_{xx} - u_{tt} + A(x)u_x + B(x)u_t = 0$$

(where subscripts denote partial derivatives), it is easy to see that as a first-order system in x,

(3) $$\frac{\partial}{\partial x}\begin{bmatrix} u \\ u_x \end{bmatrix} = \begin{bmatrix} 0 & 1 \\ (\partial_t^2 - B\partial_t) & -A \end{bmatrix}\begin{bmatrix} u \\ u_x \end{bmatrix} .$$

When the medium is homogeneous and lossless, A=B=0 and the resulting system can be diagonalized to

$$\frac{\partial}{\partial x}\begin{bmatrix} u^+ \\ u^- \end{bmatrix} = \begin{bmatrix} -\partial_t & 0 \\ 0 & \partial_t \end{bmatrix}\begin{bmatrix} u^+ \\ u^- \end{bmatrix}$$

using the change of basis

(4) $$\begin{bmatrix} u^+ \\ u^- \end{bmatrix} = \frac{1}{2}\begin{bmatrix} 1 & -\partial_t^{-1} \\ 1 & \partial_t^{-1} \end{bmatrix}\begin{bmatrix} u \\ u_x \end{bmatrix} .$$

Notice that $u^{\pm}$ are indeed right and left moving waves when A=B=0. Since the change of basis (4) is well defined even if A≠0, B≠0 the full system (3) can be equivalently described using the new basis. The system equivalent to (3) in this basis is

$$\frac{\partial}{\partial x}\begin{bmatrix} u^+ \\ u^- \end{bmatrix} = \frac{1}{2}\begin{bmatrix} -2\partial_t + B + A & B+A \\ A-B & 2\partial_t - B - A \end{bmatrix}\begin{bmatrix} u^+ \\ u^- \end{bmatrix} .$$

Notice also that the hyperbolic equation (2) has now been rewritten as a first order hyperbolic system of equations.

The pair $(u^+(a,t), (u^-(b,t))$ are the natural set of input variables for this problem and the set $(u^-(a,t),u^+(b,t))$ are the natural output pair.

It can be shown that

$$u^-(a,t) = \tilde{R}^+ u^+(a,t) = \int_0^t R(a,t-s)u^+(a,s)ds$$

which in turn implies, when substituted into the operator equation (1) for $\tilde{R}^+$, that

$$(5) \quad R_x - 2R_t = -B(x)R - \frac{1}{2}(A(x)+B(x))R*R(x,t),$$

$$(6) \quad R(x,0^+) = -\frac{1}{4}(A(x)-B(x)),$$

$$R(b,t) = 0$$

where

$$R*R(x,t) = \int_0^t R(x,s)R(x,t-s)ds.$$

The equation for the kernel R together with the behavior of R in the limit of short time (Eq. (6)) provides the basis for an algorithm to reconstruct A(x) if B(x) is known, or B(x) if A(x) is known.

As an example, consider the problem of reconstructing the electrical permittivity of a medium assuming that the conductivity is known. If a plane wave is normally incident on the slab, a transverse component, E(x,t), of the electric field satisfies

$$(7) \quad E_{xx} - c^{-2}(x)E_{tt} - b(x)E_t = 0$$

where

$$c^{-2}(x) = \varepsilon(x)\mu_o,$$

$$b(x) = \sigma(x)\mu_o$$

and $\varepsilon, \sigma, \mu_0$ are the permittivity, conductivity and (constant) magnetic permeability, respectively. Although Eq. (7) can be mapped to the form of Eq. (2), it is in fact not necessary to do so. In any event, given R(a,t) and $\sigma(x)$ [or $\varepsilon(x)$] it can be shown that the pair of equations (5) and (6) can be used to recover $\varepsilon(x)$ [or $\sigma(x)$]. Figure (1) shows the reconstruction of permittivity profiles for a single reflection kernel R(a,t) under various assumptions regarding the conductivity σ. The data R(a,t) was generated numerically assuming $\sigma_{TRUE} = 10^{-4}$ mho/m and is shown in Fig. (2). Curve C in Fig. (1) is the true permittivity profile and was reconstructed to high accuracy when σ_{TRUE} was used in the algorithm. Notice that one of the effects of assuming incorrect values for dissipation is that the boundaries between layers of the medium are not properly determined.

c. An Inverse Problem for Dispersive Media. Electromagnetic wave propagation in a dispersive medium is characterized by the fact that the phase and group velocities are functions of frequency. This causes a transient pulse to spread and change shape even if the medium is homogeneous. Since all non-vacuous media display some degree of dispersion, it is of interest to incorporate dispersive effects into inverse problems.

The physical basis for dispersion lies in the constitutive relation between the displacement field $\underset{\sim}{D}(\underset{\sim}{x},t)$ and the electric field $\underset{\sim}{E}(\underset{\sim}{x},t)$. In the simplest case this can be expressed as

$$(8) \qquad \underset{\sim}{D}(\underset{\sim}{x},t) = \varepsilon_0 \left[\underset{\sim}{E}(\underset{\sim}{x},t) + \int_0^\infty G(s)\underset{\sim}{E}(\underset{\sim}{x},t-s)ds\right]$$

where ε_0 is the permittivity of free space. The susceptibility kernel, $G(t)$, can be related to the complex permittivity, $\varepsilon(\omega)$, via

$$[\varepsilon(\omega)-\varepsilon_0]/\varepsilon_0 = \int_0^\infty G(t)e^{i\omega t}dt$$

The inverse problem considered in this section is that of determining the susceptibility kernel for a one-dimensional homogeneous isotropic medium. Thus, assume the medium occupies the region bounded by the planes $x = 0$ and $x = L$, with free space elsewhere. The magnetic permeability, μ_0, is everywhere constant. A right moving plane wave in the region $x < 0$ impinges on the medium at normal incidence at time $t = 0$, giving rise to a reflected wave. Letting $E(x,t)$ denote a transverse component of the electric field, it follows from Maxwell's equations that

$$E_{xx} - \frac{1}{c^2} E_{tt} = 0, \qquad x < 0 \quad \text{or } x > L$$

and, using Eq. (8),

$$(9) \qquad E_{xx} - \frac{1}{c^2}\left[E_{tt} + \partial_t^2 \int_0^\infty G(s)\, E(x,t-s)ds\right] = 0, \quad 0 < x < L$$

where $c^2 = (\varepsilon_0\mu_0)^{-1}$. In the region $x < 0$, E can be split into the right moving incident field, $E^+(x,t)$, and the left moving scattered field, $E^-(x,t)$, such that

$$E(x,t) = E^+(x,t) + E^-(x\ ,t)$$

where E^+ and E^- are related by

$$E^-(0,t) = \int_0^t R(0,t-s)E^+(0,s)ds$$

The inverse problem to be solved is: Given $R(0,t)$ for $0<t<T$ (for some T) find $G(t)$ for $0<t<T$. Notice that this problem involves reconstructing a function of t rather than a function of x.

Now use the splitting

$$E^\pm(x,t) = \frac{1}{2}\left[E(x,t) \pm c\partial_t^{-1}E_x(x,t)\right]$$

to rewrite (9) as

$$\partial_z \begin{bmatrix} E^+ \\ E^- \end{bmatrix} = \begin{bmatrix} \tilde{\alpha} & \tilde{\beta} \\ \tilde{\gamma} & \tilde{\delta} \end{bmatrix} \begin{bmatrix} E^+ \\ E^- \end{bmatrix}$$

where

$$\tilde{\delta} = -\tilde{\alpha} = [\partial_t + \frac{1}{2} G * \partial_t]/c$$

$$\tilde{\gamma} = -\tilde{\beta} = [\frac{1}{2} G * \partial_t]/c$$

In terms of the reflection kernel defined by

$$E^-(x,t) = \int_0^t R(x,t-s)E^+(x,s)ds$$

it follows that

(10) $2cR_x - 4R_t = G'(t) + G(0)[2R+R*R] + G'*[2R+R*R], \quad 0<x<L, \ t>0$

(11) $R(x,0^+) = -\frac{1}{4} G(0^+), \quad 0<x<L$

(12) $R(L,t) = 0 , \quad t>0 .$

These equations can be used to study both the direct and inverse problem. Figure (3) shows the domain for this system. In the direct problem, $G(t)$ is given and $R(0,t)$ is to be determined. In the inverse problem, $R(0,t)$ is given and the system (10)-(12) can be interpreted as a two point boundary value problem for a first order equation, yielding $G(t)$.

In the special case of a semi-infinite medium, R is independent of x and Eq. (10) becomes

$$4R+G+G*(2R+R*R) = 0, \quad t > 0 .$$

In this case, the solution of the inverse problem reduces to the solution of a Volterra equation of the second kind.

The more general case of a medium which is composed of a stack of homogeneous dispersive layers has also been analyzed under the simplifying assumption that the functional form of the susceptibility kernel in each layer is known [2]. The fully inhomogeneous one dimensional dispersive medium is currently being studied.

Some numerical experiments were performed to test the sensitivity of the inversion algorithm to noise in the data. A two resonance model of the electron contribution to the permittivity was used, with

$$G(t) = e^{-0.1t}\sin(2\pi t) + 0.5e^{-0.5t}\sin(5.6\pi t), \quad 0 < t < 6.$$

The reflection kernel was generated numerically, and then Gaussian noise with zero mean and .005 variance was added to it, yielding a data set with signal to noise ratio of 4.3. This data was then smoothed once using a five point linear least squares smoother. The resulting reconstruction using Eq. (10)-(12) is shown in Fig. (4).

d. Sensitivity. Independently of the method used, it is useful to be able to characterize the sensitivity of a reconstruction technique to changes in the input data, i.e., changes in $R(a,t)$. The bandlimiting of $R(a,t)$ is of particular concern in this regard. The following computations give some insight into why the reconstruction of profiles based on R with low frequency content missing are far from perfect.

The algorithm based on Eq. (5), (6) which gives $A(x)$ from $R(a,t)$ (with $B \equiv 0$) is displayed in [5]. The Jacobian, J, of the transformation that takes $R(a,t)$ into $A(x)$ is of the form

$$J = -4I + K \tag{13}$$

where I is the identity and K is due purely to the nonlinear term in Eq. (5). The singular values and several singular vectors of K for the profile given in Fig. (5) are displayed in Figs. (6)-(9). Notice that the singular vector associated with the largest singular values has only low frequency content while those associated with smaller singular values have much higher frequency content. Due to the size of the singular values modest changes in the high frequency content of R have little effect on the reconstruction of $A(x)$ while modest changes in the low frequency content have a much more pronounced effect. These attributes of the bandlimiting of R have been demonstrated in numerous numerical experiments. While this discussion does not quantify the effects of bandlimiting on reconstructed profiles, it does appear to be feasible to give a quantitative argument of profile degradation based on these ideas. A more complete discussion is found in [9].

3. Inverse Scattering with Sparse Data. In an uncontrolled environment (not under laboratory conditions), the required set of measurements needed to determine an object often cannot be obtained due to physical limitations of the surroundings, or due to the uncooperative nature of the object. Quite often the obtainable set of measurements is sparse.

Both the theoretical background of the sparse data case and a method to handle it have been investigated in [16]-[18] for the particular case of acoustic scattering as represented by the reduced wave equation

$$\Delta u + k^2 n^2(x) u = 0. \tag{14}$$

Here the host medium is taken to have index of refraction unity, and the object characterized by the unknown quantity $v(x)$, where

(15) $$v(x) = n^2(x)-1,$$

is embedded in a compact domain D. The problem is to determine $v(x)\varepsilon C_p(D)$, (piecewise continuous functions on D) from a finite set of N measurements. The ℓth measurement consists of the complex number

$u_m^s(x_\ell,k_\ell,u_\ell^i)$ which is the scattered field (produced by an incident wave u_ℓ^i) at the point x_ℓ exterior to D, and at a frequency associated with the wave number k_ℓ. Each measurement may or may not be associated with different incident waves and/or frequencies.

The inverse problem consists then of solving the system of N complex non-linear equations

(16) $$u_m^s(x_\ell,k_\ell,u_\ell) = u^s(x_\ell,k_\ell;v) \quad , \quad \ell=1,\dots,N$$

for the real variable v. The non-linear functional on the right-hand side has the explicit form

$$u^s(x_\ell,k_\ell;v) = \tilde{G}_\ell(I-\tilde{G}_\ell)^{-1}u_\ell^i(x_\ell)$$

where the operator $\tilde{G}_\ell$ is given by

$$\tilde{G}_\ell f(x) = k_\ell^2 \int_D \frac{e^{ik_\ell|x-y|}}{4\pi|x-y|} v(y)f(y)dy \ .$$

By setting $h_\ell(v)+i\ h_{\ell+N}(v) = u_m^s(x_\ell,k_\ell,u_\ell^i)-u^s(x_\ell,k_\ell;v)$ system (16) is reducible to the system of 2N real non-linear equations

(17) $$h_\ell(v) = 0 \quad , \quad \ell=1,\dots,2N$$

System (17) is non-unique. The investigation of the ill-posedness of system (17) and a method to alleviate this requires the introduction of the gradient $H_\ell(v;x)$ of $h_\ell(v)$, and the positive matrix H whose elements $H_{ij}(v)$ are defined by

(18) $$H_{ij}(v) = \int_D H_i(v;x)H_j(v;x)dx$$

The resolution cell (solution space) of system (17) can then be investigated [17]. If v_m is a particular solution, then the general solution has the form

$$V(x) = \mu\phi^+(x) + \sum_{k=1}^{2N} c_k\ H_k\ (v_m;x)$$

where ϕ^+ is any $C_p(D)$ function which is orthogonal to the subspace spanned by $\{H_k(v_m;x)\}_{k=1}^{2N}$. The coefficients μ, $c_1, \ldots, c_{2N}$ are related through a non-linear system. It can then be shown that the solution space is an infinite-dimensional manifold smooth everywhere except at possible singular points. However, the restriction of requiring the solution to lie in a prescribed 2N dimensional subspace will yield, in general, isolated point solutions. Such a restriction is natural for numerical computations. But for a sparse data set (N not too large) this restriction will not provide an optimum way of approaching the problem since (i) the solution depends critically upon the choice of subspace, (ii) the resulting solution may be unstable, (iii) the solution will leave poor resolution. For better resolution with sparse data (2N small), one should work with a subspace of dimension M where $M \gg 2N$.

An alternative approach to obtaining isolated point solutions is to impose the condition of stability. It has been shown [17] that the following system

$$\text{(19)} \qquad \underset{v \in C_p(D)}{\text{Min}} \| H^{-1}(v) \|_2$$

$$h_\ell(v) = 0, \quad \ell = 1,2,\ldots,2N$$

will yield, under certain conditions, isolated point solutions. (Another approach, to minimize the conditioning of the matrix H(v), is presently being analyzed.)

Since system (19) is equivalent to the following system

$$\underset{v}{\text{Min}} \quad \underset{\theta \in R^{2N}}{\text{Max}} \quad \sum_{i,j=1}^{2N} \theta_i \, \tilde{H}_{ij}(v) \theta_j ,$$

$$\underset{v}{\text{Min}} \sum_{i,j=1}^{2N} h_i(v) A_{ij} h_j(v)$$

where $\tilde{H}_{ij}(v)$ are the elements of H^{-1}, and A_{ij} is any positive definite matrix, the simpler system is suggested

$$\text{(20)} \qquad \underset{v}{\text{Min}} \sum_{i,j=1}^{2N} h_i(v) \, \tilde{H}_{ij}(v) \, h_j(v)$$

for practical application. (System (20) was also derived from another approach [16] based upon selecting from a "catalogue" of bodies the one whose scattered data were closest to the measured data.) The functional in (20) was minimized by using a descent-type approach.

Numerical implementation of the procedure has been carried out successfully for the one-dimensional case of a plane wave incident normally on a slab [16],[18]. Here the data consisted of measurements of either the reflected field at different frequencies or else a combination of reflected and transmitted wave at different frequencies.

The procedure has also been generalized in [1] to the case of acoustic scattering by an obstacle characterized by the Dirichlet boundary condition.

Acknowledgement. The authors thank Curt Vogel for his aid in the sensitivity study in Section 2-d.

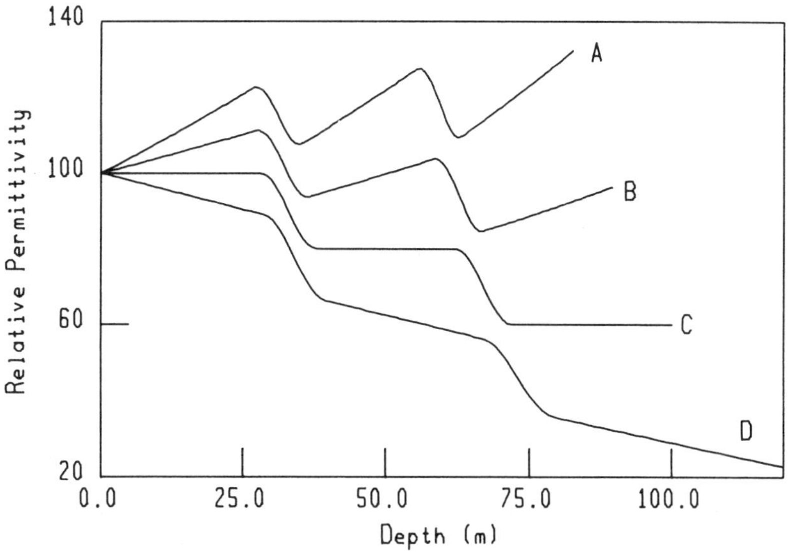

FIGURE 1. Reconstructed permittivity profiles. A. sigma = 0
B. sigma = 0.00005 C. true sigma = 0.0001
D. sigma = 0.00015

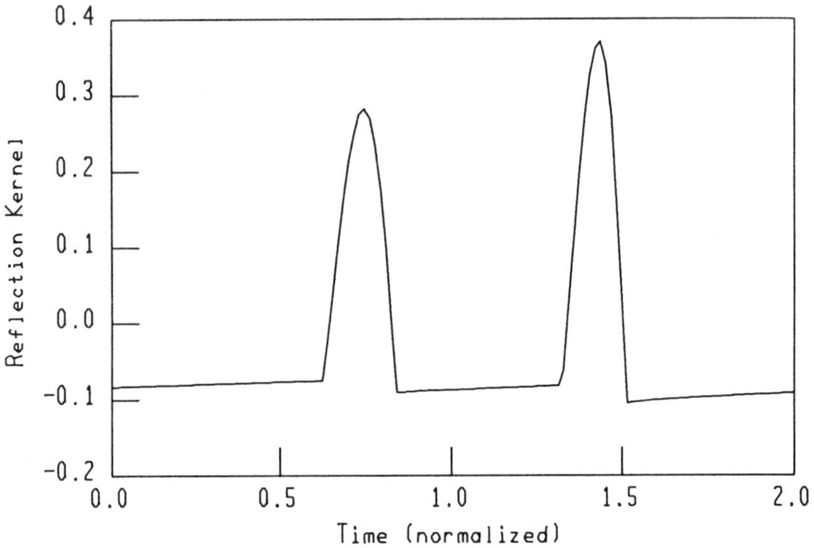

FIGURE 2. Reflection data used for the reconstructions shown in Fig. 1.

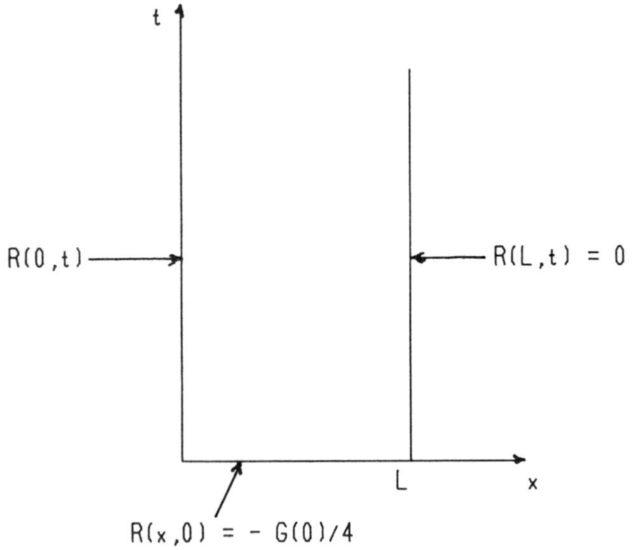

FIGURE 3. Domain for the system of equations (10) - (12).

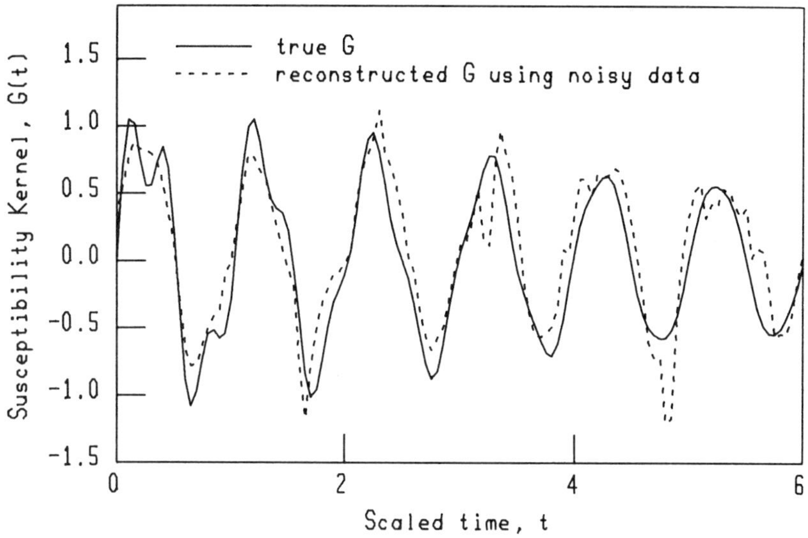

FIGURE 4. Reconstructed susceptibility kernel using smoothed noisy data with signal to noise ratio 4.3

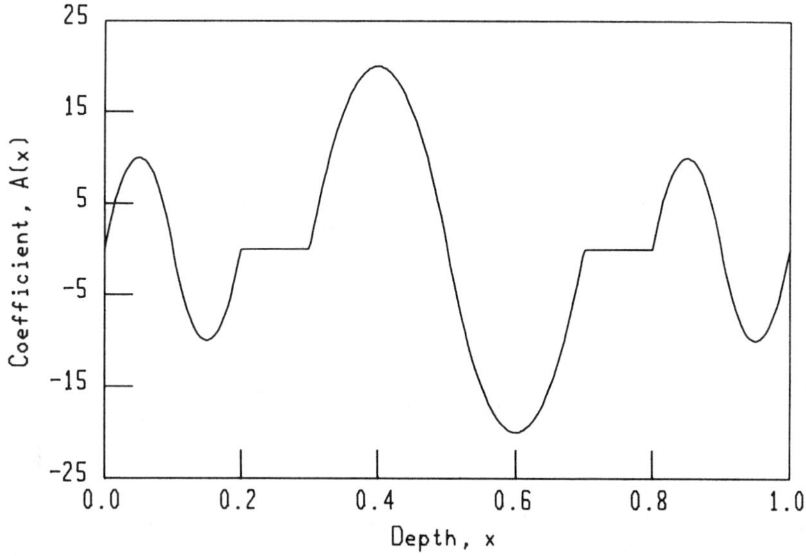

FIGURE 5. Coefficient function used in the sensitivity example in Section 2-d.

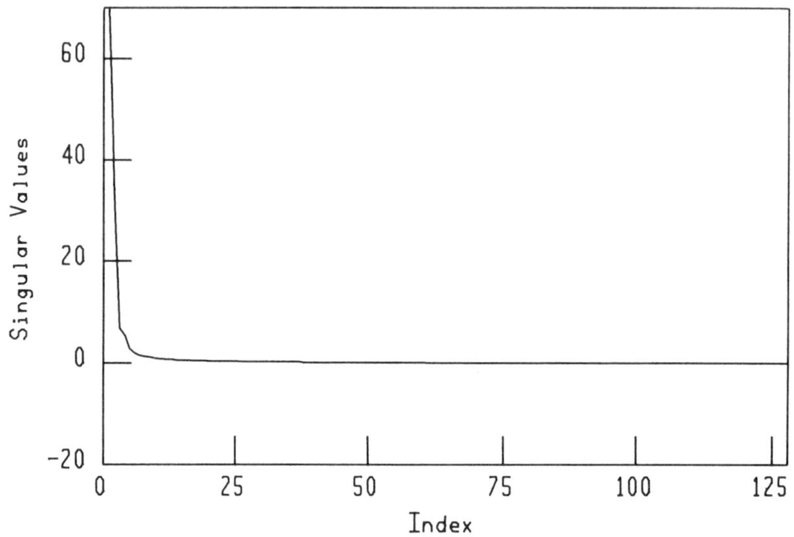

FIGURE 6. Singular values of K in Eq. (13) for A(x) given in Fig. 5.

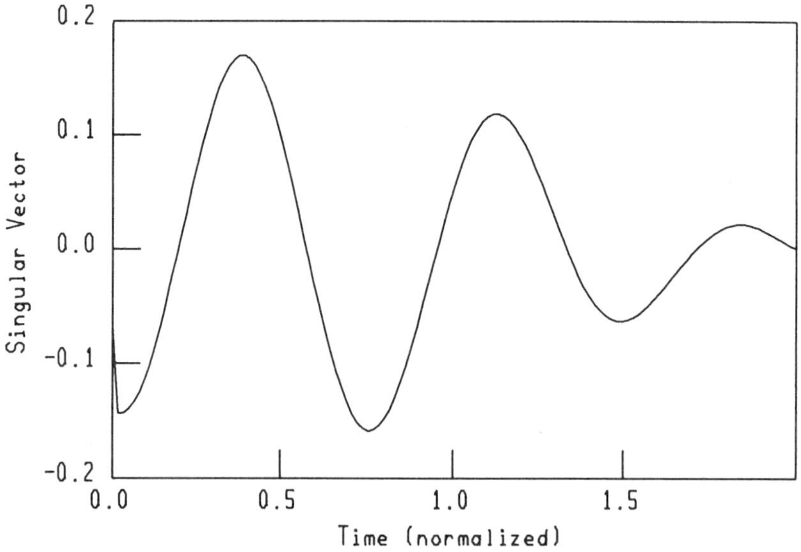

FIGURE 7. First singular vector of K.

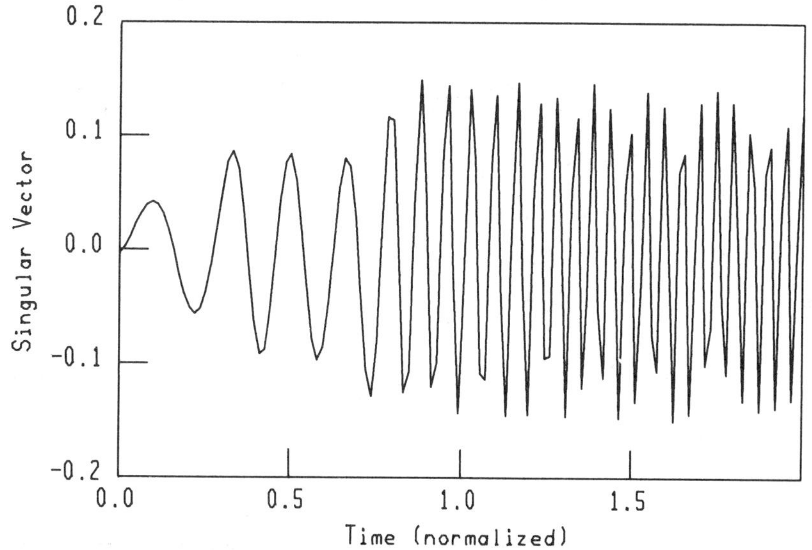

FIGURE 8. Fiftieth singular vector of K.

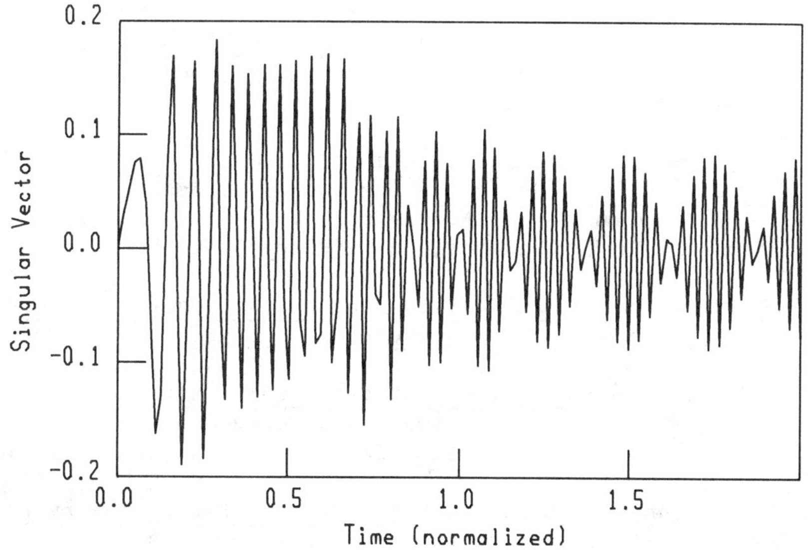

FIGURE 9. One hundredth singular vector of K.

REFERENCES

[1] M. BAUTISTA, A nonlinear approach to inverse scattering for an acoustic soft obstacle, Ph.D. thesis, Purdue U. (1984).

[2] R. S. BEEZLEY and R. J. KRUEGER, An electromagnetic inverse problem for dispersive media, preprint.

[3] M. CHENEY and J. ROSE, Three dimensional inverse scattering: high-frequency analysis of Newton's Manchenko equation, preprint.

[4] J. P. CORONES, Wave splitting and invariant imbedding in direct and inverse scattering, in Wave Propagation in Homogeneous Media and Ultrasonic Nondestructive Evaluation, AMD, v. 62, G. C. Johnson, Ed., ASME, NY, NY (1984) p. 31.

[5] J. P. CORONES, M. E. DAVISON and R. J. KRUEGER, Direct and inverse scattering in the time domain via invariant imbedding equations, J. Acoust. Soc. Am. 74 (1983) pp. 1535-1541.

[6] J. P. CORONES, M. E. DAVISON and R. J. KRUEGER, Dissipative inverse problems in the time domain, to appear in Proc. NATO-ARW-IMEI, W-M Boerner, Ed.

[7] J. P. CORONES, M. E. DAVISON and R. J. KRUEGER, Wave splittings, invariant imbedding and inverse scattering, in Inverse Optics, Proc. SPIE 413, A. J. Devaney, Ed., SPIE, Bellingham, WA (1983) pp. 102-106.

[8] J. P. CORONES, M. E. DAVISON and R. J. KRUEGER, The effects of dissipation in one-dimensional inverse problems, in Inverse Optics, Proc. SPIE 413, A. J. Devaney, Ed., SPIE, Bellingham, WA (1983) pp. 107-114.

[9] J. P. CORONES, R. J. KRUEGER and C. R. VOGEL, Noise and band-limiting in an inverse scattering model, to appear in Review of Progress in Quantitative Nondestructive Evaluation, v. 4, D. O. Thompson and D. E. Chimenti, Eds., Plenum, New York.

[10] B. DEFACIO, What do curve-fits mean on the inverse scattering theory for nondestructive evaluation?, preprint.

[11] B. DEFACIO, Ultrasonic flaw detection and sizing for quantitative nondestructive evaluation, in Inverse Optics, Proc. SPIE 413, A. J. Devaney, Ed., SPIE, Bellingham, WA (1983) pp. 48-55.

[12] B. DEFACIO and J. ROSE, Exact inverse scattering theory for the non-spherically symmetric three-dimensional plasma wave equation, preprint.

[13] J. ROSE, M. CHENEY and B. DEFACIO, The connection between time- and frequency-domain three dimensional inverse scattering methods, to appear in J. Math. Phys.

[14] J. ROSE, M. CHENEY and B. DEFACIO, Connection between time and frequency domain three dimensional inverse problems for the Schrödinger equation, to appear in Review of Progress in Quantitative Nondestructive Evaluation, v. 4, D. O. Thompson and D. E. Chimenti, Ed., Plenum, New York.

[15] J. ROSE, B. DEFACIO and M. CHENEY, Physical basis of three dimensional inverse scattering for the plasma wave equation, preprint.

[16] V. WESTON, Multi frequency inverse problem for the reduced wave equation with sparse data, J. Math. Phys. 25 (1984) pp. 1382-1390.

[17] V. WESTON, Multi frequency inverse problem for the reduced wave equation; resolution cell and stability, to appear in J. Math. Phys.

[18] V. WESTON, Recent advances in the theory of inverse scattering with sparse data, to appear in Proc. NATO-ARW-IMEI, W-M Boerner, Ed.

[19] V. WESTON, On the convergence of the Rytov approximation for the reduced wave equation, preprint.

WELL-POSED QUESTIONS AND EXPLORATION OF THE SPACE OF PARAMETERS IN LINEAR AND NON-LINEAR INVERSION

PIERRE C. SABATIER*

Abstract. Solving inverse problems is defining convenient routes, in the space of parameters, towards those called "solutions" of the problem. Thus methods for solving inverse problems are primarily related to the existence of routes in a space rather than to statistical, algebraic, or other considerations. After a brief survey of the "local" routes which yield for instance least square inversion, this paper deals with "global" routes obtained by using geometric transforms. One shows examples on how to construct these routes, how to use them, what questions can be answered which escape the range of "local routes". These methods apply at least to all inverse problems related with exact solutions of non-linear partial differential equations, and in particular to the inverse problems of wave equations. They yield efficient algorithms. The case in which localization of the desired solutions is only obtained through some extreme points of a domain is then reviewed. In this method of analysis, one replaces the ill-posed problem of finding solutions that correspond to actual measurements by a set of well-posed questions whose answers are inequalities on "theoretical measurements" of the set of solutions- and enable decision in applied problems. Several applications in engineering and geophysical research were given in the last few years.

1. Introduction. Let C be a set of parameters introduced to describe a physical system P. Let E be the set of results which may be obtained in the measurements related to P. Let M be a set of possible mappings of C into E, tentatively used to describe the

*Laboratoire de Physique Mathematique 34060 Montpellier Cedex, Equipe de Recherche associee au CNRS. This paper is part of a program sponsored by the CNRS: RCP 264 (Etudes interdisciplinaire des problemes inverses).

"direct problem". It is often possible to define the sets C and E in such a way that they are metric spaces. E contains the set of images of C by the mappings of M and "other" results (including "errors").

The "inverse problem" is solved whenever it is possible to associate to any element e of E a set of parameters C in C in such a way that:

(a) The set C' of images of C by the mappings of M has a "small" diameter in E and is close to e (existence problem).

(b) A mapping or multimapping of e to C is well-defined and continuous (stability problem).

(c) A "physical" criterion enables one to classify the elements in C and the mappings in M.

The word "physical", of course, refers to the other information on the problem.

Thus, solving inverse problems is nothing but exploring the space of parameters and writing a guidebook for this exploration, based on physical grounds. For instance, some physicists use statistical methods to write this guidebook [1]. Other first linearize the problem and use then generalized inverses or more specific methods. My own view relies on an obvious remark: ways for world exploration do not depend mainly on the guidebook you use but on the routes which can be used. This is why the same formulas appear as well in statistical inference or in algebraic inference. We show here some of these routes. Thus Section 2 is devoted to a very brief survey of the techniques of local exploration, which are well-known (but more simple than what is usually "known"). Section 3 is devoted to techniques of global exploration, especially algorithmic approaches based on geometric transforms, on which we are currently working. Section 4 is devoted to techniques in which semilocal informations are enclosed in "well-posed questions," on which we have worked in the last years. For the sake of simplicity we assume throughout that the mappings of M depend on a finite set of parameters, so that we can consider only one mapping M, and manage the additional degrees of freedom by suitably augmenting C.

2. Local Exploration of the Set of Parameters. The starting point of any local exploration is made of the following components:

(1) a priori information in C : for instance, there

exists a parameter x_0 such that $d(x,x_0)$ should not be too large, or/and x should be "smooth," or/and the a priori statistical distribution of x is given (if joint probabilities of components of x are given, a convenient definition of the statistical distribution can take these into account). All this information is "physical criterions" for classifying "solutions" x in C. Hopefully, they can be represented by one or several functionals $I_p(x)$ such that x_1 is "better" than x_2 with respect to the criterion I_p if $I_p(x_1) < I_p(x_2)$.

(2) a priori information in E : for instance, there exists a "distance" in E, or an a priori distribution of errors, which yields a "quality" criterion of a fit, i.e., a symmetric functional $I(e_1,e_2)$ defined on the product space ExE and such that e_i is a better fit of e then e_j if $I(e,e_i) = I(e_i,e)$ is smaller than $I(e,e_j)$. Now, for a given result e, we may try to solve the inverse problem by two ways.

(1) By seeking first e' in $C' = M(C)$ for which $I(e,e')$ is minimum over C' (if it is unique, this point is called the "punctum proximum" of e, and denoted as PP(e)). The reciprocal image C of e' is then determined in C, and the so-called "quasisolutions" c are obtained if $e \to c$ is continuous. They are classified by means of the criterions I_k. If there is ony one criterion, there may be "best quasisolutions".

(2) By first making a convex combination of $I(e,M(x))$ and the $I_p(x)$'s into a cost function $K(x)$, which usually depends on a set of trade-off parameters μ_p introduced as weights of the varius criterions. Then the elements of C are classified by means of K. However, in practical cases, the minimum of K over C is determined for various values of the trade-off parameters and these ones are fixed where C is reasonably stable against their (small) variations. One obtains in this way the so called "approximate solutions." The extent of ambiguities is then "appraised" by starting from the approximate solution and allowing the parameters or the values of the cost funcitonal to vary between definite bounds.

The two ways we have described give a reasonable approach to data inversion if the "best" quasisolution or the "best" approximate solution vary continuously depends on results in E. Several cases have been studied and are encompassed by the so called "regularization techniques" of Tichonov,

Miller, Morozov, etc., or by the "least square fits" of Marquardt and so many others. All these techniques of exploration essentially reduce to mathematical problems of minimizing functionals and calculating the subsets in which cost functionals lie between two bounds. The natural framework is the variational calculus and numerical methods of optimization, including linear programing. Approximate solutions usually are easier to deal with and more robust with respect to perturbations of the data or of the mappings than quasisolutions.

Let us illustrate this brief survey by an example which is simple and yet applies in many linear or non linear problems. We assume that M is represented by M functionals of the form

$$(2.1)\qquad m_i = \int_a^b G_i(t,x(t),x'(t))dt = e_i+\delta e_i \qquad i = 1,2,\ldots,M$$

where $x \in C_2[a,b]$. We set $C = C_2[a,b]$, $E = \mathbb{R}_M$,

$$(2.2)\qquad I(x)= \lambda^2 \int_a^b[x(t)-x_o(t)]^2 w_1(t)dt + \mu^2 \int_a^b[x'(t)]^2 w_2(t)dt$$

$$(2.3)\qquad I(m,e) = \sum_{i=1}^{M} p_i(m_i-e_i)^2 \ ; \qquad K(x) = I(x) + I(m(x),e)$$

where $w_1(t)$, $w_2(t)$, the p_i's, are convenient "weights". An algorithmic approach to this problem is feasible by means of numerical methods, e.g., the gradient method which defines a "route" of steepest descent. We also know an iterative approach. Writing down that the first variation of $K(x)$ vanishes as a functional [2] yields a second order differential equation for $x(t)$ whose solution $X(t,\underset{\sim}{\delta})$ depends on $\underset{\sim}{\delta}$, which must be a solution of the algebraic equation

$$(2.4)\qquad \underset{\sim}{\delta} = \int_a^b \underset{\sim}{G}[X(t,\underset{\sim}{\delta}),\ X'(t,\underset{\sim}{\delta}),\ t]dt - \underset{\sim}{e}$$

If $\underset{\sim}{G}$ is Lipschitz continuous, and the first guess $\underset{\sim}{\delta}_o$ is "small" enough, an iterative method can give $\underset{\sim}{\delta}$, and hence the desired approximate solution. Whatever the method is, for $\mu = 0$, this gives an approximate solution "a la Tichonov, Miller, Marquardt, etc.". For $\lambda = 0$, it is "a la Philips, Twomey, Morozov, etc.". Results are particularly simple in the linear case, i.e., if $G_i(x,x',t) = G_i(t)\; x(t)$, and $\mu = 0$, $w_1(t) = 1$. The result is

$$(2.5)\quad x(t) = x_o(t) - \lambda^{-2} \sum_{k=1}^{M} \delta_k\, G_k(t)$$

where $\underset{\sim}{\delta}$ is given by

$$(2.6)\quad \underset{\sim}{\delta} = [I + \lambda^{-1}\, \underset{\sim}{g}]^{-1}\, \underset{\sim}{\delta}^{(0)} \;; \qquad [\underset{\sim}{g}]_{ik} = \int_a^b G_i(t)\, G_k(t)\, dt$$

$$(2.7)\quad \delta_1^{(o)} = \int_a^b G_i(t)\, x_o(t)\, dt - e_i$$

which replace (2.4).

The reader recognizes the usual regularized result appearing as well in the Lanczos-Moore-Penrose generalized inverse as in the Marquardt least-squares inversion or the stochastic Franklin-Miller best estimator in linear inference. The δ-ness of the resolvent in (2.6) gives a feeling of the range of equivalent solutions, "a la Backus and Gilbert". It is clear that using the first or the second iterate for solving (2.4) gives an approximate value of the approximate solution around which the tangent mapping can be used as above in order to extend to the non linear case the motion of δ-ness or the usual estimates of the range of equivalent solutions. It remains to wish that the minimum is not only a "local" one.

3. Global Exploration of the Set of Parameters. "Local" methods for exploring the set of parameters enabled us to seek solutions, quasisolutions, or approximate solutions of an inverse problem in the neighborhood of a given parameter. They can be extended into "global" methods if M is a linear mapping. If M is not linear, specific methods must be used for a "global" exploration of C. The word "global" is of course understood as an opposite word to "local", but it is also reminiscent of the way we explore our natural world, the surface of the Earth.

The fist obvious way to find a parameter is by using a "complete" [3] method, e.g., the Gelfand-Levitan method. "Complete" methods can exist when there is a bijection between the set of data and a large subset of the set of parameters. However, even the Gelfand-Levitan method does not reach the set of all possible parameters in the physical problems it applies. Moreover so-called "complete" methods exist only in a few inverse problems, and giving them is never sufficient: algorithms must be devised either in addition or separately.

Exploring land can be done by means of random walks, with short random incursions to approach points of interest. Transposing this in the space of parameters is the principle of the so-called Monte Carlo and hedgehog methods, used in geophysical sciences when no other method is convenient and (much) computer time is available.

Recent approaches to global explorations of the space of parameters are of a geometrical nature. The so-called "topological" approach was introduced in our subject by W. Guttinger. In this approach, "imposing the principle of structural stability (i.e., qualitative insensitivity to slight perturbations) on the inversion process shows the following: the dominant singularities that generically occur in recorded signals, travel-time curves, surface contour maps and Fresnel-zone topographies can, together with the associated diffraction patterns, be classified into a few topological forms described by catastrophe polynomials" [4].

We shall study here a different kind of geometrical methods, which may not be of such general nature, but which apply to several great physical inverse problems, and when they do, yield the most complete answer to all questions of Inversion Theory. We suggest to call these methods "navigation methods" through the space of parameters because their counterpart in sea exploration is the captain's art of making the route of his ship. Indeed, one seldom sees a ship using in deep sea an iterative approach, sort of spiralling around and down into the harbour. Captains like to know the harbour coordinates. They enable them to follow (in deep sea) coordinate curves, or loxodromic curves defined by a fixed angle with the net of coordinate curves. Thus they get close to the port, where they begin the final approach, and a different ("local") navigation. The same navigation in C has been suggested by recent discoveries on exactly solvable non-linear problems. In each problem of this kind, there exists geometrical transforms T_i of the space of parameters which are associated with simple transforms t_i of the space of data [5] in such a way that if 0 denotes the composition law in C for the transform T, and x the composition law in E for the transforms t, $T_i \, 0 \, T_j$ corresponds to $t_i \times t_j$. Let c_o be a reference parameter, e_o the corresponding result, let

$T_i(a)$ be a transform depending on one parameter a and $t_i(a)$ the corresponding calculated result $c_0\, T_i(a)$ defines a coordinate curve in C and $e_0\, t_i(a)$ the corresponding one in E. A point of E has finitely many coordinates when it can be obtained from e_0 by means of a finite product $\prod_{i=1}^{N} t_i\,(a)$, and its coordinates in the system $e_0, t_1, t_2, \ldots, t_N$ are then $a_1, a_2, \ldots a_N$. The inverse problem for such a point has then the obvious solution obtained by following from c_0 the route $c_0 T_1(a_1) T_2(a_2) \ldots T_n(a_n)$. Admittedly, an arbitrary point e of E is never defined by a finite number of coordinates. But if, given e, we can find e_a with finitely many coordinates and such that $d_E(e, e_a)$ is small enough to justify a "local inversion" giving c from c_a, the inverse problem is solved. Moreover, as we show here for the first time, playing with the global transforms may yield explicit inversion formulas for the local inverse problem.

We began to apply these ideas to classical inverse problems last year [6,7,8]. With the local inverse formulas, the results for the Schrodinger one-dimensional problem are now complete. Results on the Zakharov-Shabat problem and on other Schrödinger problems are to be published. Some of these results are surveyed below to give thematic examples of the different steps in this method. These examples are also used to support comments on the interest, the difficulties, and the possible generalizations of this method to explore parameter spaces.

3.1 Construction of transformations T. As an example, consider the Schrödinger equation

$$(3.1)\qquad \frac{d^2f}{dx^2} + [k^2 - W(x)]f = 0$$

T must preserve the structure of (3.1), or, equivalently, that of the Riccati equation for the logarithmic derivative F of f:

$$(3.2)\qquad F' + F^2 + k^2 - W(x) = 0$$

Thus we are led to look into the Moebius group, and, taking into account the necessary vanishing of the parameter W at ∞, we are led to propose:

$$(3.3) \quad T_F(F) \equiv \tilde{F} = F + \frac{F' + \alpha'}{F + \alpha} - \frac{\beta'}{\beta}$$

Writing down explicitly the Riccati equation for $\tilde{F}$, we readily see that a necessary and sufficient condition for preserving (3.2) is that there exists a constant C such that

$$(3.4) \quad k^2 - W(x) + \alpha^2 - \alpha' = C\,\beta^2$$

The transformed equation (i.e. the equation for $\tilde{F}$) is then (3.2) with a new parameter $\tilde{W}$ given by

$$(3.5) \quad T_W(W) = \tilde{W} = W + 2\alpha' - 2\alpha\beta'/\beta + 2\beta'^2/\beta^2 - \beta''/\beta$$

Several choices of α and β are possible, but any given one corresponds to a special choice of the space where the transform T_W acts, and to a special choice of the transforms t, so that only few of them are consistent with non singular parameters and convenient transforms t. If we deal with the inverse problem on the line, W is the potential, and the best choice β is a constant, so that $-\alpha$ is itself a solution of (3.2), say, $U_0 = u_0'/u_0$, for a fixed value of k, say, k_0. Then $\tilde{W} = W - 2U_0'$;

$$(3.6) \quad \tilde{f} = f' - f\,U_0$$

whereas the reflection coefficients $R^{\pm}(k)$ are multiplied by phase factors of the form $(k \pm k_0)/(k \mp k_0)$. In this case, the tranformations T_W, which can be identified as "Darboux" transformations, are those we were seeking: they depend on one parameter (k_0), and the composition law of the t_i's is simply the multiplication. Particular choices are defined, introducing or not bound states. The "results" with finitely many coordinates are rational fractions of k. In the inverse problem at fixed energy, with a spherically symmetric potential $V(x)$, W is equal to $V(x) - \frac{1}{4}x^{-2} + \lambda^2 x^{-2}$, where $\lambda = (\ell + \frac{1}{2})$. A good choice for β is $F_{\nu^*} - F_\nu$, where ν is a complex value of λ, ν^* its conjugate. Then

$$(3.7)\quad \tilde{V} = V - 4\,\mathrm{Im}\; x^{-1} \frac{d}{dx} x^{-1}(F_{\nu *} - F_{\nu})^{-1}$$

These transforms are due to Lipperheide and Fiedeldey [9] who showed that the corresponding law for the t_i's is again a multiplication, so that points with finitely many coordinates are easily identified.

It is well-known that the stationary wave equation appearing in acoustics, seismic waves, electromagnetic wave propagation, is equivalent to a regular Schrödinger equation, if the admittance A is non-negative and twice differentiable (set $W = A^{-1/2}\, d^2/dx^2\, A^{1/2}$, $f = A^{1/2} p$) . If these assumptions fail, the equation

$$(3.8)\quad A^{-1} \frac{d}{dx} A \frac{d}{dx} p(k,x) + k^2\, p(k,x) = 0$$

must be solved with p , $A\, dp/dx$ absolutely continuous. The convenient transforms in C are then defined by means of particular solution of (3.8) for $k = \mu$, through the formula

$$(3.9)\quad \tilde{p}(k,x) = p(k,x) - p'(k,x)\, p(\mu,x)/p'(\mu,x)$$

where the primes denote the x-derivative. The transform T acts then as follows [8]:

$$(3.10)\quad T_A(A^{1/2}) = \tilde{A}^{1/2} = -A^{1/2}\, p'(\mu,x)/p(\mu,x)$$

There is not point in giving other examples. The recent developments of inverse methods in non linear partial differential equations have produced a large number of Bäcklund transformations among which many convenient transformations T_W can be picked up. Let us only say that they are as simple as the ones cited above in the case of the Zakharov-Shabat inverse problem [10] and hardly more complicated in the Jaulent-Jean problem [11].

3.2 Algorithms of global exploration. One first has to find reference points. Going back to our first example we assume for the sake of simplicity that there is no bound state. If the results $R_e^+(k)$ is a rational fraction, it can be written down as the product of a rational fraction $R^o(k)$ without pole in the upper half-plane by a finite number of phase factors $(k+k_n)/(k-k_n)$. R^o and the k_n's are the "coordinates" of the result R_e^+ in E and therefore of the corresponding parameter V in C. The transmission coefficient $T(k)$ can be constructed by using a dispersion relation and $|T(k)|$ on the real axis, which is equal to $[1 - |R^o(k)|^2]^{\frac{1}{2}}$. In the set of possible Darboux transformations T_n we shall us those which conserve $T(k)$ (the so-called "isospectral" ones) and introduce - or suppress - a phase factor. Now the potential V_o^+ which corresponds to $R^+(k) = R^o(k)$ vanishes for $x>0$. Hence the corresponding Jost solutions are trivially given $x > 0$ in terms of $R^o(k)$. Thus the transformation T_1 can be applied on the $(x > 0)$ part of V_o^+, giving altogether the $(x > 0)$ part of the potential whose $R^+(k)$ is equal to $R^o(k)$ times the first phase factor $(k+k_1)(k-k_1)^{-1}$ and the corresponding Jost solutions on $x > 0$. The process is continued to obtain R_e^+. In order to obtain $V(x)$ on $(x < 0)$, $R_e^-(k)$ is (trivially) determined from R^+_e and $T(k)$, together with its phase factor $R_e^-(k)/R^o(k)$. Setting $R^-(k) = R^o(k)$ yields a potential $V_o^-(x)$ vanishing on $(x < 0)$, from which the part of $V(x)$ for $(x < 0)$ can be determined like above. This algorithm has been given in other papers [12]. Reference parameters and algorithms work in the Zakharov-Shabat problem in a quite similar way. In the inverse problem at fixed energy, the reference potential is simply $V = 0$.

3.3 Implementing the method by local algorithms. We present the method together with the example of the Schrodinger equation on the line. The most obvious way to obtain a local inversion is managed by first linearizing the direct problem in the neighborhood of a given parameter. In our example, this yield the so-called generalized Born approximation:

$$(3.11a)\qquad 2i\ k[T(k)]^{-2}\ \delta R^+(k) = \int_{-\infty}^{+\infty} \delta V(x)\ f_-^2(k,x)\ dx$$

$$(3.11b)\qquad 2i\ k[T(k)]^{-2}\ \delta R^-(k) = \int_{-\infty}^{+\infty} V(x)\ f_+^2(k,x)\ dx$$

where δV is the shift from the parameter V and $\delta R^{\pm}$ the corresponding shift of the parameter, $f_\pm(k,x)$ denote the Jost solutions of (3.1), i.e., the solutions which are respectively asymptotic to $\exp[\pm ikx]$ as $x \to \pm\infty$, $T(k)$

is the transmission coefficient corresponding to the potential V .

This linear inverse problem is trivial if $V = 0$, since $f^{0}_{\pm}(k,x)$ reduces to $\exp[\pm ikx]$, $T^{0}(k)$ is equal to 1 , the formulas (3.11) are Fourier transforms, and inverting the yields

$$(3.12a)\quad \int_{-\infty}^{x} \delta V(t)\,dt = \pi^{-1} \int_{-\infty}^{+\infty} \delta R^{-}(k)\,[f^{0}_{-}(k,x)]^{2}\,dk$$

$$(3.12b)\quad \int_{x}^{\infty} \delta V(t)\,dt = -\pi^{-1} \int_{-\infty}^{+\infty} \delta R^{+}(k)\,[f^{0}_{+}(k,x)]^{2}\,dk$$

Setting $a(k) = 2ik[T(k)]^{-2}$, we see that the linear formula (3.11b), for instance, can be written as a scalar product in the space C :

$$(3.13)\quad a(k)\,\delta R^{-}(k) = \langle \delta V, f^{2}_{+}(k,.) \rangle_{C}$$

If there exists an inversion formula, it can be written as

$$(3.14)\quad \int_{-\infty}^{x} \delta V(t)\,dt = \langle g_{+}(x,.), \delta R^{-}(.) \rangle_{E}$$

Now it follows from (3.1) and (3.6) that T transforms f_{+} in such a way that there exists a linear transform T giving $(\tilde{f}^{2}_{+})$ from $(f_{+})^{2}$:

$$(3.15)\quad T(f^{2}_{+}) = \tilde{f}^{2}_{+} = [\tfrac{1}{4} \frac{d^{2}}{dx^{2}} - U_{0} \frac{d}{dx} + U^{2}_{o} + \tfrac{1}{2}\int_{x}^{\infty} dt\, V'(t)]f^{2}_{+}$$

Let us write down the generalized Born approximation (3.13) in the neighborhood of a transformed parameter $\tilde{V}$, with reflection coefficients $\tilde{R}^{-}$, Jost solution $\tilde{f}_{+}$. The righthand side of (3.13) can be written as well as $\langle T^{*} \delta V, \tilde{f}^{2}_{+}(k,.) \rangle$, where the adjoint operator T^{*} is equal to

$$(3.16)\quad T^{*}z = \tfrac{1}{4}\, u_{0}^{-2} \frac{d}{dx}\, u_{0}^{2} \frac{d}{dx}\, u_{0}^{-2} \frac{d}{dx}\, u_{0}^{2} \int_{-\infty}^{x} z(t)\,dt = \tau^{*}\int_{-\infty}^{x} z(t)\,dt$$

Thus, if (3.13) has an inverse of the form (3.14) for V and for $\tilde{V}$, and if T is isospectral, so that $a(k) = \tilde{a}(k)$, (and the developments below exclude bound states) the following equality must hold for any perturbation $\delta\tilde{V}$ or $\delta\tilde{R}^-$:

$$(3.17) \quad T^* \delta V = \frac{d}{dx} \langle g_+(x,.), \delta\tilde{R}^- \rangle = \tau^* \langle \tilde{g}_+(x,.), \delta\tilde{R}^-(.) \rangle$$

Hence we get a recurrence relation, $\tau^* \tilde{g}_+ = d/dx\, g_+$, which can be solved if τ^* has an unique inverse and if we know a first point in the recurrence. Now, we choose as a reference point the potential V_0 obtained by setting $R_o^-(k)$ with no pole in the upper-half plane. The corresponding potential vanishes on $(x<0)$, where $f_-(k,x)$ is $\exp[-ikx]$, and where $T(k)\, f_+(k,x)$ is a linear combination of $\exp[ikx]$ and $\exp[-ikx]$, with poles in the lower half plane. On $(k > 0)$, $f_+(k,x)$ can be constructed by applying (3.5) as many times as it is necessary to introduce the N poles k_i of $R_o^-(k)$ in the upper half-plane. The result is, for $x > 0$, of the form [7]:

$$(3.18) \quad f_+(k,x) = \exp[ikx] \sum_{i=1}^{N} (k+k_i)^{-1} f_i(x)$$

so that, for $x > 0$ also, $f_+(k,x)$ has poles in the lower half plane only. It readily follows from these remarks, by closing the contour in the upper-half k-plane, that, for the reference parameter V_0, for negative x

$$(3.19) \quad \int_{-\infty}^{+\infty} dk [f_-^o(k,x)]^2\, k^{-1} \int_{-\infty}^{+\infty} dx'\, \delta V(x')\, [f_+^o(k,x')]^2$$

$$= 2i\pi \int_{-\infty}^{x} \delta V(t)\, dt$$

so that (3.12a) is still the inverse of (3.11b). One easily checks that τ^* in the recurrent sequence conserves this property so that the index 0 can be dropped in (3.12a), which is generally the inverse of (3.11b) for negative x. The result is easily generalized. The formula (3.19) is not new, since it follows from a theorem due to Khristov [5,13]. In the Zakharov-Shabot problem, a quite similar technique holds. In the other inverse problems we have cited, again the technique can be adapted and yields new formulas. The main interest of the present approach is that it follows from (3.12) and (3.18) that only trivial calculations are necessary therein to make use of the generalized inverse Born formula, and no supplementary computer time is required. If the inverse problem is underdetermined, this ill-posedness is transformed into that of a linear operator, which is easier to study.

3.4 Proving that global algorithms are necessary. Global algorithms are necessary to study "ambiguities," i.e., cases where several different parameters correspond to the same "exact" result. Needless to say, the existence of ambiguities is related to the definition of the spaces C and E, and we consider them only when they exist for "reasonable" definitions of C and E. We shall study as a typical example the problem described by (3.8), with A twice differentiable except at finitely many points (set S) where it may have jumps, and such that the "potential" W equal to $A^{-\frac{1}{2}}\, d^2/dx^2\, A^{\frac{1}{2}}$ for $x \notin S$ belongs to the class L_2^1 of functions such that $\int_{-\infty}^{+\infty} dx(1+|x|^2)|W(x)| < \infty$. With this assumption $A^{\frac{1}{2}}$ is asymptotic to a straight line $Cx + D$ at infinite x. The "Jost solutions" $p_\pm$ of (3.8) are the solutions whose asymptotic behavior is respectively $[A(x)]^{-\frac{1}{2}} \exp[\pm ikx]$ at $x = \pm\infty$. The "result" is the reflection coefficient $\tilde{R}^+(k)$ defined by the linear relation:

$$(3.20) \quad \tilde{T}(k)\; p_-(k,x) = p_+(-k,x) + \tilde{R}^+(k)\; p_+(k,x)$$

If S is void, $\tilde{T}(k)$ and $\tilde{R}^+(k)$ are equal to the corresponding scattering coefficients of the potential W in the Schrödinger equation - we call this case the regular potential case. For any set S, if ambiguities exist, they are preserved by applying the tranformation (3.9), since $\tilde{R}^+(k)$ is then multiplied by a fixed factor. The transformation (3.9) depends on the choice of p_μ among the linear combinations $ap_-(\mu,x) + bp_-(-\mu,x)$ or $a'p_+(\mu,x) + b'p_+(-\mu,x)$. It is enough to study the case $\operatorname{Im}\mu > 0$, and $b' \neq 0$. Hence, $\tilde{R}^+(k)$ under the transformation (3.9) is multiplied by $(\mu+k)/(\mu-k)$. If $b = 0$, $\tilde{T}(k)$ remains invariant, and so is the whole discrete spectrum (eigenvalues and their normalizing constants). If $b \neq 0$, $\tilde{T}(k)$ is itself multiplied by $(\mu+k)/(\mu-k)$. Setting $\mu = i\gamma$ $(\gamma > 0)$ so as to keep real potentials or admittances, we can see than an eigenvalue appears at $k = i\gamma$, with a normalizing constant depending on a/b. It is well known that the uniqueness of W in L_2^1 is guaranteed if we choose $R^+(k)$ for example in the set R of rational fractions such that, if $|T(k)|^2 = 1-|R^+(k)|^2$ for $k \in \mathbb{R}$,

(a) $T(k)$ is holomorphic in $\operatorname{Im} k > 0$, continuous to $k \in \mathbb{R}$

(b) $T(k) - 1 = 0(|k|^{-1})$ and $0 < |k^{-1} T(k)| < \infty$ for $\operatorname{Im} k \geq 0$

(c) $R^+(k) = 0(|k|^{-1})$ and $|k^{-1}(1+R^+(k))| < \infty$ for $k \in \mathbb{R}$

(d) The Fourier transform of R^+ has a derivative in L^1_2

The condition (b) prevents the existence of a discrete spectrum in $\operatorname{Im} k > 0$. We can apply (3.9) with $\mu = i\gamma$ and let $\gamma \to 0^+$, to a regular case as well as to a discontinuous admittance. In any case, we obtain an infinity of equivalent admittances corresponding to $-R^+(k)$, and differing from each other by the limit value of a/b . One used to say that there is a zero energy bound state. To be more specific, let us introduce the even functions of μ :

(3.21a) $$\psi_\pm(\mu,x) = \tfrac{1}{2}[p_\pm(\mu,x) + p_\pm(-\mu,x)]$$

(3.21b) $$\Phi_\pm(\mu,x) = (2i\mu)^{-1}[p_\pm(\mu,x) - p_\pm(-\mu,x)]$$

so that, as x tends respectively to $\pm\infty$,

(3.22) $$\begin{cases} \psi_\pm(\mu,x) \sim [A(x)]^{-\frac{1}{2}} \cos \mu x \\ \Phi_\pm(\mu,x) \sim \pm[A(x)]^{-\frac{1}{2}} \sin \mu x \end{cases}$$

The pairs of functions Φ,ψ labelled + , or − , remain well-defined and linearly independent as $x \to 0$. The expansion coefficients of ψ_- and Φ_- (for example) in terms of ψ_+ and Φ_+ , trivially depend on $R^+(k)$ and $T(k)$, and they go to definite limits as $\mu \to 0$ for $R^+ \in R$. According to (3.10), the equivalent admittances A_0 corresponding to $-R^+$ are given in terms of the origianl admittance A by

(3.23) $$(A_o)^{\frac{1}{2}} = - \lim_{\gamma\to 0^+} A^{\frac{1}{2}} p'_{i\gamma}/p_{i\gamma}$$

$$(3.24)\quad p_\mu(x) = c\psi_-(\mu,x) + d\Phi_-(\mu,x) = c'\psi_+(\mu,x) + d'\Phi_+(\mu,x)$$

Suppose $A^{\frac{1}{2}}$ goes to β^2 as $x \to -\infty$, to $(\alpha+x)$ as $x \to +\infty$. Then it follows from (3.8), (3.21), and (3.22) that

$$(3.25)\quad \psi_+(0,x) = \int_x^\infty A^{-1}(y)\,dy\ ; \quad \Phi_+(0,x) = 1-\alpha \int_x^\infty A^{-1}(y)dy$$

$$(3.26)\quad \psi_-(0,x) = \beta^{-1}\ ; \quad \Phi_-(0,x) = \beta \int_x^\infty A^{-1}(y)dy-\gamma$$

where γ is the limit of $(\beta \int_x^\infty A^{-1}(y)\,dy + \beta^{-1} x)$ as $x \to -\infty$. Inserting (3.25) or (3.26) into (3.24) and (3.24) into (3.23) explicitly yields this set of equivalent admittances. If the original $\tilde{R}$ was in $\mathcal{R}$, this yields ambiguities for a set of results which differs from $\mathcal{R}$ by giving up the 2nd condition (c). The method works as well if we start from discontinuous admittances. As an exercise, the reader may calculate exactly <u>all</u> quantities corresponding to the admittance equal to 1 for $x < 0$, $(u-2sx/t)^2$ for $x > 0$, with $s<0, t>0$. A particular case is $u = t = 1$, for which $W(x) = -2s\,\delta(x)$. This case contains all the ambiguities [14] given by previous authors and for which $\tilde{R}^+(0) = 1$, and others. The corresponding potentials are regular except at $x = 0$ where there is a δ-function and at $|x| = \infty$ where they vanish like x^{-2} only. One can also start from admittances corresponding to potentials of L^1_2, but one easiy obtains double poles on the real axis. Asymptotic behaviors in x^{-2} and double poles on $\mathbb{R}$ are the common result when one applies the transformation twice (or more times), and let $\mu \to 0$, obtaining for instance

$$(3.27)\quad (A_{oo})^{\frac{1}{2}} = 2p_o p'_0 \lim_{\mu\to o} \frac{\mu}{p'_\mu \frac{\partial}{\partial\mu} p_\mu - p_\mu \frac{\partial}{\partial\mu} p'_\mu}$$

$$= 4p_o p'_o \left[p'_o \left(\frac{\partial^2}{\partial\mu^2} p_\mu\right)_{\mu=0} - p_o \left(\frac{\partial^2}{\partial\mu^2} p'_\mu\right)_{\mu=0}\right]^{-1}$$

this last result being obtained when p_μ is an even function of μ (it is enough that c and d are), and giving a set of ambiguities for which $\tilde{R}(0) = -1$.

3.5 The difficulties in a global exploration. There are many difficulties and we are only beginning to realize them. The most obvious ones are those of finding in general optimal approximations by points "of finitely many coordinates." When this is done, the problem of ambiguities necessarily appears. If we only know $\tilde{R}^{+}(k)$ approximately on a finite interval of $\mathbb{R}$, the best approximations by results with finitely many coordinates of $\tilde{R}^{+}$ and $\tilde{T}$ may contain an eigenvalue close to zero. Hence it is necessary either to know it by a priori assumption or to deal with instabiliites. This difficulty, and other ones, which appear in constructing the algorithms, are very deep, because they show that the natural space in which our transformations operate is not the reduced space of parameters we would like but a much larger one-probably the one in which the most general associated Riemann-Hilbert problem is defined. Even in the simplest "Schrödinger" case, the simplest Darboux transformations very naturally introduce complex potentials. In the Zakharov-Shabat case as well as in the fixed E problem, the transformations naturally introduce singular potentials, and we have seen above in the Schrödinger case the double poles of Cornille's "ghosts". Now this can also be considered as the best point of our algorithmic approach since it shows, by contrast, how much the Marchenko or Gelfand-Levitan method reduce the set C .

4. Well-Posed Questions for Ill-Posed Problems. A well-posed question is a question whose answer can be constructed and depends continuously on the data. Whenever a complete exploration of C , classifying all solution, is not possible, or very expensive, it is good to seek [14] a set of well-posed questions, with the following ultimate aim:

(1) The answer to any of them is clear-cut -- yes or no -- or an exact result -- and is independent of any assumption that is not clearly imposed by physics.

(2) The set of answers yields several "sections" of the set of solutions (weak meanings of this word).

(3) The set of answers enables "decisions" in applied problems -- e.g., those of engineering, geophysical prospecting, etc.

(4) The set of questions is optimized in the following sense: minimum computer work to obtain maximum information consistent with the prerequisites 1, 2, 3.

We began to fully realize the interest of such a set of

questions as an original approach to inverse problems when we applied linear programming [16] to linear inverse problems with a priori bounds (e.g., gravimetry). Partial extensions to linearized inverse problems with bounds (e.g., the consistency bounds necessary to justify the linearizing process) were then processed, in particular by our co-workers [17-21], and helped to clarify our view of this approach. The process can be understood in a quite general case. Suppose for example a set of measurements of an unknown function $x(t)$ $(x \in C_1(\Omega))$, is represented by the M functionals (2.1), where the δe_i's are the (unknown) measurement errors, and $\Omega = [a,b]$ if we work in $\mathbb{R}$, or a finite volume if we work in $\mathbb{R}_3$. Suppose "a priori information" on $x(t)$ is represented in either of the ways:

(a) there are two functions $x_m^o(t)$ and $x_M^o(t)$ such that for any $t \in \Omega$

$$0 \le x_m^o(t) \le x(t) \le x_M^o(t) \tag{4.1}$$

(b) there exists a reference point x_o and a series of distances d_k in C such that

$$d_k(x,x_o) \le X_k^o \tag{4.2}$$

The problem of determining the "infinite information" $x(t)$ from the finite set of measurements (2.1) and the constraints (4.1) or/and (4.2) is obviously ill-posed, and describing the set of its solutions is often awkward. On the other hand, it is clear that we have gained information if, after the measurements have been done, we can write down the inequalities (4.1) or (4.2) with new bounds that are locally better than $x_m^o(t)$, $x_M^o(t)$, X_k^o.

Let us especially consider (4.1). A first set of clear-cut questions can be obtained by covering Ω with a set of intervals $\Omega_1, \Omega_2, \ldots\ldots, \Omega_N$, and setting the questions: C being the set solutions of (2.1), x_k being the average value of $x(t)$ in Ω_k, what is the lower and upper bound of x_k over C ? Clearly, the answer depends on the size of Ω_k. Because all the functionals in (2.1) involve only integrals, if meas. $\Omega_k \to 0$, the upper bound of x_k goes up to $\sup_{\Omega_k} x_M^o(t)$, and there is no gain of information observable in Ω_k. When meas. Ω_k increases, the upper bound of x_k over C may go down and below the lower bound of $x_M^o(t)$ in Ω_k . Hence, information has been gained in Ω_k but its quality obviously depends also on meas. Ω_k . Actually, in the range where information is gained, there is a tradeoff between what can be called its two "components",

$$[\operatorname*{Sup}_{\Omega_k} x_k - \operatorname*{Inf}_{\Omega_k} x_k]$$

and meas. Ω_k . The graph of this tradeoff is usually convex towards 0 . In the highest curvature range, we have the best choices of Ω_k , which give an elementary idea of the "resolving power of the data". Hence, our first set of questions is designed for defining this resolving power of our measurements -- clearly a similar analysis would apply to more general measurements.

The average values x_k on Ω_k can be considered the simplest of the "theoretical measurements" that can be done on a set of "equivalent" solutions obtained from the "actual measurements". Clearly, there are many possible generalizations. Any functional defined on C yields a "theoretical measurement" of a parameter x . If "a priori" information yields bounds on this functional, it is sound to determine its bounds over C , i.e., after the measurements have been done, and to compare them with the previous ones. Examples are the linear moments (e.g., center-of-mass components, mass, inertia moments, etc.), which we introduced when dealing with linear inverse problems with positivity constraints, and which can be easily obtained by linear programming. One can also compare the bounds of any "theoretical measurements" defined on Ω or on a subset Ω_k before and after the actual measurements. This describes the "resolving power" or the "informative content" of the actual measurements relative to this theoretical measurement inside this domain. In some sense, one may think of this approach as a "maximal pattern recognition" of the set of solutions.

"Ideal bodies" [15] are the solutions which achieve extremal properties saturating the bounds of the theoretical measurment we study. The most typical example was introduced in gravimetry: one consider the density distribution consistent with gravity measurements and whose supremum is minimum over the set C of all possible solutions. This ideal body is a "bang-bang" solution, i.e., a solution alternatively saturating its upper and lower bound (the lower bound is 0 in this case). It can be constructed by linear programming, is unique for certain set of functionals [16], stable against small perturbations [22], and can be used as a distinguished solution to give all well-defined informations by direct comparison. The generalization to linearized problems, with consistency bounds of the form (4.1) or (4.2), is simple. There are generalizations to non linear cases [2, 15].

Systematic approaches of applied goephysical problems by a set of "theoretical measurements" and the construction of "ideal bodies" were done in the last few years by the research team of Montpellier, including two mathematical physicists and a few geophysicists. Convergence and stability of the methods and their algorithms were studied together with the application of corresponding codes to real data. This produced interesting results, of a more and more practical nature. Let us cite

(a) (in general) ideal bodies of minimum contrast, or minimizing the largest density, or other binary distribution of parameters [17, 19, 20],

(b) in the magnetism problem, research of maximal extensions of domains in which the magnetism has a defined prescribed sign. The result implies bounds for the spreading rate of oceanic crusts [18],

(c) a posteriori bounds on the focal depth of a seism-determined after linearizing the problem in the neighborhood of the clasical rays, and stating some "physical" a priori bounds [18],

(d) informative results giving parameters bound in slices of the domain, in gravity ([20], geomagnetism [20], seismology [18]. In particular, a systematic study of the content of a polymetallic mine in Neves Corvo by gravimetry, electric and electromagnetic prospecting, has been analyzed along these lines and enables one to give precise bounds for the total content and other value information [21].

This illustrates the conceptual difference between this kind of approach of inversion and the usual one. In the usual one, physicists try to have a "best estimator" of the solution, i.e., the most likely parameter, and usually do not evaluate the range of uncertainties. Here one tries to reduce a problem described by inequalities on real measurements to bounds on theoretical measurements, i.e., to a set of inequalities on ideal, well-chosen theoretical measurements. Hence the results are exact, subject to the validity of a priori assumptions and clear cut, but do not give precise values of the parameters. They are the ultimate generalization of error bars. In terms of "exploration" concept they are typically a "first localization". This method gives a "view" of the set of solutions which is in many ways comparable to the "views" of an object in industrial design, or, more precisely, sort of a convex hull of these views. From the point of view of the scientific method, replacing the usual goal of seeking solutions by seeking a "good" set of well-posed questions is a fascinating work which needs to associate the competences of scientists of several fields and is a remarkable example of interdisciplinary approach.

Concluding Remark. Statistical methods, stochastic processes, algebraic methods, and others, can inspire guide-books for exploring the space of parameters. But the fundamental concept of a route is of geometrical nature and drawing such a route is ultimatelly always a geometrical problem. We hope we have showed here that this very simple remark is essential as well in theoretical approaches as in practical problems of inversion.

References.

[1] D.D. JACKSON, Linear inverse theory with a priori data in Applied Inverse Problems, P.C. Sabatier ed., Springer, Berlin, Heidelberg, and New York, 1978, pp. 83-99.

[2] P.C. SABATIER, Application de la Theorie de l'inversion, Revue du Cethedee 76, (1983) pp. 1-18.

[3] K. CHADAN and P.C. SABATIER, Inverse Problems of Quantum Scattering Theory, Springer, New York, Heidelberg, Berlin, 1977.

[4] W. GUTTINGER, Topological approach to inverse scattering in remote sensing, to be published in Proceedings of Nato Workshop on Inverse Methods in Electromagnetic Imaging, D. Reidel, Dordrecht, Boston, London, 1984.

[5] G. CALOGERO and A. DEGASPERIS, Spectral Transform and Solitons, tome I, North Holland Pub., Cy Amsterdam, New York, Oxford, 1982.

[6] P.C. SABATIER, Rational reflection coefficients and inverse scattering on the line, Nuovo Cimento 78B (1983), pp. 235-248. Also see [12].

[7] P.C. SABATIER, Rational reflection coefficients in one dimensional inverse scattering and applications in Conference on Inverse Scattering: Theory and

Application, SIAM, Philadelphia, 1983, pp. Also see [12].

[8] P.C. SABATIER, Critical analysis of the mathematical methods used in electromagnetic inverse theories: a quest for new routes in the space of parameters, to be published in Proceedings of Nato Workshop on Inverse Methods in Electromagnetic Imaging, D. Reidel, Dordrecht, Boston, London, 1984.

[9] R. LIPPERHEIDE and H. FIEDELDEY, Inverse problem for potential scattering at fixed energy II, Z. Phys. A - Atoms and Nuclei 301, (1981) pp. 81-89.

[10] See the second tom (in preparation) of Reference [5], and A. DEGASPERIS and P.C. SABATIER, in preparation.

[11] C. LADDOMADA and G-Z. TU, Backlund transformations for the Jaulent-Miodek equations, Lett. Math. Phys 6, 1982, pp. 453-462. See also M. BOITI, C. LADDOMADA, and F. PEMPINELLI, Backlund transformations for the Schrödinger equation with a spectral dependence in the potential, to be published.

[12] This algorithm was given in references [6] and [7], where several material errors unfortunately remained: the following erratum must be applied to reference [7], whereas in reference [6] only the correction marked with an asterisk applies.

- in formula (1.8), instead of $\theta(x)$, read $\theta(-x)$
- in formula (1.16), instead of $V(y)$, read $V(x)$
- in the algorithm given in Section 3,

§8° read $-i(\mu_j + \mu_n)^{-1}$ instead of $-\frac{1}{2} i(\mu_j)^{-1}$

§9° read $-i(\mu_j + \mu_n)^{-1}$ instead of $-\frac{1}{2} i(\mu_j)^{-1}$

and $u_{n-1}(x)$ instead of $f_+^{(n-1)}(\mu_n, x)$

* §5° and 10°, read,

$$\Big[\sum_{j=1}^{n-1} (\mu_n-\mu_j)^{-1} R_+^{(n-1),j}/R_+^0(\mu_j)+(-1)^{n-1}\Big]R_+^0(\mu_n)$$

- in formula (3.19), on the left-hand side, a factor 2 is missing.
- (page 12), lines 17-18, read "Poles and residues of $R_0(k)$ are related to resonances. Poles and residues of $T(k)$ in ..."

[13] E.K. KHRISTOV, On spectral properties of operators generating KdV-type equations, in Russian, 1981.

[14] See references in the reference [8], and, for more recent ones, R.G. NEWTON, Remarks on Inverse Scattering in One Dimension, to be published in J. Math. Phys. (1984).

[15] Some general remarks on this research appeared in reference [2] and in an invited paper, entitled like the present section, in the 1983 Joint Meeting of the IEEE Geoscience and Remote Sensing Society and URSI/USNC, September 1983, San Francisco.

[16] P.C. SABATIER, Positivity Constraints in linear inverse problems I and II, Geophys. J.R. Astr. Soc. 48, (1977) pp. 415-466.

[17] R. BAYER and M. CUER, A package of routines for linear inverse problems. Cahiers Mathematiques de Montpellier, 18, (1980).

[18] J.R. GRASSO, M. CUER, and G. PASCAL, Use of two inverse techniques. Application to a local structure in the New Hebrides Island arc, Geophys. J.R. Astr. Soc. 75 (1983), pp. 437-472.

[19] R. BAYER and M. CUER, Inversion tridimensionnelle des donnees aeromagnetiques sur le masif volcanique du Mont-Dore. Implications structurales et geothermiques, Ann. Geophys. 37 (1981), pp. 347-365.

[20] R. BAYER, Interpretation des anomalies du champ de gravite et du champ geomagnetique: methodes et applications geologiques, Thesis Montpellier University (1984).

[21] V. RICHARD, Exploration par la gravimetrie des cibles minieres profondes: application de deux techniques inverses. Exemple de Neves Corvo, Thesis Montpellier University (1984).

[22] M. CUER, Thesis Montpellier University (1984).

THE SEISMIC REFLECTION INVERSE PROBLEM

ALBERT TARANTOLA*

Abstract. The problem of seismic imaging is defined as the problem of obtaining accurate images of the interior of the Earth from the observed seismograms produced by natural or artificial sources. The algebra of nonlinear operators over functional spaces, which constitutes the mathematical background for seismic imaging is introduced. The Fréchet derivatives of seismograms with respect to density, Lame's parameters, and source functions are derived. Particular attention is paid to the problem of interpreting seismic reflection data. It is found that the nonlinear, elastic, seismic reflection inverse problem can be solved by methods which are a natural generalization of the techniques and concepts of classical migration.

This paper is a condensate of two papers submitted to the Geophysical Journal of the Royal Astronomical Society.

1. Introduction. There exists as many images of the interior of the Earth as there exist, more or less, well defined descriptions of the Earth's interior. For instance, images of the Earth can be 2-D or 3-D maps of density, viscoelastic parameters, or even of "reflectivity." Seismic imaging can be defined as the set of methods for obtaining of such images using as inputs observed seismograms.

The seismic imaging inverse problem can be properly posed using the concepts of inverse problem theory. Some inverse problems in geophysics are linear (as for instance some gravity inverse problems); this is not the case for the seismic imaging inverse problem, where it is clear that displacements of the Earth's surface produced by a natural or artifical seismic source are not linearly related with Earth's parameters.

*Institut de Physique de Globe, Universite P-M Curie, 4 Place Jussieu, 75230 Paris, France.

The seismic imaging inverse problem is a large sized problem in the sense that if the three-dimensional parameters describing the Earth (or even a two-dimensional approximation) are discretized (over a given basis of functions or, more simply, over a grid of points in space) the number of discrete parameters thus defined is large enough for usual methods of resolution of inverse problems to be inapplicable with present day computers.

One of the results shown in this paper is that not only the discretization of such problems is unnecessary, but it can even prevent deep understanding of the mathematical objects manipulated and physical interpretation of the operations involved in the solution of the inverse problem. As a corollary, the functional approach can lead to savings of some orders of magnitude in computing time when compared to a naive discretized approach. The price to be paid for that is that this approach is based on the concepts of functional analysis, which are more abstract than the theory of matrix algebra, the only one needed for discretized problems. One of the sections of this paper introduces the basic concepts of functional analysis (norm, scalar product, transpose operator, adjoint operator, derivative of an operator, etc.) from a pragmatic point of view, i.e., neglecting some questions (compactness of operator, completeness of space, etc.) which are important from a mathematical point of view, but which do not arise in practical applications.

Some methods of resolution of nonlinear inverse problems are fully nonlinear, i.e., they never use any linearization of the forward problem (Press 1968, Tarantola and Valette 1982a). With present century computers we cannot hope such methods to be useful for large-sized inverse problems. The seismic imaging inverse problem has then to be set as the problem of obtaining the Earth model which optimizes some functional conveniently defined over the model space. That optimization has to be made using iterative methods, each iteration using a linearization of the forward problem in the vicinity of the "current model". In Section (2.3) we discuss some usual methods of unconstrained optimization useful for the inverse problem theory.

The seismic imaging inverse problem has, of course, some connections with the problem of medical acoustic diffraction tomography (see for instance Devaney, 1983). In fact, the problem of seismic imaging is much more complicated (and conceptually more stimulating) for two reasons: the first one is that elastic (versus acoustic) properties of the Earth cannot be neglected, and the second reason is that in geophysics the inverse model may in general be sufficiently far away from the starting model for usual linearized approximations (Born, etc.) to apply.

The method developed in this paper can be applied to any inverse problem in seismology if it is decided to use as data the whole seismogram, and not only a restricted part of information (arrival times of some phases, periods of normal modes, etc.). This supposes that routines are available which compute realistic synthetic seismograms (i.e., which realistically solve the forward problem) for the configurations under study.

Due to the economic importance of the problem of seismic prospecting, a special effort is made at the end of the paper for proposing a clear solution to the seismic reflection imaging inverse problem.

2. Large Sized Nonlinear Inverse Problems.

2.1. - Model space and data space. Basic data in seismology are seismograms, i.e., the output of seismometers. Solving the forward problem means computing the predicted seismograms when, for instance, the following are given

- density and viscoelastic parameters describing the interior of the Earth: $\rho(\underline{x})$, $\lambda(\underline{x})$, $\mu(\underline{x})$, $Q(\underline{x})$...
- initial conditions for displacement and velocity: $\underline{U}(\underline{x},0)$, $\dot{\underline{U}}(\underline{x},0)$.
- traction boundary conditions $\underline{T}(\underline{x},t)$ and volume density of external forces $\underline{f}(\underline{x},t)$, both representing a single source or a collection of sources spaced in time.
- the transfer function of seismometers, $R(t)$.
- The coordinates of the seismometers location, (x_i, y_i, z_i) .

Each hypothetical value of the whole set of the previous quantities is named a model and is denoted $\underline{m}$. We symbolically write

$$\underline{m} = \begin{bmatrix} \rho(\underline{x}) \\ \lambda(\underline{x}) \\ \cdots \\ \pi_i \end{bmatrix}$$

where π_i (i=1,2,...) represent the discrete part of the model. When solving inverse problems, some of the components of $\underline{m}$ are assumed perfectly known, and the name of model is then reserved to the set of imperfectly known (or totally unknown) functions or discrete parameters. A suitably defined space where we assume the models $\underline{m}$ take their values is named the <u>model space</u> and is denoted M .

We will need a definition of the sum of two models, $\underline{m}_1 + \underline{m}_2$, and a definition of the multiplication of a model by a scalar, $\alpha\underline{m}$. The most natural definition of the sum is "component by component":

$$\underline{m} + \underline{m}' = \begin{bmatrix} \rho(\underline{x}) \\ \lambda(\underline{x}) \\ \cdots \\ \pi_i \end{bmatrix} + \begin{bmatrix} \rho'(\underline{x}) \\ \lambda'(\underline{x}) \\ \cdots \\ \pi_i' \end{bmatrix} = \begin{bmatrix} \rho(\underline{x}) + \rho'(\underline{x}) \\ \lambda(\underline{x}) + \lambda'(\underline{x}) \\ \cdots \\ \pi_i + \pi_i' \end{bmatrix} ,$$

the multiplication by a scalar being defined by

$$\alpha\underline{m} = \begin{bmatrix} \alpha\ \rho(\underline{x}) \\ \alpha\ \lambda(\underline{x}) \\ \cdots \\ \alpha\ \pi_i \end{bmatrix} .$$

As $\rho(\underline{x})$, $\lambda(\underline{x})$,... π_i are physical quantities, the scalar α has to be assumed real (not complex).

With these definitions, we can consider M as an <u>infinite dimensional real vector space</u>.

Each hypothetical value of the set of seismograms (for the whole collection of sources) is named a <u>data set</u> and it is denoted by $\underline{d}$. We symbolically write

$$\underline{d} = \begin{bmatrix} d_1(t) \\ d_2(t) \\ \cdots \end{bmatrix} ,$$

where $d_i(t)$ represents the i'th seismogram. A suitably defined space where we assume the data sets $\underline{d}$ take their values is named the data space and is denoted by $\mathbb{D}$.

As we can naturally define the sum

$$\underline{d} + \underline{d}' = \begin{bmatrix} d_1(t) + d_1'(t) \\ d_2(t) + d_2'(t) \\ \cdots \end{bmatrix} ,$$

and the multiplication by a real number

$$\alpha\underline{d} = \begin{bmatrix} \alpha\ d_1(t) \\ \alpha\ d_2(t) \\ \cdots \end{bmatrix} ,$$

we can consider $\mathbb{D}$ as a vector space. If seismograms are considered as functions of time, $\mathbb{D}$ is infinite-dimensional. If they are considered as discretized functions of time, then $\mathbb{D}$ is finite-dimensional. As this choice is not essential for our theory, let us assume the more general (infinite dimensional) case.

For solving inverse problems, we need to give to the model and data vector spaces an additional structure. Let us first give in the next section the basic concepts of the theory of functional analysis which are useful for our problem.

2.2. - Basic concepts of functional analysis. Most of the definitions of this section are classical, some have been adapted to the purposes of Inverse Problem Theory, and some definitions which are only important from a mathematical point of view but unimportant for practical applications have been suppressed. For more details, the reader should refer to any modern treatise of functional analysis, as for instance Taylor and Lay (1980).

The model space M , the data space $\mathbb{D}$, and the real line $\mathbb{R}$ are the main spaces to be considered. Let $\mathbb{S}$ and T be two arbitrary vector spaces. $\mathbb{S}$ can represent either M or $\mathbb{D}$, while T can represent either M , $\mathbb{D}$, or $\mathbb{R}$.

An arbitrary operator from $\mathbb{S}$ into T (or into a subset of T) will be represented by $\underline{\phi}$, and

(2.1a) $\underline{t} = \underline{\phi}(\underline{s})$

where $\underline{s} \in \mathbb{S}$ and $\underline{t} \in T$. The operator $\underline{\phi}$ is named linear if for any $\underline{s}_1 \in \mathbb{S}$ and $\underline{s}_2 \in \mathbb{S}$ and for any $\alpha \in \mathbb{R}$ and $\beta \in \mathbb{R}$,

$$\underline{\phi}(\alpha \underline{s}_1 + \beta \underline{s}_2) = \alpha\, \underline{\phi}(\underline{s}_1) + \beta\, \underline{\phi}(\underline{s}_2) .$$

Let us give some examples of linear operators between Hilbert spaces. If, for instance, $\mathbb{S}$ and T were discrete spaces, (2.1a) would take the explicit form

$$t_i = \sum_\alpha \Phi_{i\alpha}\, s_\alpha ;$$

if $\mathbb{S}$ were a functional space with functions $s(u)$ as elements, then (2.1a) would take the form

$$t_i = \int dU\, \Phi_i(u)\, s(u)$$

and if $\mathbb{S}$ and T were functional spaces with functions $s(u)$ and $t(v)$, respectively, as elements, then (2.1a) would take the form

$$t(v) = \int du\, \Phi(v,u)\, s(u) .$$

The matrix $\Phi_{i\alpha}$, the column maxtrix of functions $\Phi_i(u)$, and the two-variable function $\Phi(u,v)$ are named, in each case, the kernel of the linear operator $\underline{\phi}$. In such examples, the notation

(2.1b) $\underline{t} = \underline{\Phi}\ \underline{s}$

is preferred to the notation $\underline{t} = \underline{\phi}(\underline{s})$ as introduced in (2.1a).

When $\mathbb{S}$ is a functional space, the "functions $\Phi_i(u)$ or $\Phi(v, u)$ can, in fact, be distributions. For instance, the linear differential equation

$$t(v) = \left[\frac{d^2 s(u)}{du^2} \right]_{u=v}$$

can formally be written as the integral equation

$$t(v) = \int dU \; \delta''(v-u) \; s(u) \; ,$$

where the kernel δ'' represents the second derivative of the Dirac's 'function'.

To simplify the discussion, all through the rest of this section, the vector space $\mathbb{S}$ is considered composed of functions $s(u)$ and the vector space T is considered composed of functions $t(v)$. We have seen in the previous section that the model space M and the data space $\mathbb{D}$ have, in fact, a more complicated structure, but the reader will easily generalize the examples given in this section.

A norm over $\mathbb{S}$ is an operator from $\mathbb{S}$ into the positive real line, notated $|| \; ||$, and verifying the following properties:

$$||s_1 + s_2|| \leq ||\underline{s}_1|| + ||\underline{s}_2||$$

$$||\alpha \, \underline{s}|| = |\alpha| \; ||\underline{s}||$$

$$||\underline{s}|| \neq 0 \quad \text{if} \quad \underline{s} \neq 0 \; .$$

Let $Q(\underline{u})$ be an arbitrary positive function. When it is defined, the quantity

$$(2.2) \qquad ||\underline{s}|| = \{ \int dU \; Q(u) \; |s(u)|^p \}^{1/p}$$

verifies the properties of a norm. It is named the weighted L^p-norm of the function $s(u)$. The set of functions for which (2.2) is defined is named a weighted L^p-space. If a vector space $\mathbb{S}$ furnished with a norm is complete (i.e., each absolutely convergent series of elements of $\mathbb{S}$ convergences to an element of $\mathbb{S}$) it is named Banach Space.

The weighting in (2.2) is unsatisfactory for the purposes of inverse problem theory because it is not general enough: it does not allow to take into account, in the weighting, possible correlations between the values of $\underline{s}$ at different points u and $u' \neq u$. As we will see in section (2.4) this is essential when approaching inverse problems from a functional point of view.

Let $W_s(u,u')$ be a weighting function over $\mathbb{S}$ (we will later see the relation existing between a weighting function and a covariance function over $\mathbb{S}$). The <u>weighted</u> L^2-<u>norm</u> of a function $s(u)$ is defined by

$$(2.3) \qquad ||\underline{s}|| = \{ \int du \int du' \; s(u) \; W_s(u,u') \; s(u') \}^{\frac{1}{2}} \; .$$

With such a definition, the theory of inverse problems can be developed, and in fact, all functional (i.e., not discretized) theories of inverse problems are based on such a norm (Franklin 1970, Tarantola and Valettte 1982b, δ-ness criterion of Backus and Gilbert 1970, etc.). The trouble is that the L^2-norm is known for its lack of robustness, i.e., for its strong sensitivity to a small number of largely erratic data. If such erratic data are suspected in a data set, the L^1-norm should be preferred to the L^2-norm, as shown by Claerbout and Muir (1973) for discrete problems.

As we are not able to give to the definition (2.2) of the L^p-norm (for $p \neq 2$) the same level of generality as the definition (2.3) of the L^2-norm, we have to limit the scope of this paper to methods of inversion over L^2-spaces. On the other hand, one enormous advantage of this restriction is that a L^2-norm necessarily derives from a scalar product, so that we can always define a scalar product over a L^2-space. This fact considerably simplifies the conceptual statement and the practical resolution of inverse problems. If a Banach (i.e., normed, complete) space admits a scalar product, it is then named a <u>Hilbert space</u>. In what follows we consider only Hilbert spaces.

A <u>scalar product</u> is an operator from $\mathbb{S}x\mathbb{S}$ into $\mathbb{R}$, denoted by $< \, , \, >$ and characterized by the following properties

$$<\underline{s},\underline{s}_1+\underline{s}_2> \; = \; <\underline{s},\underline{s}_1> + <\underline{s},\underline{s}_2>$$

$$<\underline{s}_1,\underline{s}_2> \; = \; <\underline{s}_2,\underline{s}_1>$$

$$<\alpha \, \underline{s}_1,\underline{s}_2> \; = \; \alpha <\underline{s}_1,\underline{s}_2>$$

$$<\underline{s},\underline{s}> \; \geq 0 \quad \text{and} \quad <\underline{s},\underline{s}> \; \neq 0 \quad \text{if} \quad \underline{s} \neq 0 \; .$$

Given the weighting function $W_S(u,u')$, the scalar product over $\mathbb{S}$ can be simply defined by

$$(2.4) \quad \langle \underline{s}_1, \underline{s}_2 \rangle = \int du \int du' \; s_1(u) \; W_s(u,u') \; s_2(u')$$

and the L^2-norm (2.3) is obtained as

$$(2.5) \quad ||\underline{s}|| = \langle \underline{s}, \underline{s} \rangle^{1/2} \; .$$

An operator mapping $\mathbb{S}$ into the real line $\mathbb{R}$ is named a <u>form</u> over $\mathbb{S}$. Of particular interest are the <u>linear forms</u> over $\mathbb{S}$. It can be shown that they constitute a vectorial space, which is named the <u>dual</u> of $\mathbb{S}$ and is denoted by $\hat{\mathbb{S}}$. The results of the action of a linear form $\underline{\sigma} \in \hat{\mathbb{S}}$ over an arbitrary $\underline{s} \in \mathbb{S}$ is denoted by $\underline{\sigma}[\underline{s}]$. For a given $\underline{s}_1 \in \mathbb{S}$, the scalar project $\langle \underline{s}_1, \underline{s}_2 \rangle$ clearly defines, for arbitrary $\underline{s}_2 \in \mathbb{S}$, a linear form over $\mathbb{S}$. This means that to any $\underline{s}_1 \in \mathbb{S}$ we can associate an element $\underline{\sigma} \in \hat{\mathbb{S}}$, which will be denoted $\hat{\underline{s}}_1$ and which is defined by

$$(2.6) \quad \hat{\underline{s}}_1[\underline{s}_2] = \langle \underline{s}_1, \underline{s}_2 \rangle$$

Defining

$$(2.7) \quad \hat{s}_1(u) = \int du' \; W_s(u,u') \; s_1(u') \; ,$$

using the symmetry of $W_s(u,u')$, and the definition of the scalar product (2.4), we have

$$(2.8) \quad \hat{\underline{s}}_1[\underline{s}_2] = \int du \; \hat{s}_1(u) \; s_2(u) \; .$$

Thus, we see that if the space $\mathbb{S}$ is composed of functions $s(u)$, the space $\hat{\mathbb{S}}$, dual of $\mathbb{S}$, can be identified to a space of functions $s(u)$, defined for the same variable u, and related with the functions $s(u)$ through the equation (2.7).

It is useful to interpret intuitively $s(u)$ and $\hat{s}(u)$ as infinite dimensional column matrices. The notation $\hat{\underline{s}}_1^T \underline{s}_2$ is then preferred to the notation $\hat{\underline{s}}_1[\underline{s}_2]$:

$$(2.9) \quad \hat{\underline{s}}_1^T \underline{s}_2 = \hat{\underline{s}}_1[\underline{s}_2] = \int du \; \hat{s}_1(u) \; s_2(u) \; .$$

The quantity $\underline{s}_2^T \hat{\underline{s}}_1$ can also naturally be defined thus giving

(2.10) $$\underline{s}_2^T \hat{\underline{s}}_1 = \hat{\underline{s}}_1^T \underline{s}_2 .$$

If a Hilbert space $\mathbb{S}$ is furnished with a scalar product $< , >_S$, it is useful to furnish $\hat{\mathbb{S}}$ with the associated scalar product.

(2.11) $$<\hat{\underline{s}}_1,\hat{\underline{s}}_2>_{\hat{\mathbb{S}}} = <\underline{s}_1,\underline{s}_2>_{\mathbb{S}} .$$

We accept here without demonstration, that the space $\hat{\hat{\mathbb{S}}}$, the dual of $\hat{\mathbb{S}}$, can be identified with $\mathbb{S}$.

Let $\underline{\Phi}$ be a linear operator mapping $\mathbb{S}$ into $\mathbb{T}$, let $\underline{\Lambda}$ be a linear operator mapping $\hat{\mathbb{T}}$ into $\hat{\mathbb{S}}$. $\underline{\Lambda}$ is the <u>transpose</u> of $\underline{\Phi}$, and is denoted $\underline{\Phi}^T$ if

(2.12) $$<\underline{\Phi}\underline{s},\underline{t}>_{\mathbb{T}} = <\hat{\underline{s}},\underline{\Phi}^T\hat{\underline{t}}>_{\hat{\mathbb{S}}} .$$

The following properties can be demonstrated

(2.13) $$\begin{cases} (\underline{\Phi}_1 \underline{\Phi}_2)^T = \underline{\Phi}_2^T \underline{\Phi}_1^T \\ (\underline{\Phi}^T)^T = \underline{\Phi} \\ (\underline{\Phi}\underline{s})^T = \underline{s}^T \underline{\Phi}^T \end{cases}$$

thus showing that the symbol $(\ )^T$ can be used as in classical matricial analysis.

If $\mathbb{T} = \hat{\mathbb{S}}$, i.e., if $\underline{\Phi}$ maps $\mathbb{S}$ into its dual $\hat{\mathbb{S}}$, then, by definition, $\underline{\Phi}^T$ also maps $\mathbb{S}$ into $\hat{\mathbb{S}}$. If, in this case, $\underline{\Phi}^T = \underline{\Phi}$, then the operator $\underline{\Phi}$ is <u>symmetric</u>. If $\underline{\Phi}$ is symmetric and, if for any $\underline{s} \neq \underline{0}$ we have

$$\underline{s}^T \underline{\Phi} \underline{s} > 0 ,$$

the operator $\underline{\Phi}$ is positive definite.

Let $\underline{\Phi}$ be a linear operator mapping $\mathbb{S}$ into T , and let $\underline{\Lambda}$ be a linear operator mapping T into $\mathbb{S}$. The adjoint and the inverse of $\underline{\Phi}$ are defined as follows:

$\underline{\Lambda}$ is the adjoint of $\underline{\Phi}$, and is denoted $\underline{\Phi}^*$ if

$$(2.14) \qquad \langle \underline{t}, \underline{\Phi}\,\underline{s} \rangle_T = \langle \underline{\Phi}^* \underline{t}, \underline{s} \rangle_{\mathbb{S}} \ .$$

The following properties can be demonstrated

$$(2.15a) \qquad (\underline{\Phi}_1 \underline{\Phi}_2)^* = \underline{\Phi}_2^* \, \underline{\Phi}_1^* \qquad (\underline{\Phi}^*)^* = \underline{\Phi} \ .$$

If $T = \mathbb{S}$, i.e., if $\underline{\Phi}$ maps $\mathbb{S}$ into itself, then by definition $\underline{\Phi}^*$ also maps $\mathbb{S}$ into itself. If, in that case, $\underline{\Phi}^* = \underline{\Phi}$, then the operator $\underline{\Phi}$ is self-adjoint. If $\underline{\Phi}$ is self-adjoint and, for any $\underline{s} \neq 0$ we have

$$\langle \underline{s}, \underline{\Phi}\,\underline{s} \rangle_{\mathbb{S}} > 0 \ ,$$

the operator $\underline{\Phi}$ is positive definite.

$\underline{\Lambda}$ is the inverse of $\underline{\Phi}$, and is denoted $\underline{\Phi}^{-1}$ if

$$(2.16) \qquad \underline{\Phi}^{-1} \, \underline{\Phi} \, \underline{s} = \underline{s} \quad (\underline{s} \in \mathbb{S}) \qquad \underline{\Phi} \, \underline{\Phi}^{-1} \, \underline{t} = \underline{t} \quad (\underline{t} \in T) \ .$$

Neglecting some mathematical subtleties, we can crudely say that, given a linear operator $\underline{\Phi}$ mapping $\mathbb{S}$ into T , the transpose of $\underline{\Phi}$, $\underline{\Phi}^T$, and the adjoint of $\underline{\Phi}$, $\underline{\Phi}^*$, always exist. For the purposes of our theory, as we will see, we only need to define the inverse of positive definite operators. It also always exists.

A linear, symmetric, positive definite operator $\underline{C}_S$ mapping $\hat{\mathbb{S}}$ into $\mathbb{S}$ is named a covariance operator over $\mathbb{S}$.

As $\underline{C}_s$, is assumed positive definite, its inverse $\underline{C}_s^{-1}$ exists, and maps $\mathbb{S}$ into $\hat{\mathbb{S}}$.

If the function $W_s(u,u')$ that was used for defining the scalar product in $\mathbb{S}$ (eq. 2.4) is identified with the kernel of $\underline{C}_s^{-1}$, the corresponding scalar product is called the <u>natural</u> scalar product associated to the covariance operator $\underline{C}_s$.

Equation (2.7) can then be written

$$\hat{\underline{s}} = \underline{C}_s^{-1} \underline{s} . \tag{2.17}$$

In that case, the space $\hat{\mathbb{S}}$ itself can be identified with the image of $\mathbb{S}$ through $\underline{C}_s^{-1}$:

$$\hat{\mathbb{S}} = \underline{C}_s^{-1} \mathbb{S} .$$

Of course, we also have the inverse equations

$$\begin{aligned} \underline{s} &= \underline{C}_s \hat{\underline{s}} \\ \mathbb{S} &= \underline{C}_s \hat{\mathbb{S}} . \end{aligned} \tag{2.18}$$

The scalar product can then be written, using (2.6), (2.9), (2.10), and (2.17),

$$\langle \underline{s}_1, \underline{s}_2 \rangle_{\mathbb{S}} = \hat{\underline{s}}_1^T \underline{s}_2 = \underline{s}_1^T \underline{C}_s^{-1} \underline{s}_2 . \tag{2.19}$$

Properties (2.15a) of the symbol $(\ )^*$ are similar to properties (2.13) of the symbol $(\ )^T$, but the equality

$$(\underline{\Phi} \underline{s})^* = \underline{s}^* \underline{\Phi}^* \tag{2.15b}$$

is missing in (2.15a). This last equation makes sense if we define

$$\underline{s}^* = \hat{\underline{s}}^T = \underline{s}^T \underline{C}_s^{-1} . \tag{2.20}$$

We can now give the final notations for the scalar product (2.19) on the space $\mathbb{S}$

$$\langle \underline{s}_1, \underline{s}_2 \rangle_{\mathbb{S}} = \underline{s}_1^* \underline{s}_2 = \underline{s}_1^T \underline{C}_s^{-1} \underline{s}_2 . \tag{2.21}$$

The norm on $\mathbb{S}$ is then given by

$$(2.22) \quad ||\underline{s}||^2_{\mathbb{S}} = \langle \underline{s},\underline{s}\rangle = \underline{s}^* \underline{s} = \underline{s}^T \underline{C}_s^{-1} \underline{s} .$$

The reader will easily verify that the scalar product over $\hat{\mathbb{S}}$ can be written

$$\langle \hat{\underline{s}}_1, \hat{\underline{s}}_2 \rangle_{\hat{\mathbb{S}}} = \hat{\underline{s}}_1^T \underline{C}_s \hat{\underline{s}}_2 ,$$

thus showing that if the inverse of the covariance operator defines the scalar product in $\mathbb{S}$ (equation 2.21), it is the covariance operator $\underline{C}_S$ itself which defines the scalar product on $\hat{\mathbb{S}}$.

We can now derive the relationship between the adjoint and the transpose of an operator. Let $\underline{\Phi}$ be a linear operator from a Hilbert space $\mathbb{S}$ into a Hilbert space T, let $\underline{C}_S$ be the covariance operator defining the scalar product in $\mathbb{S}$ and let $\underline{C}_T$ be the one in T. We successively have

$$\langle \underline{t},\underline{\Phi}\underline{s}\rangle_T = \underline{t}^T\underline{C}_T^{-1}\underline{\Phi}\underline{s} = \underline{t}^T\underline{C}_T^{-1}\underline{\Phi}\underline{C}_s\underline{C}_s^{-1}\underline{s}$$

$$= (\underline{C}_s\underline{\Phi}^T\underline{C}_T^{-1}\underline{t})^T\underline{C}_s^{-1}\underline{s} = \langle \underline{C}_s\underline{\Phi}^T\underline{C}_T^{-1}\underline{t},\underline{s}\rangle_{\mathbb{S}}$$

and, by comparison with (2.18) we deduce the formula

$$(2.23) \quad \underline{\Phi}^* = \underline{C}_s\underline{\Phi}^T\underline{C}_T^{-1} .$$

Let us denote by $C_S(u,u')$ the kernel of the covariance operator $\underline{C}_S$ (such a kernel is named a <u>covariance function</u>) and let us denote by $W_S(u,u')$ the kernel of $\underline{C}_S^{-1}$ (which is named a <u>weighting function</u>). The equation

$$\underline{C}_s \underline{C}_s^{-1} = \underline{I}$$

is written, in terms of kernels,

$$(2.24) \quad \int du' \, C_s(u,u') \, W_s(u',u'') = \delta(u-u'') .$$

The scalar product (2.21) is written

$$(2.25) \quad \langle \underline{s}_1, \underline{s}_2 \rangle_{\mathbb{S}} = \int du \int du' \; s_1(u) \; W_s(u,u') \; s_2(u') \; ,$$

equation identical to (2.4). The covariance function $C_S(u,u')$ can be chosen arbitrarily, but it happens that in usual applications covariance functions are, in general, smoothing functions (Pugachev, 1965), i.e., very regular functions. This means in particular that covariance operators are well behaved integral operators. Their inverses, $\underline{C}_S^{-1}$, are then differential operators, i.e., the kernel $W_S(u,u')$ is a distribution. The more $C_S(u,u')$ is regular, the more $W_S(u,u')$ is pathological. For usual covariance functions, weighting functions are, in fact, rather complicated distributions. This means that, in practice, we will not be able to use (2.25) for effectively computing a scalar product. Instead, we shall proceed as follows. We first write

$$\langle \underline{s}_1, \underline{s}_2 \rangle_{\mathbb{S}} = \underline{s}_1^T \; \hat{\underline{s}}_2$$

where

$$\hat{\underline{s}}_2 = \underline{C}_s^{-1} \; \underline{s}_2$$

i.e.,

$$\underline{C}_s \; \hat{\underline{s}}_2 = \underline{s}_2 \; .$$

Explicitly

$$(2.26a) \quad \langle \underline{s}_1, \underline{s}_2 \rangle_{\mathbb{S}} = \int du \; \underline{s}_1(u) \; \hat{s}_2(u)$$

$$(2.26b) \quad \int du' \; C_s(u,u') \; \hat{s}_2(u') = s_2(u) \; .$$

Given $C_S(u,u')$ and $s_2(u)$, equation (2.26b) can be solved using any numerical method as, for instance, by discretization of the variables u and u', which transforms the integral equation into a matrix equation. The same discretization used for solving (2.26b) can be used for evaluating the integral in (2.26a). With such a method, the scalar product can, in principle, be computed with arbitrary accuracy, in spite of the intrinsic instability of the solution of (2.26b).

We turn now to the characterization of the transpose, $\underline{\Phi}^T$, of a linear operator $\underline{\Phi}$. Let

(2.27a) $\underline{t}_1 = \underline{\Phi}\ \underline{s}_1$

represent a linear operator from $\mathbb{S}$ into $\mathbb{T}$, and

(2.27b) $t_1(v) = \int du\ \Phi(v,u)\ s_1(u)$

its integral representation, where $\Phi(v,u)$ is the kernel of $\underline{\Phi}$. By definition, $\underline{\Phi}^T$ is a linear operator from $\hat{\mathbb{T}}$ into $\hat{\mathbb{S}}$. Let

(2.28a) $\hat{\underline{s}}_2 = \underline{\Phi}^T\ \hat{\underline{t}}_2$

represent such an operator, and

(2.28b) $\hat{s}_2(u) = \int dv\ \underline{\Phi}^T(u,v)\ \hat{t}_2(v)$

its integral representation, where $\Phi^T(u,v)$ is the kernel of $\underline{\Phi}^T$. We have to seek the relation between $\Phi(v,u)$ and $\Phi^T(u,v)$. Using for instance the properties (2.13) we have, for arbitrary $\underline{s} \in \mathbb{S}$ and $\hat{\underline{t}} \in \hat{\mathbb{T}}$,

$$\underline{s}^T\ \underline{\Phi}^T\hat{\underline{t}} = \hat{\underline{t}}^T\underline{\Phi}\,\underline{s}$$

or, in explicit form

$$\int du\ s(u) \int dv\ \Phi^T(u,v)\ \hat{t}(v) = \int dv\ \hat{t}(v) \int du\ \Phi(v,u)\ s(u) \ ,$$

and from this equation we deduce

(2.29) $\Phi^T(u,v) = \Phi(v,u)$.

So, if the integral representation of (2.27a) is (2.27b),

the integral representation of (2.28a) is

$$\hat{s}_2(u) = \int dv\ \Phi(v,u)\ \hat{t}_2(v) \quad . \tag{2.28c}$$

This result simply generalizes to the functional case the well known result that in discrete analysis there is, in fact, no difference between the kernel of an operator (a matrix) and the kernel of the transpose operator (the transposed matrix).

To end this section we have to introduce the concept of the derivative of a nonlinear operator. Let $\underline{\phi}$ represent an arbitrary operator from $\mathbb{S}$ into $\mathbb{T}$. If a given point $\underline{s} \in \mathbb{S}$ and for an arbitrary $\delta\underline{s} \in \mathbb{S}$ we can write

$$\underline{\phi}(\underline{s}_o + \delta\underline{s}) = \underline{\phi}(\underline{s}_o) + \underline{\psi}(\underline{s}_o,\delta\underline{s}) + \underline{o}(||\delta\underline{s}||^2_{\mathbb{S}}) \tag{2.30}$$

where $\underline{o}(||\delta s||^2_{\mathbb{S}})$ represents a term whose norm tends to vanish more rapidly than the norm of $\delta\underline{s}$ when this last tends to zero

$$\lim_{||\delta\underline{s}||_{\mathbb{S}}\to 0} \frac{||\underline{o}(||\delta\underline{s}||^2_{\mathbb{S}})||_{\mathbb{T}}}{||\delta\underline{s}||_{\mathbb{S}}} = 0$$

and where, for fixed $\underline{s}_o$, $\underline{\psi}$ is linear in $\delta\underline{s}$, then the operator $\underline{\phi}$ is said differentiable at $\underline{s} = \underline{s}_o$, while $\underline{\psi}$ is named the <u>Frechet derivative of $\underline{\phi}$ with respect to $\underline{s}$ at the point</u> $\underline{s} = \underline{s}_o$.

Sometimes, the notation

$$\underline{\psi} = \frac{\partial\underline{\phi}}{\partial\underline{s}}$$

is used to recall that $\underline{\psi}$ is the derivative of $\underline{\phi}$ with respect to $\underline{s}$. Also, to recall that, for given $\underline{s}_o$, $\underline{\psi}$ is linear in $\delta\underline{s}$, the notation $\underline{\Psi}_o\ \delta\underline{s}$ should be preferred to the notation $\underline{\Psi}_o(\delta\underline{s})$ thus giving for (2.30),

$$\phi(\underline{s}_o + \delta\underline{s}) = \underline{\phi}(\underline{s}_o) + \underline{\Psi}_o\ \delta\underline{s} + \underline{o}(||\delta\underline{s}||^2_{s}) \quad . \tag{2.31}$$

It is easy to demonstrate that if $\underline{\phi}$ is a linear operator, then, for any $\underline{s} \in \mathbb{S}$,

$$\underline{\phi}(\underline{s}) = \underline{\Psi}\ \underline{s} ,$$

where $\underline{\Psi}$ is independent of $\underline{s}$.

For a later use, we should recall that, if $\underline{\phi}$ represents an application from $\mathbb{S}$ into T , its derivative $\underline{\Psi}_0$ at a given point $\underline{s}_0$ is a linear operator also from $\mathbb{S}$ into T .

In the particular case where we consider a nonlinear form over $\mathbb{S}$ denoted $\phi(\underline{s})$, equation (2.31) is rewritten

$$(2.32) \quad \phi(\underline{s}_0 + \delta\underline{s}) = \phi(\underline{s}_0) + \langle \underline{\gamma}_0, \delta\underline{s} \rangle_S + \underline{o}(||\delta\underline{s}||^2_{\mathbb{S}}) .$$

thus defining $\underline{\gamma}_0$, the <u>gradient</u> of ϕ at the point $\underline{s}_0$. It is clear that, for given $\underline{s}_0$, $\underline{\gamma}_0$ is an element of $\mathbb{S}$. For variable $\underline{s}_0$, the gradient $\underline{\gamma}_0$ has to be considered as a (in general nonlinear) function of $\underline{s}_0$. For a differentiable $\underline{\gamma}$, the <u>Hessian</u> of ϕ at the point $\underline{s}_0$ is defined by

$$(2.33) \quad \underline{h}_0 = \left(\frac{\partial \underline{\gamma}}{\partial \underline{s}}\right)_{\underline{s}=\underline{s}_0}$$

Here, $\underline{\gamma}$ is clearly considered as an application from $\mathbb{S}$ into $\mathbb{S}$, so that, for given $\underline{s}_0$, $\underline{h}_0$ is a linear application also from $\mathbb{S}$ into $\mathbb{S}$. The notation $\underline{s}' = \underline{H}_0\underline{s}$ is then preferred to the notation $\underline{s}' = \underline{h}_0(\underline{s})$.

It can be shown that the Hessian is a self-adjoint operator:

$$\underline{H} = \underline{H}^* .$$

<u>2.3. - Basic concepts of optimization theory</u>. As previously discussed, any realistic method for solving large sized inverse problems has to be based on the optimization of some real functional (nonlinear form) $S(\underline{m})$ defined over M . For instance, $S(\underline{m})$ can be a measure of discrepancy between the values of some measurable quantities, observed during a physical experiment (the data set), and the values predicted (using a physical theory) for the model $\underline{m}$. As geophysical imaging problems are always underdetermined

(Backus and Gilbert, 1968), the function $S(\underline{m})$ will contain, in addition to this measure of discrepancy between observed and predicted values of data, a measure of discrepancy between $\underline{m}$ and some a priori assumptions on the model (smoothness, etc.). The model $\underline{m}$ which minimizes $S(\underline{m})$ is then named the best model (with respect to $S(\underline{m})$).

Alternatively, $S(\underline{m})$ may represent the likelihood of $\underline{m}$, and can be conveniently defined from statistics in the data set and a priori statistics of the model (Tarantola and Valette, 1982a). In that case the model $\underline{m}$ maximizing $S(\underline{m})$ is named maximum likelihood model.

The model space M is assumed to be a Hilbert space with a scalar product represented by $<\underline{m}_1,\underline{m}_2>$. Let us try to recall some usual rules for minimizing a given nonlinear functional $S(\underline{m})$ defined over M. We assume here that $S(\underline{m})$ is twice differentiable (i.e., that second order Frechet derivative of S exists). If instead of a minimization, we have a maximization problem, the changes to be made to the following discussion are trivial.

Let $\underline{\gamma}_n \in M$ represent the gradient of $S(\underline{m})$ at an arbitrary point $\underline{m}_n$. We accept here without demonstration that $\underline{\gamma}_n$ defines the direction of steepest ascent at the point $\underline{m}_n$.

The condition:

(2.34a) S minimum

at an arbitrary point $\underline{m}_n$ implies

(2.34b) $\underline{\gamma}_n = \underline{0}$.

Of course, we have not the equivalence ($\underline{\gamma}_n = \underline{0} \not\Rightarrow S$ minimum), unless $S(\underline{m})$ is strictly convex. As we will deal with nonlinear inverse problems, this condition will never be warranted in advance, so in each case if a point $\underline{m}_n$ has been obtained such that $\underline{\gamma}_n = \underline{0}$, a special analysis should be made to ensure that we are not in a secondary minima of S (we will later see that maxima and saddle points are easily eliminated).

We say that a method for solving (2.34) is "iterative" if at any current point $\underline{m}_n$ it can furnish a point $\underline{m}_{n+1}$ which is better than $\underline{m}_n$ (in the sense that $S(\underline{m}_{n+1}) < S(\underline{m}_n)$). Starting at an arbitrary point $\underline{m}_0$, such a method furnishes a series of points $\underline{m}_0, \underline{m}_1, \ldots$ that, if it converges, converges necessarily to a (local) minimum of S.

A wide set of such iterative methods are known, as for instance Monte-Carlo (random) methods (Press, 1968, Keilis-Borok and Yanovskaya, 1967). Nevertheless, for large sized problems we have to use methods which use the properties of the function $S(\underline{m})$ at the vicinity of $\underline{m}_n$ (tangent space, curvature of S, etc.) in order to estimate $\underline{m}_{n+1}$. The most popular of these are Newton methods and gradient methods.

A Newton method chooses for $\underline{m}_{n+1}$ the point at which the first order expansion of $\underline{\gamma}$ around $\underline{m}_n$ vanishes:

$$\underline{\gamma}_{n+1} = \underline{\gamma}_n + \left(\frac{\partial \underline{\gamma}}{\partial \underline{m}}\right)_n (\underline{m}_{n+1} - \underline{m}_n) = \underline{0}$$

i.e.,

$$\underline{m}_{n+1} = \underline{m}_n - \left(\frac{\partial \underline{\gamma}}{\partial \underline{m}}\right)_n^{-1} \underline{\gamma}$$

or, introducing the Hessian (2.33),

$$(2.35) \quad \underline{m}_{n+1} = \underline{m}_n - \underline{H}_n^{-1} \underline{\gamma}_n .$$

The last is named the Newton formula. The Hessian $\underline{H}_n$ is a self-adjoint operator. We will later see that in inverse problem theory it is always positive definite, so its inverse $\underline{H}_n^{-1}$ always exists.

For nonlinear inverse problems, we have no warranty that the point $\underline{m}_{n+1}$ given by the Newton formula will verify $S(\underline{m}_{n+1}) < S(\underline{m}_n)$. If this last condition is not verified, we can replace (2.35) by

$$(2.36) \quad \underline{m}_{n+1} = \underline{m}_n - \varepsilon\, \underline{H}_n^{-1} \underline{\gamma}_n ,$$

where ε is a real number $0 < \varepsilon < 1$. As $-\underline{\gamma}_n$ is a direction of descent, for a sufficiently small value of ε the point $\underline{m}_{n+1}$ will be better that the point $\underline{m}_n$. In practical applications, the value $\varepsilon = 1$ is first considered. If it is not acceptable, then the value $\varepsilon = \frac{1}{2}$ proves generally to be good.

The operator $\underline{H}_n^{-1}$ of (2.35) can, in fact, be replaced by an arbitrary self-adjoint positive definite operator:

(2.37) $\underline{m}_{n+1} = \underline{m}_n - \underline{Q}\,\underline{\gamma}_n$.

It is clear that; provided that (2.37) is convergent, it converges to a stationary point of $S(\underline{\gamma}_\infty = \underline{0})$. A usual choice for $\underline{Q}$ is

$$\underline{Q} = \underline{H}_0^{-1}$$

or even an (astute) approximation of $\underline{H}_0^{-1}$. In general, the more $\underline{Q}$ is different from $\underline{H}_n^{-1}$ the slowest is the convergence of the algorithm. As $\underline{H}_n^{-1}$ is in general too complicated to be used, the right choice for $\underline{Q}$ results from the trade-off between the complexity of $\underline{Q}$ and the number of iterations needed to obtain an accurate enough solution of our problem. Equation (2.37) defines a <u>Quasi-Newton method</u>.

Gradient methods are based on the following idea: let $\underline{m}_n$ be the current point, and let $\underline{d}_n$ be an arbitrary direction at $\underline{m}_n$. A generic point $\underline{m}$ on the direction $\underline{d}_n$ is writen

(2.38) $\underline{m} = \underline{m}_n + \alpha_n\,\underline{d}_n$,

where α_n is an arbitrary real number. If we choose α_n so that $S(\underline{m})$ is minimized (along that diretion), and we iterate the process, we will certainly converge to some minimum of S . We have

$$S(\underline{m}_{n+1}) = S(\underline{m}_n + \alpha_n \underline{d}_n)$$

For given $\underline{m}_n$ and $\underline{d}_n$, the optimal value of α_n is obtained by the condition

$$\left(\frac{dS}{d\alpha_n}\right) = 0$$

which gives

(2.39) $\langle \underline{\gamma}_{n+1}, \underline{d}_n \rangle = 0$.

This equation means that, at the mimimum of S along the direction $\underline{d}_n$, the gradient of S (i.e., $\underline{\gamma}$) is <u>perpendicular</u> to $\underline{d}_n$. Equivalently, we can say that at the minimum of S the 'level lines' of $S(\underline{m})$ are <u>tangent</u> to $\underline{d}_n$ (because the gradient is itself perpendicular to the level lines). To the first order in $\underline{m}_{n+1} - \underline{m}_n$ we have

(2.40) $\underline{\gamma}_{n+1} \simeq \underline{\gamma}_n + \left(\frac{\partial \underline{\gamma}}{\partial \underline{m}}\right)_n (\underline{m}_{n+1} - \underline{m}_n)$

i.e.,

$$\underline{\gamma}_{n+1} \simeq \underline{\gamma}_n + \underline{H}_n \, (\underline{m}_{n+1} - \underline{m}_n) \; .$$

using (2.38) and (2.39) we obtain

$$<\underline{d}_n , \; (\underline{\gamma}_n + \alpha_n \underline{H}_n \underline{d}_n)> \; \simeq 0$$

i.e.,

$$(2.41) \qquad \alpha_n \simeq - \frac{<\underline{d}_n , \; \underline{\gamma}_n>}{<\underline{d}_n , \; \underline{H}_n \underline{d}_n>} \; .$$

If $S(\underline{m})$ is quadratic in $\underline{m}$ (as will happen for linear or linearized inverse problems), the first order development (2.40) is exact, and we can replace the symbol $\simeq$ in (2.41) by equality. The value of α_n thus obtained, perfectly minimizes $S(\underline{m})$ along the direction $\underline{d}_n$. As when $S(\underline{m})$ is quadratic, its minimum is unique, in this case, any gradient method is convergent, and converges to the unique minimum.

For nonlinear inverse problems, $S(\underline{m})$ will not be quadratic, so the value α_n given by equation (2.41) is only a linear approximation to the optimal value. The simplest strategy for still ensuring convergence in that case is to compute the value $S_{n+1} = S(\underline{m}_{n+1})$ at each iteration; if S_{n+1} is not better than S_n this geometrically means that we have gone too far along the direction $\underline{d}_n$ and crossed the level line $S = S(\underline{m}_n)$ "at the other side of the valley". It is clear that we have then to change α_n to an α'_n such that S'_{n+1} is better than S_n (for instance, the value $\alpha'_n = \alpha_n/2$ proves often to be adequate). With this strategy we can warrant the convergence of the algorithm whatever the functional $S(\underline{m})$ can be. Of course, in the general case, the extremum of S may not be unique, so we can converge to secondary minima. The only practical strategy for checking the existence of such secondary solutions is to initialize the iterative sequence at different points $\underline{m}_0$ and to verify that the convergence point $\underline{m}_\infty$ is always the same. If it is not, the problem has to be reconsidered and, methods other than least squares may prove more appropriate.

Until now, the direction $\underline{d}_n$ remains arbitrary. Let us give some examples of choice for $\underline{d}_n$:

Steepest decent. The simplest choice for $\underline{d}_n$ is the direction of steepest descent

$$\underline{d}_n = - \underline{\gamma}_n \, .$$

Using (2.38) and (2.41) this gives the algorithm

$$(2.42a) \begin{cases} \alpha_n = \dfrac{\langle \underline{\gamma}_n \, , \, \underline{\gamma}_n \rangle}{\langle \underline{\gamma}_n , \, \underline{H}_n \underline{\gamma}_n \rangle} \\ \underline{m}_{n+1} = \underline{m}_n - \alpha_n \, \underline{\gamma}_n \end{cases}$$

Preconditioned steepest descent. This choice consists in taking

$$\underline{d}_n = - \underline{Q}_n \, \underline{\gamma}_n$$

where $\underline{Q}_n$ is an arbitrary self-adjoint positive definite operator, suitably chosen for accelerating the convergence of the algorithm. Using (2.38), (2.40), and (2.41), this gives the algorithm

$$(2.42b) \begin{cases} \underline{\phi}_n = \underline{Q}_n \, \underline{\gamma}_n \\ \alpha_n = \dfrac{\langle \underline{\phi}_n , \underline{\gamma}_n \rangle}{\langle \underline{\phi}_n , \underline{H}_n \underline{\phi}_n \rangle} \\ \underline{m}_{n+1} = \underline{m}_n - \alpha_n \, \underline{\phi}_n \end{cases}$$

Taking

$$\underline{Q} = \underline{I}$$

in (2.42b) gives (2.42a), i.e., the steepest descent algorithm. Should we take $\underline{Q}$ variable with the iteration number, the optimum choice will clearly be

$$\underline{Q}_n = \underline{H}_n^{-1}$$

because this gives $\alpha_n = 1$ and

$$\underline{m}_{n+1} = \underline{m}_n - \underline{H}_n^{-1} \, \underline{\gamma}_n$$

which exactly correspond to the Newton algorithm (2.35). In practical applications, as $\underline{H}_n^{-1}$ is too complicated to be useful, a crude approximation of $\underline{H}_0^{-1}$ proves in general to be adequate for Q.

Davidon, Fletcher, Power (D.F.P.) method. In this method, the preconditioning operator Q is updated at each iteration:

$$(2.43)\quad \begin{cases} \underline{\phi}_n = \underline{Q}_n \, \underline{\gamma}_n \\[2mm] \alpha_n = \dfrac{\langle \underline{\phi}_n, \underline{\gamma}_n \rangle}{\langle \underline{\phi}_n, \underline{H}_n \underline{\phi}_n \rangle} \\[2mm] \underline{m}_{n+1} = \underline{m}_n - \alpha_n \underline{\phi}_n \\[2mm] \underline{\beta}_n = \underline{\gamma}_{n+1} - \underline{\gamma}_n \\[2mm] \underline{\mu}_n = \underline{Q}_n \, \underline{\beta}_n \\[2mm] \underline{\theta}_{n+1} = \underline{\theta}_n + \dfrac{\langle \underline{\phi}_n, \cdot \rangle}{\langle \underline{\phi}_n, \underline{H}_n \underline{\phi}_n \rangle} \, \underline{\phi}_n - \dfrac{\langle \underline{\mu}_n, \cdot \rangle}{\langle \underline{\mu}_n, \underline{\beta}_n \rangle} \, \underline{\mu}_n \end{cases}$$

and has the property

$$\underline{Q}_n \to \underline{H}_n^{-1} \qquad \text{(see Walsh, 1975).}$$

The D.F.P. iteration method is usually initialized at $\underline{Q}_0 = \underline{I}$, so at the first iteration it behaves like the steepest descent method. When we approach the solution, the method behaves like the Newton method, with its fast convergence. We will later see that the fact that $\underline{Q}_n$ approaches the inverse of the Hessian, $\underline{H}_n^{-1}$, is of great importance for studying error and resolution when solving inverse problems.

Conjugate directions. Equation (2.39) shows that two successive directions in an algorithm of steepest descent are orthogonal. This means that the steepest descent direction, although corresponding to the simple choice of direction, is probably not the optimal. Fletcher and Reeves (1964) suggest that successive directions have to be chosen conjugate (in the usual sense in conics geometry) rather than orthogonal.

They obtain

$$\underline{d}_0 = - \underline{\gamma}_0$$

$$\underline{d}_n = - \underline{\gamma}_n + \frac{\langle \underline{\gamma}_n, \underline{\gamma}_n \rangle}{\langle \underline{\gamma}_{n-1}, \underline{\gamma}_{n-1} \rangle} \underline{d}_{n-1} \qquad (n \geq 1)$$

This gives the algorithm

$$(2.45) \quad \begin{cases} \omega_n = \langle \underline{\gamma}_n, \underline{\gamma}_n \rangle \\ \underline{\phi}_n = \underline{\gamma}_n + \dfrac{\omega_n}{\omega_{n-1}} \underline{\phi}_{n-1} \qquad (\underline{\phi}_o = \underline{\gamma}_o) \\ \alpha_n = \dfrac{\omega_n}{\langle \underline{\phi}_n, \underline{H}_n \underline{\phi}_n \rangle} \\ \underline{m}_{n+1} = \underline{m}_n - \alpha_n \underline{\phi}_n , \end{cases}$$

where we have used the property

$$\langle \underline{\phi}_n, \underline{\gamma}_n \rangle = \langle \underline{\gamma}_n, \underline{\gamma}_n \rangle .$$

<u>Preconditioned conjugated directions</u>. As was the case for the steepest descent algorithm, the conjugate directions algorithm can be preconditioned. Letting

$$\underline{Q}_n \simeq \underline{H}_n^{-1}$$

we have the algorithm

$$(2.46) \quad \begin{cases} \underline{\lambda}_n = \underline{Q}_n \underline{\gamma}_n \\ \omega_n = \langle \lambda_n, \underline{\gamma}_n \rangle \\ \underline{\phi}_n = \underline{\lambda}_n + \dfrac{\omega_n}{\omega_{n-1}} \underline{\phi}_{n-1} \qquad (\underline{\phi}_0 = \underline{\lambda}_0) \end{cases}$$

$$\left| \begin{array}{l} \alpha_n = \dfrac{\omega_n}{<\underline{\phi}_n, \underline{H}_n \underline{\phi}_n>} \\ \\ \underline{m}_{n+1} = \underline{m}_n - \alpha_n \underline{\phi}_n \ , \end{array} \right.$$

where we have used the property

$$<\underline{\phi}_n, \underline{\gamma}_n> = <\underline{\lambda}_n, \underline{\gamma}_n> .$$

If Q_n is conveniently chosen, this algorithm often proves to be best of all simple gradient algorithms.

2.4. - The generalized nonlinear least-squares criterion. Let M be the model space and $\mathbb{D}$ be the data space. Solving the forward problem means to use a physical theory to predict error free values of data that would correspond to a given model. We symbolically represent by

(2.47) $\underline{d} = \underline{f}(\underline{m})$

the solution of the forward problem. $\underline{f}$ represents a (generally nonlinear) operator from M into $\mathbb{D}$.

In our seismological problem, the operator $\underline{f}$ essentially represents the solution of the elastodynamic wave equation, with given initial and boundary conditions, and the convolution of the computed displacement at the receiver locations by the instrument response. The operator $\underline{f}$ is clearly nonlinear.

As we will see, it will be the only nonlinear operator to be explicitly considered for solving inverse problems.

We assume that a physical experiment has furnished some information about some measurable quantities, and that this information can conveniently be described using the vector of observed values, $\underline{d}_{OBS}$, and the covariance operator $\underline{C}_D$.

The vector $\underline{d}_{OBS}$ generally represent the direct output of some instruments (seismometers plus recording system, in our case). Less trivial is the meaning of $\underline{C}_D$. As $\underline{d}_{OBS}$ results from a single (although possibly complex) experiment, no statistics are available for estimating errors. Instead, $\underline{C}_D$ results from a more or less subjective estimation. Often correlations between errors are neglected and $\underline{C}_D$ is assumed to be a diagonal operator (see Section 5.1 for an example).

Given $\underline{C}_D$, the only natural response to the question of how much close a given data vector $\underline{d}$ is to observed values $\underline{d}_{OBS}$ is given by the norm

$$||\underline{d}-\underline{d}_{OBS}||_D = \{(\underline{d}-\underline{d}_{OBS})^T \underline{C}_D^{-1} (\underline{d}-\underline{d}_{OBS})\}^{\frac{1}{2}} .$$

This norm clearly derives from the scalar product in the data space

$$(2.48) \quad \langle \underline{d}_1, \underline{d}_2 \rangle_D = \underline{d}_1^T \underline{C}_D^{-1} \underline{d}_2$$

by the definition

$$(2.49) \quad ||\underline{d}||_D = \langle \underline{d}, \underline{d} \rangle_D^{\frac{1}{2}} .$$

In what follows we assume our data space $\mathbb{D}$ furnished with the scalar product (2.48) to be a Hilbert space.

We assume now that we have some a priori information about the model, and that this a priori information can conveniently be described using the a priori model $\underline{m}_{PRIOR}$ and the covariance operator in the model space $\underline{C}_M$. Intuitively, $\underline{m}_{PRIOR}$ represents a model from which we do not wish our inverse solution to differ too much. $\underline{C}_M$ thus represents how much we accept our solution to differ from $\underline{m}_{PRIOR}$. For more discussion, the reader should refer to Jackson (1972) for discrete problems, or to Tarantola and Nercessian (1984) for problems involving functions. We give in Section 5.1 an example of choice for $\underline{m}_{PRIOR}$ and $\underline{C}_m$.

As it was the case in the data space, the only natural measure of discrepancy between the a priori model $\underline{m}_{PRIOR}$ and an arbitrary model $\underline{m}$ is given by the norm

$$||\underline{m} - \underline{m}_{PRIOR}||_M = \{(\underline{m} - \underline{m}_{PRIOR})^T \underline{C}_M^{-1} (\underline{m} - \underline{m}_{PRIOR})\}^{\frac{1}{2}}$$

This gives the scalar product in the model space

$$(2.50) \quad \langle \underline{m}_1, \underline{m}_2 \rangle_M = \underline{m}_1^T \underline{C}_M^{-1} \underline{m}_2$$

related with the norm by the usual formula

$$(2.51) \quad ||\underline{m}||_M = \langle \underline{m}, \underline{m} \rangle^{\frac{1}{2}} .$$

Given the observed values $\underline{d}_{OBS}$, the covariance operator $\underline{C}_D$ in the data space, the a priori model $\underline{m}_{PRIOR}$, the covariance operator $\underline{C}_M$ in the model space, and the nonlinear operator $\underline{f}(\underline{m})$ representing the solution of the forward problem, we can state the inverse problem as the problem of obtaining the model which minimizes the expression

(2.52)

$$S(\underline{m}) = \frac{1}{2} \{ ||\underline{f}(\underline{m}) - \underline{d}_{OBS}||_D^2 + ||\underline{m} - \underline{m}_{PRIOR}||_M^2 \}$$

$$= \frac{1}{2} \{ (\underline{f}(\underline{m}) - \underline{d}_{OBS})^T \underline{C}_D^{-1} (\underline{f}(\underline{m}) - \underline{d}_{OBS}) + (\underline{m} - \underline{m}_{PRIOR})^T \underline{C}_M^{-1} (\underline{m} - \underline{m}_{PRIOR}) \}$$

where the factor $\frac{1}{2}$ stands for subsequent simplifications. The solution of the problem is notated $\underline{m}_{EST}$, and is named the least-squares estimator.

We see that $\underline{m}_{EST}$ will be close to the a priori model $\underline{m}_{PRIOR}$ because of the term $||\underline{m} - \underline{m}_{PRIOR}||_M$ in (2.52), and the values of data corresponding to that model, $\underline{d} = \underline{f}(\underline{m}_{EST})$, will be close to the observed values $\underline{d}_{OBS}$, because of the term $||\underline{f}(\underline{m}) - \underline{d}_{OBS}||_D$. The relative trade-off between these two terms is automatically optimized because these terms have the right dependence on the estimated errors in $\underline{d}_{OBS}$ and $\underline{m}_{PRIOR}$ through $\underline{C}_D$ and $\underline{C}_M$ respectively.

The justification of the minimization of $S(\underline{m})$ as given by (2.52) can be obtained as follows. Considering the space $\mathbb{D} \times M$, we should define the best point $(\underline{d}, \underline{m})$ as the point minimizing the expression

$$(2.53) \quad S(\underline{d},\underline{m}) = \frac{1}{2} \begin{bmatrix} \underline{d} - \underline{d}_{OBS} \\ \underline{m} - \underline{m}_{PRIOR} \end{bmatrix}^T \begin{bmatrix} \underline{C}_D & \underline{C}_{DM} \\ \underline{C}_{MD} & \underline{C}_M \end{bmatrix}^{-1} \begin{bmatrix} \underline{d} - \underline{d}_{OBS} \\ \underline{m} - \underline{m}_{PRIOR} \end{bmatrix}$$

with the constraint

(2.54) $\underline{d} = \underline{f}(\underline{m})$,

where $\underline{C}_{DM} = \underline{C}_{MD}^T$ is a covariance operator representing the correlations between errors in $\underline{d}_{OBS}$ and errors in $\underline{m}_{PRIOR}$. In Tarantola and Valette (1982) it is shown that the problem of minimization of (2.53) arises, in a probabilistic context, as equivalent to the problem of maximizing the likelihood of the solution.

If $\underline{m}_{PRIOR}$ is truly an a priori model, this means that it is independent from the observed values of data, $\underline{d}_{OBS}$. Errors in $\underline{m}_{PRIOR}$ are thus independent from errors in $\underline{d}_{OBS}$, so that $\underline{C}_{DM} = \underline{C}_{MD}^T = \underline{0}$. The problem of minimizing (2.53) with the constraint (2.54) then reduces to the problem of the unconstrained minimization of (2.52).

In Section 2.2 we assumed that covariance operators were positive definite. Let us point here that this means in particular that $\underline{C}_D$ or $\underline{C}_M$ cannot contain null or infinite variances or perfect correlations.

We turn now to the computation of the gradient of S . We have

$$S(\underline{m}_n+\delta\underline{m}) = \frac{1}{2}\left\{(\underline{f}(\underline{m}_n+\delta\underline{m})-\underline{d}_{OBS})^T \underline{C}_D^{-1}(\underline{f}(\underline{m}_n+\delta\underline{m})-\underline{d}_{OBS}) + (\underline{m}_n+\delta\underline{m}-\underline{m}_{PRIOR})^T \underline{C}_M^{-1}(\underline{m}_n+\delta\underline{m}-\underline{m}_{PRIOR})\right\}$$

Introducing the Frechet derivative of $\underline{f}(\underline{m})$ at point $\underline{m}_n$ (c.f. Section 2.2):

(2.55) $$\underline{F}_n = \left(\frac{\partial \underline{f}}{\partial \underline{m}}\right)_{\underline{m}_n}$$

we have

(2.56) $$\underline{f}(\underline{m}_n+\delta\underline{m}) = \underline{f}(\underline{m}_n) + \underline{F}_n\delta\underline{m} + \underline{o}(||\delta\underline{m}||_M^2)$$

and

$$S(\underline{m}_n+\delta\underline{m}) - S(\underline{m}_n) = [\underline{F}_n^T\underline{C}_o^{-1}(\underline{f}(\underline{m}_n)-\underline{d}_{OBS}) +$$

$$\underline{C}_M^{-1}(\underline{m}_n - \underline{m}_{PRIOR})]^T\delta\underline{m} + o(||\delta\underline{m}||_M^2) .$$

Denoting by $\underline{\gamma}$ the gradient of S we have, by definition (equation 2.32),

$$S(\underline{m}_n+\delta\underline{m}) = S(\underline{m}_n) + \langle\underline{\gamma}_n, \delta\underline{m}\rangle_M + o(||\delta\underline{M}||_M^2)$$

i.e.,

$$S(\underline{m}_n+\delta\underline{m}) = S(\underline{m}_n) + \underline{\gamma}_n^T \underline{C}_M^{-1} \delta\underline{m} + o(||\delta\underline{m}||_M^2) ,$$

thus giving

$$(2.57) \quad \underline{\gamma}_n = \underline{C}_M\underline{F}_n^T\underline{C}_D^{-1}(\underline{f}(\underline{m}_n)-\underline{d}_{OBS}) + (\underline{m}_n-\underline{m}_{PRIOR})$$

Similarly, we obtain the Hessian

$$(2.58) \quad \underline{H}_n = \left(\frac{\partial\underline{\gamma}}{\partial\underline{m}}\right)_{\underline{m}_n} = \underline{I} + \underline{C}_M\underline{F}_n^T\underline{C}_D^{-1}\underline{F}_n + \underline{C}_M\underline{K}_n^T\underline{C}_D^{-1}(f(\underline{m}_n)-\underline{d}_{OBS}) ,$$

where

$$\underline{K}_n = \left(\frac{\partial\underline{F}}{\partial\underline{m}}\right)_{\underline{m}_n} .$$

If $\underline{d} = f(\underline{m})$ represents a linear operator, $\underline{K}_n$ identically vanishes. For usual nonlinear problems, the last term in (2.58) although not identically null is, in general, small, and as we have seen in Section 2.3 that there is no need for the Hessian to be known with high accuracy, this term is always neglected thus giving

$$(2.59) \quad \underline{H}_n = \underline{I} + \underline{C}_M\underline{F}_n^T\underline{C}_D^{-1}\underline{F}_n$$

Let us point out that, using the notations

$$\underline{m}^* = \underline{m}^T \underline{C}_M^{-1}$$

$$\underline{d}^* = \underline{d}^T \underline{C}_D^{-1}$$

introduced in Section 2.2, and using the operator $\underline{F}_n^*$, adjoint of $\underline{F}_n$, equations (2.52), (2.57), and (2.59) can be written respectively,

$$S(\underline{m}) = \frac{1}{2} \{ (\underline{f}(\underline{m})-\underline{d}_{OBS})^* (\underline{f}(\underline{m})-\underline{d}_{OBS}) + (\underline{m}-\underline{m}_{PRIOR})^* (\underline{m}-\underline{m}_{PRIOR}) \}$$

$$\underline{\gamma}_n = \underline{F}_n^* (\underline{f}(\underline{m}_n) - \underline{d}_{OBS}) + (\underline{m}_n - \underline{m}_{PRIOR})$$

$$\underline{H}_n = \underline{I} + \underline{F}_n^* \underline{F}_n$$

where we have used the property (2.23):

$$\underline{F}^* = \underline{C}_M \underline{F}^T \underline{C}_D^{-1} .$$

2.5. - Nonlinear inverse problems. In this section we focus our attention on the preconditioned steepest descent algorithm (and on two of its special cases, the ordinary steepest descent algorithm and the Newton algorithm).

Letting $\underline{F}_n$ be the Frechet derivative of $\underline{f}$, the operator representing the solution of the forward problem,

$$\underline{F}_n = \left(\frac{\partial \underline{f}}{\partial \underline{m}}\right)_{\underline{m}_n}$$

and using the results of the previous section, we obtain the following algorithms for the resolution of nonlinear inverse problems:

a) Preconditioned steepest descent. Letting $\underline{Q}_n$ be an arbitrary self-adjoint positive definite operator presumably close to

$$\underline{Q}_n \simeq (\underline{I} + \underline{C}_M \underline{F}_n^T \underline{C}_D^{-1} \underline{F}_n)^{-1}$$

we obtain

$$(2.60)\quad \begin{cases} \underline{d}_n = \underline{f}(\underline{m}_n) \\ \underline{\gamma}_n = \underline{C}_M\, \underline{F}_n^T\, \underline{C}_D^{-1}\, (\underline{d}_n - \underline{d}_{OBS}) + (\underline{m}_n - \underline{m}_{PRIOR}) \\ \underline{\phi}_n = \underline{Q}_n\, \underline{\gamma}_n \\ \underline{b}_n = \underline{F}_n\, \underline{\phi}_n \\ \alpha_n = \underline{\phi}_n^T\, \underline{C}_M^{-1}\, \underline{\gamma}_n / (\underline{\phi}_n^T\, \underline{C}_M^{-1}\, \underline{\phi}_n + \underline{b}_n^T\, \underline{C}_D^{-1}\, \underline{b}_n) \\ \underline{m}_{n+1} = \underline{m}_n - \alpha_n\, \underline{\phi}_n \end{cases}$$

b) <u>Steepest descent</u>. Taking the roughest approximation for $\underline{Q}_n$:

$$\underline{Q}_n = \underline{I} \ ,$$

we obtain

$$(2.61)\quad \begin{cases} \underline{d}_n = \underline{f}(\underline{m}_n) \\ \underline{\gamma}_n = \underline{C}_M \underline{F}_n^T \underline{C}_D^{-1} (\underline{d}_n - \underline{d}_{OBS}) + (\underline{m}_n - \underline{m}_{PRIOR}) \\ \underline{b}_n = \underline{F}_n\, \underline{\gamma}_n \\ \alpha_n = 1/(1 + \underline{b}_n^T\, \underline{C}_D^{-1}\, \underline{b}_n / \underline{\gamma}_n^T\, \underline{C}_M^{-1}\, \underline{\gamma}_n) \\ \underline{m}_{n+1} = \underline{m}_n - \alpha_n\, \underline{\gamma}_n \end{cases}$$

C) <u>Newton algorithm</u>. Taking for $\underline{Q}_n$ the approximation

$$\underline{Q}_n = (\underline{I} + \underline{C}_M\, \underline{F}_n^T\, \underline{C}_D^{-1}\, \underline{F}_n)^{-1}$$

we obtain

$$(2.62)\quad \begin{cases} \underline{d}_n = \underline{f}(\underline{m}_n) \\ \underline{m}_{n+1} = \underline{m}_n - (\underline{I}+\underline{C}_M\underline{F}_n^T\underline{C}_D^{-1}\underline{F}_n)^{-1}(\underline{C}_M\underline{F}_n^T\underline{C}_D^{-1}(\underline{d}_n-\underline{d}_{OBS}) \\ \qquad\qquad + (\underline{m}_n-\underline{m}_{PRIOR})) \end{cases}$$

This last expression can also be written in the more symmetric form

$$\underline{m}_{n+1} = \underline{m}_n - (\underline{F}_n^T\underline{C}_D^{-1}\underline{F}_n+\underline{C}_M^{-1})^{-1}(\underline{F}_n^T\underline{C}_D^{-1}(\underline{d}_n-\underline{d}_{OBS})+\underline{C}_M^{-1}(\underline{m}_n-\underline{m}_{PRIOR})$$

where we recognize the usual formula used for the resolution of discrete nonlinear inverse problems (Tarantola and Valette, 1982b). Introducing the operator $\underline{F}_k^*$, adjoint of $\underline{F}_k$, we have (equation 2.23):

$$\underline{F}^* = \underline{C}_M \, \underline{F}_n^T \, \underline{C}_D^{-1} \, ,$$

and using the identities

$$\underline{F}^*(\underline{I} + \underline{F}\underline{F}^*)^{-1} = (\underline{I} + \underline{F}^*\underline{F})^{-1}\underline{F}^*$$

$$(\underline{I} + \underline{F}^*\underline{F})^{-1} = \underline{I} - \underline{F}^*(\underline{I} + \underline{F}\,\underline{F}^*)^{-1}\,\underline{F}$$

we have the following equivalent equations:

$$\underline{m}_{n+1} = \underline{m}_n - (\underline{I} + \underline{F}_n^*\underline{F}_n)^{-1}(\underline{F}_n^*(\underline{d}_n - \underline{d}_{OBS}) + (\underline{m}_n - \underline{m}_{PRIOR}))$$

$$\underline{m}_{n+1} = \underline{m}_{PRIOR} - (\underline{I} + \underline{F}^*\underline{F}_n)^{-1}\,\underline{F}_n^*(\underline{d}_n - \underline{d}_{OBS} - \underline{F}_n(\underline{m}_n - m_{PRIOR}))$$

$$\underline{m}_{n+1} = \underline{m}_{PRIOR} - \underline{F}_n^*(\underline{I} + \underline{F}_n\underline{F}_n^*)^{-1}(\underline{d}_n - \underline{d}_{OBS} - \underline{F}_n(\underline{m}_n - \underline{m}_{PRIOR}))$$

Let us make some remarks about these results.

First, if we are able to exhibit an operator $\underline{Q}_n$ close enough to $(\underline{I} + \underline{C}_M\underline{F}_n^*\underline{C}_D^{-1}\underline{F}_n)^{-1}$ and such that the computation of $\underline{\phi}_n = \underline{Q}_n\underline{\gamma}_n$ is inexpensive, algorithm (2.60) should be preferred to (2.61) or (2.62).

Second, the computation of $\underline{\gamma}_n$ has to be made with the maximum of accuracy if we wish our final model $\underline{m}_\infty$ to be the true solution of the problem. This is not the case for the computation of $\underline{b}_n = \underline{F}_n \underline{\phi}_n$, which is only useful for evaluating the constant α_n which has to be known only approximately. For that reason, we can sometimes take advantage of the approximation.

$$\underline{b}_n \simeq \underline{F}_o \underline{\phi}_n .$$

For some problems the value α_n does not vary very much from iteration to iteration, and the approximation

$$\alpha_n = \alpha_o$$

proves to be good enough, saving a lot of computing time.

When using the Newton algorithm (2.62), at each iteration the linear equation

$$(\underline{I} + \underline{C}_M \underline{F}_n^T \underline{C}_D^{-1} \underline{F}_n)(\underline{m}_{n+1} - \underline{m}_n) = \underline{C}_M \underline{F}_n^T \underline{C}_D^{-1} (\underline{d}_n - \underline{d}_{OBS}) + (\underline{m}_n - \underline{m}_{PRIOR})$$

has to be solved in order to obtain the value $\underline{m}_{n+1}$. For large-sized problems, this equation has to be solved iteratively, and the resulting algorithm will be very similar to the gradient algorithms (2.60) or (2.61), so , from a practical point of view, there is no difference between Newton and gradient algorithms.

The last remark is that these iterative algorithms have to be initialized at an arbitrary point $\underline{m}_n = \underline{m}_o$. Of course, a reasonable choice is

$$\underline{m}_o = \underline{m}_{PRIOR}$$

but for checking the possible existence of secondary minima, the only safe strategy is to initialize at different points $\underline{m}_o \neq \underline{m}_{PRIOR}$ and to verify that the solution $\underline{m}_\infty$ is independent of $\underline{m}_o$. If not, least squares methods should be dropped, and more general methods should be used, as Monte-Carlo methods (Press, 1968) or probabilistic methods (Tarantola and Valette, 1982a).

d) Davidon-Fletcher-Powell. We have, using the results of (2.3), the following algorithm

$$\begin{cases} \underline{d}_n = \underline{f}(\underline{m}_n) \\ \underline{\gamma}_n = \underline{C}_M \underline{F}_n^T \underline{C}_D^{-1} (\underline{d}_n - \underline{d}_{OBS}) + (\underline{m}_n - \underline{m}_{PRIOR}) \end{cases}$$

$$\left\{\begin{aligned}
\underline{\phi}_n &= \underline{\underline{Q}}_n \, \underline{\gamma}_n \\
\underline{b}_n &= \underline{F}_n \, \underline{\phi}_n \\
\alpha_n &= \underline{\phi}_n^T \, \underline{C}_M^{-1} \, \underline{\gamma}_n / (\underline{\phi}_n^T \, \underline{C}_M^{-1} \, \underline{\phi}_n + \underline{b}_n^T \, \underline{C}_D^{-1} \, \underline{b}_n) \\
\underline{m}_{n+1} &= \underline{m}_n - \alpha_n \, \underline{\phi}_n \\
\underline{\beta}_n &= \underline{\gamma}_{n+1} - \underline{\gamma}_n \\
\underline{\mu}_n &= \underline{\underline{Q}}_n \, \underline{\beta}_n \\
\underline{\underline{Q}}_{n+1} &= \underline{\underline{Q}}_n + \frac{\underline{\phi}_n \, \underline{\phi}_n^T \, \underline{C}_M^{-1}}{\underline{\phi}_n^T \, \underline{C}_M^{-1} \, \underline{\phi}_n + \underline{b}_n^T \, \underline{C}_D^{-1} \, \underline{b}_n} - \frac{\underline{\mu}_n \, \underline{\mu}_n^T \, C_M^{-1}}{\mu_n^T \, C_M^{-1} \, \beta_n}
\end{aligned}\right.$$

and we have

$$\underline{\underline{Q}}_n \to (\underline{\underline{I}} + \underline{C}_M \, \underline{F}_n^T \, \underline{C}_D^{-1} \, \underline{F}_n)^{-1} .$$

Equivalently,

$$\left\{\begin{aligned}
\underline{d}_n &= \underline{f}(\underline{m}_n) \\
\hat{\underline{\gamma}}_n &= \underline{F}_n^T \, \underline{C}_D^{-1} (\underline{d}_n - \underline{d}_{OBS}) + \underline{C}_M^{-1} (\underline{m}_n - m_{PRIOR}) \\
\underline{\phi}_n &= \check{\underline{\underline{Q}}}_n \, \hat{\underline{\gamma}}_n \\
\underline{b}_n &= \underline{F}_n \, \underline{\phi}_n \\
\alpha_n &= \underline{\phi}_n^T \, \hat{\underline{\gamma}}_n / (\underline{\phi}_n^T \, \underline{C}_M^{-1} \, \underline{\phi}_n + \underline{b}_n^T \, \underline{C}_D^{-1} \, \underline{b}_n)
\end{aligned}\right.$$

$$\left|\begin{array}{l} \underline{m}_{n+1} = \underline{m}_n - \alpha_n \underline{\phi}_n \\ \\ \hat{\underline{\beta}}_n = \hat{\underline{\gamma}}_{n+1} - \hat{\underline{\gamma}}_n \\ \\ \underline{\mu}_n = \underline{\check{Q}}_n \hat{\underline{\beta}}_n \\ \\ \underline{\check{Q}}_{n+1} = \underline{\check{Q}}_n + \dfrac{\underline{\phi}_n \underline{\phi}_n^T}{\underline{\phi}_n^T \underline{C}_M^{-1} \underline{\phi}_n + \underline{b}_n^T \underline{C}_D^{-1} \underline{b}_n} - \dfrac{\underline{\mu}_n \underline{\mu}_n^T}{\underline{\mu}_n^T \hat{\underline{\beta}}_n} \end{array}\right.$$

and

$$\underline{\check{Q}}_n \to (\underline{F}_n^T \underline{C}_D^{-1} \underline{F}_n + \underline{C}_M^{-1})^{-1} .$$

See Section (2.7) for the interpretation of the limit of $\underline{\check{Q}}_n$ as the posteriori covariance in the model space.

2.6. - Linearized inverse problems. The Backus and Gilbert's solution as a special case. In the iterative methods of the previous section, the forward problem

(2.63a) $\underline{d}_n = \underline{f}(\underline{m}_n)$

was assumed to be solved exactly (i.e., in its fully nonlinear form) at each iteration, but, for given $\underline{m}_n$ and $\underline{d}_n$, the value $\underline{m}_{n+1}$ was estimated using the local linearization

(2.64a) $\underline{f}(\underline{m}_{n+1}) \simeq \underline{f}(\underline{m}_n) + \underline{F}_n(\underline{m}_{n+1} - \underline{m}_n)$.

When the algorithm converges, the differences $\underline{m}_{n+1} - \underline{m}_n$ become smaller and smaller, so that the linearization (2.64a) tends to be exact. The solution $\underline{m}_\infty$ thus obtained is then the true solution of the nonlienar problem (it exactly verifies $\underline{\gamma}_\infty = \underline{0}$).

If the starting point $\underline{m}_0$ is close enough to the final model $\underline{m}_\infty$ it may happen that the approximation

(2.64b) $f(\underline{m}_{n+1}) \simeq \underline{f}(\underline{m}_0) + \underline{F}_0(\underline{m}_{n+1} - \underline{m}_0)$

holds with sufficient accuracy for any $\underline{m}_{n+1}$. In particular, equation (2.63a) can be replaced by

(2.63b) $\underline{d}_n \simeq \underline{f}(\underline{m}_o) + \underline{F}_o(\underline{m} - \underline{m}_o)$

Such a problem is named <u>linearizable</u>. If equations (2.63a) and (2.64a) are actually replaced by the approximations (2.63b) and (2.64b), we say that the problem has been <u>linearized</u>. Note that although the starting point $\underline{m}_o$ is arbitrary, it is usually identified with the <u>a priori</u> model:

$$\underline{m}_o = \underline{m}_{PRIOR} \qquad \text{(in general)}.$$

The only criterion we need for deciding if a problem is linearizable, is that the data predicted using (2.63b) have to be realistic enough (errors introduced by the approximation (2.63b) are of the same order of magnitude as observational errors in $\underline{d}_{OBS}$). Of course, the solution of a linearized problem does not equal the solution of the original nonlinear problem, but can be very close to it (if (2.63b) is a good approximation). See also the remark about linearized problems at the end of this section.

Some problems are strictly linear, as for instance the gravitation field versus density problem. Notice that for a linear problem

$$\underline{f}(\underline{m}) = \underline{F}\ \underline{m}$$

with $\underline{F}$ independent from $\underline{m}$, equation (2.63b) simplifies to

(2.63c) $\underline{d}_n = \underline{F}\ \underline{m}_n$.

The results corresponding to linear problems will easily be obtained from the results given in this section using (2.63c) instead of using (2.63b).

From the results of the previous section, we obtain, for linearized problems, the following algorithms:

a) <u>Preconditioned steepest descent</u>. We have

$$(2.65)\quad \begin{cases} \underline{d}_n = \underline{f}(\underline{m}_o) + \underline{F}_o(\underline{m}_n - \underline{m}_o) \\ \underline{\gamma}_n = \underline{C}_M\ \underline{F}_o\ \underline{C}_D^{-1}\ (\underline{d}_n - \underline{d}_{OBS}) + (\underline{m}_n - \underline{m}_o) \\ \underline{\phi}_n = \underline{Q}\ \underline{\gamma}_n \\ \underline{b}_n = \underline{F}_o\ \underline{\phi}_n \end{cases}$$

$$\begin{cases} \alpha_n = \underline{\phi}_n^T \, \underline{C}_M^{-1} \, \underline{\gamma}_n / (\underline{\phi}_n^T \, \underline{C}_M^{-1} \, \underline{\phi}_n + \underline{b}_n^T \, \underline{C}_D^{-1} \, \underline{b}_n) \\ \underline{m}_{n+1} = \underline{m}_n - \alpha_n \, \underline{\phi}_n \end{cases}$$

where $\underline{Q}$ is an arbitrary self adjoint positive definite operator, presumably not too far from

$$\underline{Q} \simeq (\underline{I} + \underline{C}_M \, \underline{F}_o^T \, \underline{C}_D^{-1} \, \underline{F}_o)^{-1}$$

B) Steepest descent. Taking

$$\underline{Q} = \underline{I}$$

we obtain

$$(2.66) \begin{cases} \underline{d}_n = \underline{f}(\underline{m}_o) + \underline{F}_o(\underline{m}_n - \underline{m}_o) \\ \underline{\gamma}_n = \underline{C}_M \, \underline{F}_o^T \, \underline{C}_D^{-1} (\underline{d}_n - \underline{d}_{OBS}) + (\underline{m}_n - \underline{m}_o) \\ \underline{b}_n = \underline{F}_o \, \underline{\gamma}_n \\ \alpha_n = 1/(1 + \underline{b}_n^T \, \underline{C}_D^{-1} \, \underline{b}_n / \underline{\gamma}_n^T \, \underline{C}_M^{-1} \, \underline{\gamma}_n) \\ \underline{m}_{n+1} = \underline{m}_n - \alpha_n \, \underline{\gamma}_n \end{cases}$$

c) Newton algorithm. Taking for $\underline{Q}$ the exact value

$$\underline{Q} = (\underline{I} + \underline{C}_M \, \underline{F}_o^T \, \underline{C}_D^{-1} \, \underline{F}_o)^{-1}$$

we find the important result that the Newton algorithm converges in only one iteration, so that it is no more an iterative but an exact algorithm:

$$(2.67a) \quad \underline{m}_{EST} = \underline{m}_o + (\underline{I}+\underline{C}_M \, \underline{F}_o^T \, \underline{C}_D^{-1} \, \underline{F}_o)^{-1} \, \underline{C}_M \, \underline{F}_o^T \, \underline{C}_D^{-1} (\underline{d}_{OBS} - \underline{f}(\underline{m}_o))$$

where the starting point $\underline{m}_o$ has been identified with the *a priori* model $\underline{m}_{PRIOR}$. Using the identity

$$(\underline{I}+\underline{C}_M\underline{F}_o^T\underline{C}_D^{-1}\underline{F}_o)^{-1}\underline{C}_M\underline{F}_o^T\underline{C}_D^{-1} = \underline{C}_M\underline{F}_o^T\underline{C}_D^{-1}(\underline{I}+\underline{F}_o\underline{C}_M\underline{F}_o^T\underline{C}_D^{-1})^{-1}$$

we can write

(2.67b) $$\underline{m}_{EST} = \underline{m}_o + \underline{C}_M\,\underline{F}_o^T\,\underline{C}_D^{-1}(\underline{I}+\underline{F}_o\,\underline{C}_M\,\underline{F}_o^T\,\underline{C}_D^{-1})^{-1}(\underline{d}_{OBS}-\underline{f}(\underline{m}_o))$$

Equations (2.67a) and (2.67b) can also be written, in symmetric form,

(2.67c) $$\underline{m}_{EST} = \underline{m}_o + (\underline{F}_o^T\,\underline{C}_D^{-1}\,\underline{F}_o + \underline{C}_M^{-1})^{-1}\underline{F}_o^T\,\underline{C}_D^{-1}(\underline{d}_{OBS}-\underline{f}(\underline{m}_o))$$

(2.67d) $$\underline{m}_{EST} = \underline{m}_o + \underline{C}_M\,\underline{F}_o^T(\underline{F}_o\,\underline{C}_M\,\underline{F}_o^T + \underline{C}_D)^{-1}(\underline{d}_{OBS}-\underline{f}(\underline{m}_o))\ .$$

Let us here briefly discuss the Backus and Gilbert approach for the resolution of linearized problems, following for instance their 1971 paper.

Let $\underline{m}_{TRUE}$ be the element of the model space representing the (unknown) true Earth, and let $\underline{d}_{OBS}$ be the element of the data space representing the observations, which are assumed to be error free. In the Backus and Gilbert paper the data space is assumed discrete and finite, but this is, from our point of view, not essential for the discussion.

They assume the problem linearizable around a given "starting point" $\underline{m}_o$:

(2.68) $$\underline{d} = \underline{f}(\underline{m}) = \underline{f}(\underline{m}_o) + \underline{F}_o(\underline{m} - \underline{m}_o)$$

As $\underline{d}_{OBS}$ is assumed error free, they have

(2.69) $$\underline{d}_{OBS} = \underline{f}(\underline{m}_o) + \underline{F}_o(\underline{m}_{TRUE} - \underline{m}_o)$$

Introducing the data residuals

$$\Delta\underline{d}_{OBS} = \underline{d}_{OBS} - \underline{f}(\underline{m}_o)$$

and the model corrections

$$\Delta\underline{m}_{TRUE} = \underline{m}_{TRUE} - \underline{m}_o$$

equation (2.69) is rewritten

(2.69b) $\Delta \underline{d}_{OBS} = \underline{F}_o \, \Delta \underline{m}_{TRUE}$.

In this equation, the vector $\Delta \underline{d}_{OBS}$ and the linear operator $\underline{F}_o$ are known, while the vector $\Delta \underline{m}_{TRUE}$ is unknown.

The operator $\underline{F}_o$ has, of course, no inverse in general. As (2.69b) is a linear equation Backus and Gilbert suggest to define a vector $\Delta \underline{m}_{EST}$, estimate of $\Delta \underline{m}_{TRUE}$ as depending linearly on $\Delta \underline{d}_{OBS}$:

(2.70) $\Delta \underline{m}_{EST} = \underline{L} \, \Delta \underline{d}_{OBS}$,

where $\underline{L}$ represents an arbitrary linear operator to be defined using an adequate criterion. Inserting (2.69b) in (2.70) we have

(2.71) $\Delta \underline{m}_{EST} = \underline{A} \, \Delta \underline{m}_{TRUE}$

where

(2.72) $\underline{A} = \underline{L} \, \underline{F}_o$.

As $\underline{L}$ is arbitrary, so it is $\underline{A}$. Backus and Gilbert suggest to choose for $\underline{L}$ the operator which makes $\underline{A}$ as close as possible to the identity operator, i.e., which minimizes a conveniently defined norm $||\underline{A} - \underline{I}||$. The reason for imposing $\underline{A} \simeq \underline{I}$ is that, in that case, equation (2.71) gives $\Delta \underline{m}_{EST} \simeq \Delta \underline{m}_{TRUE}$.

When the norm $||\underline{A} - \underline{I}||$ is an L^2-norm they obtain (Backus and Gilbert, 1971)

$$\underline{L} = \underline{F}_o^T (\underline{F}_o \underline{F}_o^T)^{-1} .$$

(as they assume discrete, independent observations, the operator $\underline{F}_o \underline{F}_o^T$ is a regular matrix, so the inverse matrix exists). This leads to the Backus and Gilbert's solution

(2.73) $\Delta \underline{m}_{EST} = \underline{F}_o^T (\underline{F}_o \, \underline{F}_o^T)^{-1} \, \Delta \underline{d}_{OBS}$.

We can now summarize their philosophy as follows: the operator $\underline{A}$ is somewhat arbitrary (defined by minimization of one arbitrary norm $||\underline{A} - \underline{I}||$), but it is perfectly known. By equation (2.71), we see that $\Delta \underline{m}_{EST}$ is a filtered version of $\Delta \underline{m}_{TRUE}$ with the known filter $\underline{A}$. If we change the filter, we change $\Delta \underline{m}_{EST}$. Inverse problem theory is unable to obtain $\Delta \underline{m}_{TRUE}$ (no miracles in physics), its only job is, for suitably chosen filters $\underline{A}$, to exhibit the corresponding $\Delta \underline{m}_{EST}$.

Of course, I agree with that point of view. Except for one detail: when appropriate description of observational errors is made (via the covariance operator $\underline{C}_D$) and appropriate description of a priori information on the model is made (via $\underline{C}_M$), the probability theory gives arguments to choose one particular filter (see Tarantola and Valette, 1982a).

In practical applications, geophysicists need an estimation, $\underline{m}_{EST}$, of the true Earth model, which is usually defined by

$$\underline{m}_{EST} = \underline{m}_o + \Delta \underline{m}_{EST}$$

Using the Backus and Gilbert solution (2.73) and the definition (2.70) this gives

$$(2.74)\quad \underline{m}_{EST} = \underline{m}_o + \underline{F}_o^T (\underline{F}_o \underline{F}_o^T)^{-1} (\underline{d}_{OBS} - \underline{f}(\underline{m}_o)) \ .$$

We can now compare this equation with out solution (2.67d) for linearized problems:

$$(2.67\text{b again})\quad \underline{m}_{EST} = \underline{m}_o + \underline{C}_M \underline{F}_o^T (\underline{F}_o \underline{C}_M \underline{F}_o^T + \underline{C}_D)^{-1} (\underline{d}_{OBS} - \underline{f}(\underline{m}_o))$$

We clearly see that (2.74) is a special case of (2.67b) if we assume the a priori information on the model to be nonexistent (large variances in $\underline{C}_M$).

The reader should pay attention to the fact that the expression $\underline{F}_o \underline{F}_o^T$ makes only sense if all the components of the model vector have homogeneous physical dimensions. If for instance, the vector $\underline{m}$ contains as components the functions

$$\underline{m} = \begin{bmatrix} \rho(\underline{x}) \\ \lambda(\underline{x}) \\ \\ \cdots \\ \pi_i \end{bmatrix}$$

introduced in Section 2.1, with their own physical dimensions, $\underline{F}_o \underline{C}_m \underline{F}_o^T$ makes sense (it is homogeneous to a covariance operator in the data space), while $\underline{F}_o \underline{F}_o^T$ does not.

We can conclude that, neglecting some details, the Backus and Gilbert solution for linearized problems is equivalent to our linearized solution (when they choose an L^2-norm for minimizing $||\underline{A} - \underline{I}||$).

I only disagree with the Backus and Gilbert point of view if it is applied to the solution of nonlinear problems (i.e., problems in which the Backus and Gilbert procedure has to be iterated, taking as starting point for the n+1-th iteration the solution of the n-th iteration).

In Section (2.5) we gave iterative solutions for nonlinear problems, and in this section we have given iterative and direct solutions for linearized problems. It happens that iteratively using direct solutions for linearized problems is not equivalent to iteratively solving nonlinear problems. We claim that the right approach is the nonlinear one. Unfortunately, in the Backus and Gilbert point of view, the concept of linearization is essential in the theory, so it cannot be naturally extended to nonlinear problems.

The difference between the two approaches can best be seen as follows. When starting a nonlinear inverse problem we explicitly ask for our estimate $\underline{m}_{EST}$ $\underline{m}_{EST} = \underline{m}_\infty$ to remain close to the a priori point $\underline{m}_O$ (assuming here $\underline{m}_{PRIOR} = \underline{m}_O$). Backus and Gilbert ask in fact for each computed correction $(\Delta\underline{m}_{EST})_n$ to be the smallest of all possible corrections compatible with the data set. This is equivalent to saying that they ask for each model $\underline{m}_n$ to remain close to the previous model $\underline{m}_{n-1}$. In short, while we ask to $||\underline{m}_\infty - \underline{m}_O||$ to be small, Backus and Gilbert ask each $||\underline{m}_{n+1} - \underline{m}_n||$ to be small. Backus and Gilbert only envisaged purely underdetermined problems; for such problems, our solution clearly depends on $\underline{m}_O$, thus showing that $\underline{m}_O$ is for us more than a simple "starting point:" it is clearly an "a priori model." The trouble is that the Backus and Gilbert's solution also depends strongly on $\underline{m}_O$, so, even in their theory, $\underline{m}_O$ is something more than a simple starting point.

Our conclusion for that discussion is that if a problem is linearizable (in the precise sense defined at the beginning of this section) the Backus and Gilbert approach can be used. If the problem is not linearizable, the popular use of the Backus and Gilbert approach is inconsistent.

2.7. - Analysis of error and resolution. Using the methods described in Tarantola and Valette (1982a) it is possible to define the covariance operator $\underline{C}'_M$ describing errors and resolution on the computed model $\underline{m}_{EST}$. For a linearizable problem, the following relation can be shown (see for instance Franklin 1970),

$$\underline{C}'_M = (\underline{I} + \underline{C}_M \underline{F}_o^T \underline{C}_D^{-1} \underline{F}_o)^{-1} \underline{C}_M$$

$$= (\underline{F}_o^T \underline{C}_D^{-1} \underline{F}_o + \underline{C}_M^{-1})^{-1}$$

$$= \underline{C}_M - \underline{C}_M \underline{F}_o^T (\underline{F}_o \underline{C}_M \underline{F}_o^T + \underline{C}_D)^{-1} \underline{F}_o \underline{C}_M$$

$$= [\underline{I} - \underline{C}_M \underline{F}_o^T \underline{C}_D^{-1} (\underline{I} + \underline{F}_o \underline{C}_M \underline{F}_o^T \underline{C}_D^{-1})^{-1} \underline{F}_o] \underline{C}_M \ .$$

The diagonal values of $\underline{C}'_M$ clearly give the (a posteriori) variances in $\underline{m}_{EST}$, while off-diagonal values give information on a posteriori error correlation (and, in particular, about spatial resolution). For more discussion, the reader is referred to Jackson (1979) or Tarantola and Valette (1982b).

If the inverse problem is not large sized, the operator $\underline{C}'_M$ can be estimated by direct computation, using, for instance, a Newton method of resolution of the minimization problem. If it is large-sized, and we have to use a gradient method, then, the only method which gives the solution $\underline{m}_\infty$ and the a posteriori convariance operator $\underline{C}'_M$ is the D.F.P. method, where we have (see Section 2.6-d)

$$\check{\underline{Q}}_\infty = (\underline{F}^T \underline{C}_D^{-1} F + \underline{C}_M^{-1})^{-1} = \underline{C}'_M$$

We point out that, when using this method, the kernel of the operator $\check{\underline{Q}}_n$ (a matrix for the computer) is never build. We only keep in a memory of the computer the series of values $\underline{\phi}_0, \phi_1, \ldots, \underline{\mu}_0, \underline{\mu}_1, \ldots$ which allow the reconstruction of $\check{\underline{Q}}_n$ at any time we wish.

3. The Elastodynamic Forward Problem.

3.1. - Notations. I will use a general curvilinear system of coordinates (not necessarily orthogonal), $\underline{x} = (x^1, x^2, x^3)$, and the associate natural basis (not necessarily normalized) for the vectorial space attached to each point. We must then distinguish between contravariant (upper indexes) and covariant (lower indexes) components of tensors. The implicit summation convention over repeated indexes is assumed everywhere unless explicitly stated. The reader who prefers to avoid the complication of using a general system of coordinates may assume a cartesian system, so that contravariant and covariant components of tensors are identical and all the indexes can be lowered (or raised). For details on tensorial calculus, see for instance Lichnerowicz (1960).

I consider a volume $\mathcal{V}$ bounded by a surface $\mathcal{S}$. The volume $\mathcal{V}$ represents an heterogeneous elastic Earth, and is described by its density $\rho(\underline{x})$, and its elastic parameters $c^{ijk\ell}(\underline{x})$. The surface $\mathcal{S}$ represents the surface of the Earth, and the tractions on it, $T^i(\underline{x},t)$, are assumed to be given. For instance, if we consider it as a free surface, the tractions are assumed to be identically zero. Alternatively, the function $T^i(\underline{x},t)$ can be used for describing a surface artifical source of seismic waves (as for instance a Vibroseis source). Finally, we allow the existence of a volume density of external forces, $f^i(\underline{x},t)$, representing any deep source (artificial or natural) of seismic waves. Both $T^i(\underline{x},t)$ and $f^i(\underline{x},t)$ are assumed to be identically null before a time t_o:

$$T^i(\underline{x},t) = 0 \qquad \text{for} \quad t < t_o$$

$$f^i(\underline{x},t) = 0 \qquad \text{for} \quad t < t_o \; .$$

The initial conditions are homogeneous (null displacements and velocities for $t < t_o$).

<u>3.2. - Solution in terms of Green's functions</u>. The elastodynamic forward problem consists in predicting the <u>recorded output of some seismometers</u>, given the sources of seismic waves ($f^i(\underline{x},t)$ and/or $T^i(\underline{x},t)$) and given a description of the Earth's interior ($\rho(\underline{x})$ and $c^{ijk\ell}(\underline{x})$)

If $u^i(\underline{x},t)$ represents the i-th component of the <u>particle displacement</u> at the point $\underline{x}$ and at the time t, the tensor of <u>infinitesimal strain</u>, is defined by

$$\text{(3.1)} \quad u_{ij}(\underline{x},t) = \frac{1}{2}\left(\frac{Du^i}{Dx^j}(\underline{x},t) + \frac{Du^j}{Dx^i}(\underline{x},t)\right)$$

where D/Dx^i stands for the covariant derivative (which becomes identical to the partial derivative $\partial/\partial x^i$ for cartesian coordinates). Let $n^i(\underline{x})$ represent the <u>unit normal</u> to an arbitrary surface passing through $\underline{x}$. If the <u>stress vector</u> across the surface at that point is noted

$T^i(\underline{x},t)$, the <u>stress tensor</u> $\sigma^{ij}(\underline{x},t)$ is defined by

(3.2) $T^i(\underline{x},t) = \sigma^{ij}(\underline{x},t)n_j(\underline{x})$

The fundamental axiom of the dynamics of a continuous medium is written

(3.3) $\rho(\underline{x})\ \ddot{u}^i(\underline{x},t) = \dfrac{D\sigma^{ij}}{Dx^j}(\underline{x},t) + f^i(\underline{x},t)$.

where a dot denotes differentiation with respect to time. A linearly elastic medium is defined by the stress-strain relation

(3.4) $\sigma^{ij}(\underline{x},t) = c^{ijk\ell}(\underline{x})\ u_{k\ell}(\underline{x},t)$

where the tensor of elastic parameters has the fundamental symmetries

$$c^{ijk\ell} = c^{jik\ell} = c^{ij\ell k} = c^{k\ell ij}$$

In the particular case of an isotropic medium, $c^{ijk\ell}$ reduces to

(3.5) $c^{ijk\ell} = \lambda\delta^{ij}\delta^{k\ell} + \mu(\delta^{ik}\delta^{j\ell} + \delta^{i\ell}\delta^{jk})$

where $\lambda(\underline{x})$ and $\mu(\underline{x})$ are the <u>Lame's parameters</u>, and where δ^{ij} represents the <u>Kronecker's (metric) tensor</u>.

Using equations (3.1)-(3.4), we can obtain the set of equations defining the displacement field everywhere in the medium:

$$(3.6)\quad \begin{cases} \rho(\underline{x})\ \ddot{u}^i(\underline{x},t) - \dfrac{D}{Dx^j}[c^{ijk\ell}(\underline{x})\ u_{k\ell}(\underline{x},t)] = f^i(\underline{x},t) \\ c^{ijk\ell}(\underline{x})\ u_{k\ell}(\underline{x},t)\ n_j(\underline{x}) = T^i(\underline{x},t) \qquad (\underline{x} \in S) \\ u^i(\underline{x},t) = 0 \quad \text{for} \quad t < t_o \\ \dot{u}^i(\underline{x},t) = 0 \quad \text{for} \quad t < t_o\ . \end{cases}$$

The Green's function $G^i_j(\underline{x},t;\underline{x}',t')$ is usually introduced as the i'th component of the displacement at $(\underline{x},t)$ for an impulsive source in the j direction at $(\underline{x}',t')$, and satisfying homogeneous initial and boundary conditions:

$$(3.7)\quad \begin{cases} \rho(\underline{x})\ddot{G}^i_m(\underline{x},t;\underline{x}',t') - \dfrac{D}{Dx^j}\left[c^{ijk\ell}(\underline{x})G_{k\ell m}(\underline{x},t;\underline{x}',t')\right] = \\ \qquad = \delta^i_m\,\delta(\underline{x}-\underline{x}')\,\delta(t-t') \\ c^{ijk\ell}(\underline{x})G_{k\ell m}(\underline{x},t;\underline{x}',t')n_j(\underline{x}) = 0 \qquad (\underline{x}\in S) \\ G^i_m(\underline{x},t;\underline{x}',t') = 0 \qquad \text{for } t < t' \\ \dot{G}^i_m(\underline{x},t;\underline{x}',t') = 0 \qquad \text{for } t < t' . \end{cases}$$

where

$$\dot{G}^i_j(\underline{x},t;\underline{x}',t') = \frac{\partial G^i_j}{\partial t}(\underline{x},t;\underline{x}',t')$$

and

$$(3.8)\quad G_{ijk}(\underline{x},t;\underline{x}',t') = \frac{1}{2}\left(\frac{DG_{ik}}{Dx^j}(\underline{x},t;\underline{x}',t') + \frac{DG_{jk}}{Dx^i}(\underline{x},t;\underline{x}',t')\right).$$

It is clear that $G_{ijk}(\underline{x},t;\underline{x}',t')$ represents the (i,j) component of the infinitesimal strain at $(\underline{x},t)$ for an unit source in the k-direction at $(\underline{x}',t')$.

An easy generalization of the demonstrations of Aki and Richards (1980) to curvilinear coordinates gives the three following important results:

a) the Green's function is invariant by translation over the time:

$$(3.9)\quad G_{ij}(\underline{x},t;\underline{x}',t') = G_{ij}(\underline{x},t-t';\underline{x}',0) ;$$

b) it verifies the reciprocity theorem:

$$(3.10)\quad G_{ij}(\underline{x},t;\underline{x}',t') = G_{ji}(\underline{x}',t;\underline{x},t') ;$$

c) the solution of equations (3.6) admits the integral representation

$$(3.11)\quad u^i(\underline{x},t) = \int_V dV(\underline{x}') \int dt' G^i_j(\underline{x},t;\underline{x}',t') f^j(\underline{x}',t')$$

$$+ \int_S dS(\underline{x}') \int dt' G^i_j(\underline{x},t;\underline{x}',t') T^j(\underline{x}',t') .$$

Using (3.9) we can also write

$$(3.12)\quad u^i(\underline{x},t) = \int_V dV(\underline{x}') G^i_j(\underline{x},t;\underline{x}',0) * f^j(\underline{x}',t)$$

$$+ \int_S dS(\underline{x}') G^i_j(\underline{x},t;\underline{x}',0) * T^j(\underline{x}',t) ,$$

where the symbol $*$ stands for the time convolution

$$f(t)*g(t) = \int dt' f(t-t') g(t') = \int dt' g(t-t') f(t') = g(t)*f(t) .$$

Let $\underline{x}_R$ denote a generic observation point (the index R stands for receiver). By $u^i(\underline{x}_R,t)$ I denote the displacement at the point $\underline{x}_R$, and by $v^i(\underline{x}_R,t)$ the seismogram at that point (i.e., the output of the seismometer plus the recording system). In general, v^i is a voltage, and, roughly speaking, is proportional to a band limited version of the particle velocity at $\underline{x}_R$. More precisely, we can assume that modern seismometers and recording systems are linear, so that the relation between $v^i(\underline{x}_R,t)$ and $u^i(\underline{x}_R,t)$ can be written

$$(3.13)\quad v^i(\underline{x}_R,t) = u^i(\underline{x}_R,t) * R(t) ,$$

where $R(t)$ is the transfer function (or impulse response) of the system. Once the seismometer and the recording system have been specified, the function $R(t)$ can be assumed perfectly known.

From (3.12) we have

$$u^i(\underline{x}_R,t) = \int_V dV(\underline{x})\; G^i_j(\underline{x}_R,t;\underline{x},0) * f^j(\underline{x},t)$$

$$+ \int_S dS(\underline{x})\; G^i_j(\underline{x}_R,t;\underline{x},0) * T^j(\underline{x},t) \; .$$

I will assume, all through this paper, that we are able to use a numerical method (finite differencing, ray method, discrete wavenumber, etc.) which allows us to obtain reasonably accurate estimations of the Green's functions $G^i_j(\underline{x},t;\underline{x}',0)$ for arbitrary point $\underline{x}$ and $\underline{x}'$, and for an arbitrary Earth model ($\rho(\underline{x})$ and $c^{ijk\ell}(\underline{x})$). Then, for given source functions, $f^i(\underline{x},t)$ and $T^i(\underline{x},t)$, the forward problem is solved using (3.13) and (3.14).

4. The Frechet Derivatives of Seismograms

4.1. - Frechet derivatives and derivative operator. Let $\rho(\underline{x})$, $\lambda(\underline{x})$, and $\mu(\underline{x})$, represent an heterogeneous, elastic, isotropic Earth, and let $u^i(\underline{x}_R, t)$ represent the computed displacement at the receiver point $\underline{x}_R$ for given volume density of external force, $f^i(\underline{x},t)$, and given surface tractions $T^i(\underline{x},t)$ $(\underline{x} \in S)$. An arbitrary perturbation of these quantities

$$\rho(\underline{x}) \to \rho(\underline{x}) + \delta\rho(\underline{x})$$

$$\lambda(\underline{x}) \to \lambda(\underline{x}) + \delta\lambda(\underline{x})$$

$$\mu(\underline{x}) \to \mu(\underline{x}) + \delta\mu(\underline{x})$$

$$f^i(\underline{x},t) \to f^i(\underline{x},t) + \delta f^i(\underline{x},t)$$

$$T^i(\underline{x},t) \to T^i(\underline{x},t) + \delta T^i(\underline{x},t)$$

will produce a perturbation f the displacement field:

$$u^i(\underline{x},t) \to u^i(\underline{x},t) + \Delta u^i(\underline{x},t) \; .$$

By definition, the Frechet derivatives of the displacements $\underline{u}$ with respect to $\underline{\rho}$, $\underline{\lambda}$, $\underline{\mu}$, $\underline{f}$, and $\underline{T}$, are linear operators which will be denoted by

$$\frac{\partial \underline{u}}{\partial \underline{\rho}} , \quad \frac{\partial \underline{u}}{\partial \underline{\lambda}} , \quad \frac{\partial \underline{u}}{\partial \underline{\mu}} , \quad \frac{\partial \underline{u}}{\partial \underline{f}} , \quad \text{and} \quad \frac{\partial \underline{u}}{\partial \underline{T}} ,$$

and which are defined by (cf. Section 2.2.):

$$(4.1) \quad \Delta \underline{u} = \frac{\partial \underline{u}}{\partial \underline{\rho}} \delta \underline{\rho} + \frac{\partial \underline{u}}{\partial \underline{\lambda}} \delta \underline{\lambda} + \frac{\partial \underline{u}}{\partial \underline{\mu}} \delta \underline{\mu} + \frac{\partial \underline{u}}{\partial \underline{f}} \delta \underline{f} + \frac{\partial \underline{u}}{\partial \underline{T}} \delta \underline{T} + \underline{o}(\delta \underline{\rho}, \delta \underline{\lambda}, \delta \underline{\mu}, \delta \underline{f}, \delta \underline{T})^2 .$$

Explicitly, we will have

$$\begin{aligned}(4.2) \quad \Delta u_i(\underline{x}_R, t) = {} & \int_V dV(\underline{x}) A_i(\underline{x}_R, t; \underline{x}) \delta\rho(\underline{x}) + \int_V dV(\underline{x}) B_i(\underline{x}_R, t; \underline{x}) \delta\lambda(\underline{x}) \\ & + \int_V dV(\underline{x}) \; C_i(\underline{x}_R, t; \underline{x}) \; \delta\mu(\underline{x}) \\ & + \int_V dV(\underline{x}) \int dt' \; D_i^j(\underline{x}_R, t; \underline{x}, t') \; \delta f_j(\underline{x}, t') \\ & + \int_S dS(\underline{x}) \int dt' \; E_i^j(\underline{x}_R, t; \underline{x}, t') \; \delta T_j(\underline{x}, t') \\ & + o_i(\delta \underline{\rho}, \delta \underline{\lambda}, \delta \underline{\mu}, \delta \underline{f}, \delta \underline{T})^2 .\end{aligned}$$

The functions $A_i(\underline{x}_R, t; \underline{x})$, $B_i(\underline{x}_R, t; \underline{x})$, $C_i(\underline{x}_R, t; \underline{x})$, $D_i^j(\underline{x}_R, t; \underline{x}, t')$ and $E_i^j(\underline{x}_R, t; \underline{x}, t')$ are the <u>Kernels</u> of the Frechet derivatives. All through this paper we will use the <u>notation</u>

$$\begin{cases} A_i(\underline{x}_R, t; \underline{x}) = \dfrac{\partial u_i(\underline{x}_R, t)}{\partial \rho(\underline{x})} \\[2ex] B_i(\underline{x}_R, t'; \underline{x}) = \dfrac{\partial u_i(\underline{x}_R, t)}{\partial \lambda(\underline{x})} \end{cases}$$

$$(4.3)\begin{cases} C_i(\underline{x}_R,t;\underline{x}) = \dfrac{\partial u_i(\underline{x}_R,t)}{\partial \mu(\underline{x})} \\[2ex] D_i^j(\underline{x}_R,t;\underline{x},t') = \dfrac{\partial u_i(\underline{x}_R,t)}{\partial f_j(\underline{x},t')} \\[2ex] E_i^j(\underline{x}_R,t;\underline{x},t') = \dfrac{\partial u_i(\underline{x}_R,t)}{\partial T_j(\underline{x},t')} , \end{cases}$$

so that equations (4.2) will be written

$$\begin{aligned} \Delta u_i(\underline{x}_R,t) = & \int_V dV(\underline{x}) \frac{\partial u_i(\underline{x}_R,t)}{\partial \rho(\underline{x})} \delta\rho(\underline{x}) \\ & + \int_V dV(\underline{x}) \frac{\partial u_i(\underline{x}_R,t)}{\partial \lambda(\underline{x})} \delta\lambda(\underline{x}) \\ & + \int_V dV(\underline{x}) \frac{\partial u_i(\underline{x}_R,t)}{\partial \mu(\underline{x})} \delta\mu(\underline{x}) \\ & + \int_V dV(\underline{x}) \int dt' \frac{\partial u_i(\underline{x}_R,t)}{\partial f_j(\underline{x},t')} \delta f_j(\underline{x},t') \\ & + \int_S dS(\underline{x}) \int dt' \frac{\partial u_i(\underline{x}_R,t)}{\partial T_j(\underline{x},t')} \delta T_j(\underline{x},t') \\ & + o_i(\delta\underline{\rho}, \delta\underline{\lambda}, \delta\underline{\mu}, \delta\underline{f}, \delta\underline{T})^2 . \end{aligned}$$

For more compactness of subsequent formulae let us define

$$(4.4) \quad \delta c^{ijk\ell} = \delta\lambda\, \delta^{ij}\, \delta^{k\ell} + \delta\mu(\delta^{i\ell}\delta^{jk} + \delta^{ik}\delta^{j\ell}) .$$

Using the elastodynamic wave equation (3.6) we have

$$(\rho+\delta\rho)(\underline{x})(\ddot{u}^i+\Delta\ddot{u}^i)(\underline{x},t) - \frac{D}{Dx^j}[(c^{ijk\ell}+\delta c^{ijk\ell})(\underline{x})$$

$$(u_{k\ell}+\Delta u_{k\ell})(\underline{x},t)] = (f^i+\delta f^i)(\underline{x},t)$$

$$(c^{ijk\ell}+\delta c^{ijk\ell})(\underline{x})(u_{k\ell}+u_{k\ell})(\underline{x},t)n_j(\underline{x}) =$$

$$(T^i+\delta T^i)(\underline{x},t) \qquad (\underline{x}\in S)$$

$$(u^i+\Delta u^i)(\underline{x},t) = 0 \qquad \text{for} \quad t < t_o$$

$$(\dot{u}^i+\Delta\dot{u}^i)(\underline{x},t) = 0 \qquad \text{for} \quad t < t_o .$$

Using (3.6) again and reordering we have

$$(4.5)\begin{cases} \rho(\underline{x})\Delta\ddot{u}^i(\underline{x},t) - \frac{D}{Dx^j}[c^{ijk\ell}(\underline{x})\Delta u_{k\ell}(\underline{x},t)] = \Delta f^i(\underline{x},t) \\ c^{ijk\ell}(\underline{x})\Delta u_{k\ell}(\underline{x},t)n_j(\underline{x}) = \Delta T^i(\underline{x},t) \qquad (\underline{x}\in S) \\ \Delta u^i(\underline{x},t) = 0 \qquad \text{for} \quad t < t_o \\ \Delta\dot{u}^i(\underline{x},t) = 0 \qquad \text{for} \quad t < t_o . \end{cases}$$

where

$$\Delta f^i(\underline{x},t) = \delta f^i(\underline{x},t) - \delta\rho(\underline{x})\ddot{u}^i(\underline{x},t)$$

$$+ \frac{D}{Dx^j}[\delta c^{ijk\ell}(\underline{x})u_{k\ell}(\underline{x},t)] + o^i(\delta\underline{\rho},\ \delta\underline{c},\ \delta\underline{f},\ \delta\underline{T})^2$$

$$\Delta T^i(\underline{x},t) = \delta T^i(\underline{x},t) - \delta c^{ijk\ell}(\underline{x}) u_{k\ell}(\underline{x},t) n_j(\underline{x})$$

$$+ o^i(\delta\underline{\rho}, \delta\underline{c}, \delta\underline{f}, \delta\underline{T})^2 \qquad (\underline{x} \in S) .$$

Equation (4.5) is easily interpreted: it corresponds to the well known fact that the perturbation $\Delta u^i(\underline{x},t)$ due to perturbations $\delta\rho(\underline{x})$, $\delta c^{ijk\ell}(\underline{x})$, $\delta f^i(\underline{x},t)$, and $\delta T^i(\underline{x},t)$ can be interpreted as the field propagating in the unperturbed medium (i.e., with density $\rho(\underline{x})$ and elastic parameters $c^{ijk\ell}(\underline{x})$) due to the "secondary volume sources" $\Delta f^i(\underline{x},t)$ and to the "secondary surface sources" $\Delta T^i(\underline{x},t)$. Neglecting the term $o^i(\delta\underline{\rho}, \delta\underline{c}, \delta\underline{f}, \delta\underline{T})^2$ in (4.5) would correspond to the "Born approximation".

As $\Delta u^i(\underline{x},t)$ is propagating in the unperturbed medium, we have to use the unperturbed Green's functions for obtaining the solution of (4.5). Using (3.15) we have

$$(4.6) \quad \Delta u^i(\underline{x}_R,t) = \int_V dV(\underline{x})\, G^i_j(\underline{x}_R,t;\underline{x},0) * \Delta f^j(\underline{x},t)$$

$$+ \int_S dS(\underline{x})\, G^i_j(\underline{x}_R,t;\underline{x},0) * \Delta T^j(\underline{x},t) .$$

After some computations (see Appendix) we obtain, using (4.4),

$$(4.7) \quad \Delta u_i(\underline{x}_R,t) = - \int_V dV(\underline{x}) \dot{G}^j_i(\underline{x},t;\underline{x}_R,0) * \dot{u}_j(\underline{x},t)\,\delta\rho(\underline{x})$$

$$- \int_V dV(\underline{x})\, G^\ell_{\ell i}(\underline{x},t;\underline{x}_R,0) * u^m_m(\underline{x},t)\,\delta\lambda(\underline{x})$$

$$- 2\int_V dV(\underline{x})\, G^\ell_{ki}(\underline{x},t;\underline{x}_R,0) * u^k_\ell(\underline{x},t)\,\delta\mu(\underline{x})$$

$$+ \int_V dV(\underline{x}) \int dt'\, G^j_i(\underline{x},t;\underline{x}_R,0)\, \delta f_j(\underline{x},t')$$

$$+ \int_S dS(\underline{x}) \int dt'\, G^j_i(\underline{x},t;\underline{x}_R,0)\, \delta T_j(\underline{x},t')$$

$$+ o_i(\delta\underline{\rho}, \delta\underline{\lambda}, \delta\underline{\mu}, \delta\underline{f}, \delta\underline{T})^2 .$$

and by comparison with (4.3) we deduce

$$
(4.8)\left\{
\begin{aligned}
\frac{\partial u_i(\underline{x}_R,t)}{\partial \rho(\underline{x})} &= -\dot{G}_i^j(\underline{x},t;\underline{x}_R,0) * \dot{u}_j(\underline{x},t) \\
\frac{\partial u_i(\underline{x}_R,t)}{\partial \lambda(\underline{x})} &= -\, G^{\ell}_{\ell i}(\underline{x},t;\underline{x}_R,0) * u^m_m(\underline{x},t) \\
\frac{\partial u_i(\underline{x}_R,t)}{\partial \mu(\underline{x})} &= -2G^{\ell}_{ki}(\underline{x},t;\underline{x}_R,0) * u^k_{\ell}(\underline{x},t) \\
\frac{\partial u_i(\underline{x}_R,t)}{\partial f_j(\underline{x},t')} &= G_i^j(\underline{x},t-t';\underline{x}_R,o) \\
\frac{\partial u_i(\underline{x}_R,t)}{\partial T_j(\underline{x},t')} &= G_i^j(\underline{x},t-t';\underline{x}_R,o)
\end{aligned}
\right.
$$

Let me, for instance, interprete the first of these questions. For computing the partial derivative of the displacement $u_i(\underline{x}_R,t)$ with respect to the density at an arbitrary point $\underline{x}$, we have first to solve the wave equation in order to obtain $u_j(\underline{x},t)$. We have then to solve again the wave equation, with an impulsive source at the receiver location $\underline{x}_R$ in order to obtain the Green's functions $G_i^j(\underline{x},t;\underline{x}_R,0)$. The derivative we are looking for is then easily obtained by time convolution. Remark that with two resolutions of the wave equation we are able to obtain the derivative of $u_i(\underline{x}_R,t)$ with respect to $\rho(\underline{x})$ <u>everywhere in the medium</u> (i.e., for any $\underline{x}$).

Equations (4.8) give the Frechet derivatives of the displacements. For solving inverse problems we would need the Frechet derivatives of seismograms. Using (3.13) we easily obtain

$$\frac{\partial v_i(\underline{x}_R,t)}{\partial \rho(\underline{x})} = \frac{\partial u_i(\underline{x}_R,t)}{\partial \rho(\underline{x})} * R(t)$$

$$\frac{\partial v_i(\underline{x}_R,t)}{\partial \lambda(\underline{x})} = \frac{\partial u_i(\underline{x}_R,t)}{\partial \lambda(\underline{x})} * R(t)$$

$$(4.9)\quad \frac{\partial v_i(\underline{x}_R,t)}{\partial \mu(\underline{x})} = \frac{\partial u_i(\underline{x}_R,t)}{\partial \mu(\underline{x})} * R(t)$$

$$\frac{\partial v_i(\underline{x}_R,t)}{\partial f_j(\underline{x},t)} = \frac{\partial u_i(\underline{x}_R,t)}{\partial f_j(\underline{x},t)} * R(t)$$

$$\frac{\partial v_i(\underline{x}_R,t)}{\partial T_j(\underline{x},t)} = \frac{\partial u_i(\underline{x}_R,t)}{\partial T_j(\underline{x},t)} * R(t) .$$

Let

$$\underline{m} = \begin{bmatrix} \rho(\underline{x}) \\ \lambda(\underline{x}) \\ \mu(\underline{x}) \\ f^i(\underline{x},t) \\ T^i(\underline{x},t) \end{bmatrix}$$

represent an arbitrary model of Earth and of seismic sources, which will be named the "current model". Let

$$\underline{v} = [v^i(\underline{x}_R,t)]$$

represent the computed values of seismograms corresponding to the current model. We can symbolically write

$$\underline{v} = \underline{\phi}(\underline{m}) = \underline{\phi}(\underline{\rho}, \underline{\lambda}, \underline{\mu}, \underline{f}, \underline{T}) ,$$

where the nonlinear operator $\underline{\phi}$ represents the resolution of the forward problem.

For given $(\underline{\rho}, \underline{\lambda}, \underline{\mu}, \underline{f}, \underline{T})$ we have just computed the Kernels of the derivative operators

$$(4.10)\quad \frac{\partial \underline{v}}{\partial \underline{m}} = \left(\frac{\partial \underline{v}}{\partial \underline{\rho}}\ \frac{\partial \underline{v}}{\partial \underline{\lambda}}\ \frac{\partial \underline{v}}{\partial \underline{\mu}}\ \frac{\partial \underline{v}}{\partial \underline{f}}\ \frac{\partial \underline{v}}{\partial \underline{T}}\right) .$$

Let

$$(4.11)\qquad \delta\underline{m} = \begin{bmatrix} \partial\rho(\underline{x}) \\ \partial\lambda(\underline{x}) \\ \partial\mu(\underline{x}) \\ \partial f^{i}(\underline{x},t) \\ \partial T^{i}(\underline{x},t) \end{bmatrix}$$

be an arbitrary model perturbation. When solving inverse problems we will find expressions like

$$(4.12)\qquad \delta\underline{v} = \left(\frac{\partial\underline{v}}{\partial\underline{m}}\right)\delta\underline{m}$$

$$= \left(\frac{\partial\underline{v}}{\partial\underline{\rho}}\right)\delta\underline{\rho} + \left(\frac{\partial\underline{v}}{\partial\underline{\lambda}}\right)\delta\underline{\lambda} + \left(\frac{\partial\underline{v}}{\partial\underline{\mu}}\right)\delta\underline{\mu} + \left(\frac{\partial\underline{v}}{\partial\underline{f}}\right)\delta\underline{f} + \left(\frac{\partial\underline{v}}{\partial\underline{T}}\right)\delta\underline{T} .$$

Let me briefly discuss here its meaning.

Introducing the Kernels of the derivative operators and using the notations defined in (4.3), equation (4.12) can be written

$$\delta v_i(\underline{x}_R,t) = \int_{\mathcal{V}} dV(\underline{x}) \frac{\partial v_i(\underline{x}_R,t)}{\partial\rho(\underline{x})} \delta\rho(\underline{x})$$

$$+ \int_{\mathcal{V}} dV(\underline{x}) \frac{\partial v_i(\underline{x}_R,t)}{\partial\lambda(\underline{x})} \delta\lambda(\underline{x})$$

$$+ \int_{\mathcal{V}} dV(\underline{x}) \frac{\partial v_i(\underline{x}_R,t)}{\partial\mu(\underline{x})} \delta\mu(\underline{x})$$

$$+ \int_{\mathcal{V}} dV(x) \int dt' \frac{\partial v_i(\underline{x}_R,t)}{\partial f_j(\underline{x}_R,t')} \delta f_j(\underline{x},t')$$

$$+ \int_{S} dS(\underline{x}) \int dt' \frac{\partial v_i(\underline{x}_R,t)}{\partial T_j(\underline{x}_R,t')} \delta T_j(\underline{x},t') .$$

Using (4.8) and (4.9) we have

$$(4.13)\quad \delta v^i(\underline{x}_R,t) = \delta u^i(\underline{x}_R,t) * R(t)$$

where

$$\delta u_i(\underline{x}_R,t) = -\int_V dV(\underline{x})\ \dot{G}_i^j(\underline{x},t;\underline{x}_R,0) * \dot{u}_j(\underline{x},t)\ \delta\rho(\underline{x})$$

$$-\int_V dV(\underline{x})\ G_{\ell i}^{\ell}(\underline{x},t;\underline{x}_R,0) * u_m^m(\underline{x},t)\ \delta\lambda(\underline{x})$$

$$-\ 2\int_V dV(\underline{x})\ G_{ki}^{\ell}(\underline{x},t;\underline{x}_R,0) * u_\ell^k(\underline{x},t)\ \delta\mu(\underline{x})$$

$$+\int_V dV(\underline{x}) \int dt'\ G_i^j(\underline{x},t-t';\underline{x}_R,0)\ \delta f_j(\underline{x},t')$$

$$+\int_S dS(\underline{x}) \int dt'\ G_i^j(\underline{x},t-t';\underline{x}_R,0)\ \delta T_j(\underline{x},t')\ .$$

The last equation is of no simple direct use. Running backwards all the way we did between equation (4.6) and (4.7) we finally obtain

$$(4.14)\quad \delta u^i(\underline{x}_R,t) = \int_V dV(\underline{x})\ G_j^i(\underline{x}_R,t;\underline{x},0) * \Delta f^j(\underline{x},t)$$

$$+\int_S dS(\underline{x})\ G_j^i(\underline{x}_R,t;\underline{x},0) * \Delta T^j(\underline{x},t)$$

where

$$\Delta f^i(\underline{x},t) = \delta f^i(\underline{x},t) - \delta\rho(\underline{x})\ddot{u}^i(\underline{x},t)$$

$$+\frac{D}{Dx_i}\left[\delta\lambda(\underline{x})\ u_k^k(\underline{x},t)\right] + 2\frac{D}{Dx_j}\left[\delta\mu(\underline{x})u_j^i(\underline{x},t)\right]$$

and

$$\Delta T^i(\underline{x},t) = \delta T^i(\underline{x},t) - \delta\lambda(\underline{x}) n^i(\underline{x}) u^\ell_\ell(\underline{x},t)$$

$$- 2\delta\mu(\underline{x}) n^j(\underline{x}) u^i_j(\underline{x},t) \qquad (\underline{x} \in S) .$$

As we could expect, $\delta v^i(\underline{x}_R,t)$ represents the convolution by the instrument (equation 4.13) of the displacement originated in the current model $(\rho(\underline{x}), \lambda(\underline{x}), \mu(\underline{x}), f^i(\underline{x},t), T^i(\underline{x},t))$ by the Born's secondary sources $\delta\rho(\underline{x})$, $\delta\lambda(\underline{x})$, $\delta\mu(\underline{x})$, $\delta f^i(\underline{x},t)$, $\delta T^i(\underline{x},t)$ (equation (4.14)). We see thus that the computation of $\delta\underline{v}$ as defined in (4.12) (i.e., the application of the derivative operator) corresponds to the resolution of one forward problem, over the current model, for a complex distribution of volume and surface sources.

Much less evident is the meaning of the application of the transpose of the derivative operator. The following section is devoted to that problem.

4.2. - The transpose derivative operator. The derivative operator $\partial\underline{v}/\partial\underline{m}$ acts on the model space and gives a data vector (equation 4.10). By definition, the transpose operator $(\partial\underline{v}/\partial\underline{m})^T$ acts on the dual of the data space and gives an element of the dual of the model space (Section 2.2). We have seen that the dual of a space can be identified with the space weighted by the corresponding covariance operator (equations 2.17-2.18). Let $\delta\hat{\underline{v}}$ be an arbitrary element of the space of weighted data. The application of $(\partial\underline{v}/\partial\underline{m})^T$ on $\delta\hat{\underline{v}}$ will give an element of the space of weighted models which will be denoted by $\delta\hat{\underline{m}}$:

$$(4.15) \qquad \delta\hat{\underline{m}} = \left(\frac{\partial\underline{v}}{\partial\underline{m}}\right)^T \delta\hat{\underline{v}} .$$

Using the notation (4.10) we can write

$$(\partial\underline{v}/\partial\underline{m})^T = \begin{bmatrix} (\partial\underline{v}/\partial\rho)^T \\ (\partial\underline{v}/\partial\lambda)^T \\ (\partial\underline{v}/\partial\mu)^T \\ (\partial\underline{v}/\partial\underline{f})^T \\ (\partial\underline{v}/\partial\underline{T})^T \end{bmatrix}$$

So that equation (4.15) becomes

$$
\begin{aligned}
\delta\hat{\underline{\rho}} &= \left(\frac{\partial \underline{v}}{\partial \underline{\rho}}\right)^T \delta\hat{\underline{v}} \\
\delta\hat{\underline{\lambda}} &= \left(\frac{\partial \underline{v}}{\partial \underline{\lambda}}\right)^T \delta\hat{\underline{v}} \\
\partial\hat{\underline{\mu}} &= \left(\frac{\partial \underline{v}}{\partial \underline{\mu}}\right)^T \delta\hat{\underline{v}} \\
\delta\hat{\underline{f}} &= \left(\frac{\partial \underline{v}}{\partial \underline{f}}\right)^T \delta\hat{\underline{v}} \\
\delta\hat{\underline{T}} &= \left(\frac{\partial \underline{v}}{\partial \underline{T}}\right)^T \delta\hat{\underline{v}} \ .
\end{aligned}
\tag{4.16}
$$

In Section (2.2) we have seen that in fact there is no difference between the Kernel of an operator and the Kernel of its transpose (equation 2.29). Equation (4.16) can then be written, explicitly,

$$
\begin{aligned}
\delta\hat{\rho}(\underline{x}) &= \int dt \sum_R \frac{\partial v_i(\underline{x}_R,t)}{\partial\rho(\underline{x})} \delta\hat{v}^i(\underline{x}_R,t) \\
\delta\hat{\lambda}(\underline{x}) &= \int dt \sum_R \frac{\partial v_i(\underline{x}_R,t)}{\partial\lambda(\underline{x})} \delta\hat{v}^i(\underline{x}_R,t) \\
\delta\hat{\mu}(\underline{x}) &= \int dt \sum_R \frac{\partial v_i(\underline{x}_R,t)}{\partial\mu(\underline{x})} \delta\hat{v}^i(\underline{x}_R,t) \\
\delta\hat{f}^j(\underline{x},t') &= \int dt \sum_R \frac{\partial v_i(\underline{x}_R,t)}{\partial f_j(\underline{x},t)} \delta\hat{v}^i(\underline{x}_R,t) \\
\delta\hat{T}^j(\underline{x},t') &= \int dt \sum_R \frac{\partial v_i(\underline{x}_R,t)}{\partial T_j(\underline{x},t')} \delta\hat{v}^i(\underline{x}_R,t) \qquad (\underline{x} \in S) \ .
\end{aligned}
$$

Using (4.8), (4.9), and the property

$$\int dt\ f(t) * g(t)\ h(t) = \int dt\ f(t)\ g(t) * h(-t)$$

we obtain

$$(4.17)\begin{cases} \delta\hat{\rho}(\underline{x}) = -\int dt\ \dot{\psi}^i(\underline{x},t)\quad \dot{u}_i(\underline{x},t) \\ \delta\hat{\lambda}(\underline{x}) = -\int dt\ \psi^\ell_\ell(\underline{x},t)\quad u^m_m(\underline{x},t) \\ \delta\hat{\mu}(\underline{x}) = -2\int dt\ \psi^m_\ell(\underline{x},t)\quad u^\ell_m(\underline{x},t) \\ \delta\hat{f}^i(\underline{x},t) = \psi^i(\underline{x},t) \\ \delta\hat{T}^i(\underline{x},t) = \psi^i(\underline{x},t) \qquad (\underline{x}\ \varepsilon\ )S\ , \end{cases}$$

where

$$(4.18)\quad \psi^i(\underline{x},t) = \sum_R G^i_j(\underline{x},o;\underline{x}_R,t) * \delta w^j(\underline{x}_R,t)$$

and

$$(4.19)\quad \delta w_j(\underline{x}_R,t) = \delta\hat{v}^j(\underline{x}_R,t) * R(-t)\ .$$

As $\delta\hat{v}(\underline{x}_R,t)$ represents a weighted data vector (a collection of seismograms, weighted by estimated errors), $\delta w^i(\underline{x}_R,t)$ represents the same seismograms cross-correlated by the impulse response of the seismometer, i.e., the "deconvolved" seismograms (equation 4.19). $\psi^i(\underline{x},t)$ can be interpreted as the field obtained by propagation backwards in time of a collection of "sources": one at each receiver location $\underline{x}_R$, and radiating the values $\delta w^i(\underline{x}_R,t)$ (equation 4.18). (The propagation is backwards in time because of the presence of $G^i_j(\underline{x},o;\underline{x}_R,t)$ in (4.13) instead of $G^i_j(\underline{x},t;\underline{x}_R,o)$. The field $\psi^i(\underline{x},t)$ represents then the field which, propagated forward in time, would give, at the observation points $\underline{x}_R$, the values $\delta w^i(\underline{x}_R,t)$. Once the field $\psi^i(\underline{x},t)$ has been obtained, the values $\delta\hat{\rho}(\underline{x})$, $\delta\hat{\lambda}(\underline{x})$, $\delta\hat{\mu}(\underline{x})$, $\delta f^i(\underline{x},t)$, and $\delta T^i(\underline{x},t)$ are readily computed using (4.17). These equations will be interpreted in the next section.

We remark here that the application of the transpose operator $(\partial \underline{v}/\partial \underline{m})^T$ on a weighted data vector $\delta \hat{\underline{v}}$ corresponds to the following operations:

- Deconvolution of $\delta \hat{\underline{v}}$ using (4.19).
- Propagation backwards in time of a composite source, one point source per receiver point, radiating the deconvolved values of the previous step.
- Computation of (4.17).

As we see, the mathematical operations involved in the computation of $(\partial \underline{v}/\partial \underline{m})^T \delta \hat{\underline{v}}$ are completely different of those involved in the computation of $(\partial \underline{v}/\partial \underline{m})\ \delta \underline{m}$. When solving inverse problems using a traditional discretized approach, the computation of $(\partial \underline{v}/\partial \underline{m})\ \delta \underline{m}$ corresponds to the multiplication of the matrix representing $(\partial \underline{v}/\partial \underline{m})$ by the discrete vector representing $\delta \underline{m}$; the computation of $(\partial \underline{v}/\partial \underline{m})^T \delta \hat{\underline{v}}$ corresponds then to the multiplication of the transposed matrix by the discrete vector representing $\delta \underline{v}$. If the discretization is properly made, the discrete approach and the continuous approach will of course give equivalent results (the "continuous" approach will also discretize something, somewhere, for the purposes of using analogic computers). The important difference between a functional approach and a naive discretized approach is that the matrix multiplication is a blind operation, which in fact corresponds to operations easily interpreted in terms of deconvolution, wave propagation, time correlation, etc., as shown above.

5. An Example of Inversion: The Problem of Interpreting Seismic Reflection Data

5.1. - Introduction. The forward problem. The observations. Let me consider here a seismic reflection experiment on land, where vibrators are used as sources of seismic waves.

An idealized seismic reflection experiment can be described as follows. A vibrator occupies, respectively, the "shot points" $\underline{x}_s$ (s = 1,2,...) . For each shot point, the soil displacement is monitored at some receiver locations $\underline{x}_R$ (R = 1,2,...) . In present day experiments, and for economic reasons, the source and receiver locations are usually limited to the immediate viccinity of a line (the survey line). The number of shot points, can be arbitrarily large, depending on the length of the survey line. The number of seismic traces recorded per shot point ranges from a few tens to some hundreds.

In fact, an actual experiment is slightly more complex. On one hand, and in order to increase the energy radiated into the Earth at each vibration, an array of vibrations is used instead of a single one. On the other hand, in order to attenuate the surface waves (for the purposes of an optimal use of the dynamic range of geophones), each seismic trace is obtained as the sum of the outputs of an array of geophones spread over a distance approximately equal to the mean wavelength of the surface waves.

As the wavelengths radiated by an individual vibrator are large compared with the size of the vibrating plate, a vibrator source can conveniently be modelled by a traction T^i acting on a point $\underline{x}_S$ of the surface of the Earth:

$$T^i(\underline{x},t) = \delta(\underline{x} - \underline{x}_S)\; S^i(t)$$

where S^i represents the <u>source time function</u>. For a vertical vibrator ("P-wave experiment") only the vertical component of the vector S^i is non-vanishing. For more compactness in our notations, let us for the moment do not make this assumption.

More precisely, if for a given short number S we denote by $\underline{x}_S^\alpha$ $(\alpha = 1,2,\ldots)$ the surface location of the α-th vibrator of the array, the total source function is clearly given by

$$(5.1) \qquad T^i(\underline{x},t;S) = \sum_\alpha \delta(\underline{x} - \underline{x}_S^\alpha)\; S^i(t)$$

where it is assumed that vibrators vibrate in phase.

For given shot point S, let use denote by $v^i(R,t;S)$ the i-th component of the seismic trace number R. In ordinary P-wave experiments only the vertical component of the soil displacement is monitored, but let us keep for the moment more general assumptions. If we denote by $\underline{x}_R^\beta$ the location of the β-th geophone of the array, and as geophones of an array are simply mounted in series, the total output (voltage) is the sum of the outputs of each geophone of the array:

$$(5.2) \qquad v^i(R,t;S) = \sum_\beta v^i(\underline{x}_R^\beta,t;S) \; .$$

For given a Earth model (i.e., for given $\rho(\underline{x})$, $\lambda(\underline{x})$, and $\mu(\underline{x})$, the predicted displacement at a point $\underline{x}_R^\beta$ can readily be obtained using equation (3.12):

$$u^i(\underline{x}_R^\beta, t; S) = \int_S dS(\underline{x}) \; G_j^i(\underline{x}_R^\beta, t; \underline{x}, 0) * T^j(\underline{x}, t; S)$$

where, as we assume that there are no other sources of seismic waves than vibrators, the volume density of external forces, $f^i(\underline{x},t)$, is taken identically null. Using (5.1) we obtain:

$$(5.3) \quad u^i(\underline{x}_R^\beta, t; S) = \sum_\alpha G_j^i(\underline{x}_R^\beta, t; \underline{x}_s^\alpha, 0) * S^j(t) \; .$$

Introducing the impulse response of the seismometer, $R(t)$, (see Section 3.1) we have, for the computed output of a geophone at $\underline{x}_R^\beta$:

$$(5.4) \quad v^i(\underline{x}_R^\beta, t; s) = u^i(\underline{x}_R^\beta, t; S) * R(t) \; .$$

Using (5.2) we obtain the total output the trace number R :

$$(5.5) \quad v^i(R, t; S) = \sum_\beta \sum_\alpha G_j^i(\underline{x}_R^\beta, t; \underline{x}_s^\alpha, 0) * S^j(t) * R(t) \; .$$

If we are able to obtain the Green's functions $G_j^i(\underline{x}, t; \underline{x}', 0)$ by using any numerical method (finite differencing, ray theory, discrete wave-number, etc.), equation (5.5) completely solves the forward problem.

I point out here that, as we explicitly consider that the source and receive arrays are composite, concepts such as the "radiation pattern of an array of vibrators," or the "admission pattern of an array of geophones," have not to be introduced.

The recorded ("observed") values of our seismic reflection experiment will be denoted by

$$v^i(R, t; S)_{OBS}$$

where I use for the time variable the notation t , instead of t_n which would more properly correspond to a digitally recorded data set. A quite general description of experimetnal incertitudes in the data set can be achieved by defining a covariance operator $\underline{C}_V$. As our data set is discrete, the Kernel of $\underline{C}_V$ is a matrix, whose elements can be denoted by

$$C_V^{ii'}(R,t;S|R',t';S') \ .$$

We make here the most trivial assumption for the statistics of errors, i.e., completely uncorrelated errors (along each trace and between different traces):

$$(5.6) \quad C_V^{ii'}(R,t;s|R',t';s') = \sigma_V^2 \ \delta^{ii'} \delta_{RR'} \delta_{tt'} \delta_{ss'}$$

where σ_V^2 represents the estimated value of the variance of the noise.

5.2. - The a priori model. I assume here that the source time function of the vibrator, $s^1(t)$, is prefectly known. A detailed description of how the source function itself can be integrated as an unknown in the inverse problem can be found in Tarantola (1984).

What was named model in Section (2.1) is then here composed of the three functions $\rho(\underline{x})$, $\lambda(\underline{x})$, and $\mu(\underline{x})$:

$$\underline{m} = \begin{bmatrix} \underline{\rho} \\ \underline{\lambda} \\ \underline{\mu} \end{bmatrix}$$

The a priori model will then be composed by the three functions

$$\underline{m}_{PRIOR} = \begin{bmatrix} \underline{\rho}_{PRIOR} \\ \underline{\lambda}_{PRIOR} \\ \underline{\mu}_{PRIOR} \end{bmatrix}$$

and the covariance operator representing our confidence in $\underline{m}_{PRIOR}$ will then have the partitioned form

$$\underline{C}_M = \begin{bmatrix} \underline{C}_{\rho\rho} & \underline{C}_{\rho\lambda} & \underline{C}_{\rho\mu} \\ \underline{C}_{\lambda\rho} & \underline{C}_{\lambda\lambda} & \underline{C}_{\lambda\mu} \\ \underline{C}_{\mu\rho} & \underline{C}_{\mu\lambda} & \underline{C}_{\mu\mu} \end{bmatrix} .$$

From a rigorous point of view, $\underline{m}_{PRIOR}$ and $\underline{C}_M$ should be obtained from statistical studies (see for instance Kinoshita, 1981). More easily, we can choose for $\underline{m}_{PRIOR}$ and $\underline{C}_M$ some reasonable values, as for instance

$$\rho(\underline{x})_{PRIOR} = \text{const.} = 2.4 \text{ g cm}^{-3}$$

$$\lambda(\underline{x})_{PRIOR} = \text{const.} = 0.8\ 10^4 \text{ MPa}$$

$$\mu(\underline{x})_{PRIOR} = \text{const.} = 0.6\ 10^4 \text{ MPa}$$

with

$$\underline{C}_M = \begin{bmatrix} \underline{C}_{\rho\rho} & 0 & 0 \\ 0 & \underline{C}_{\lambda\lambda} & 0 \\ 0 & 0 & \underline{C}_{\mu\mu} \end{bmatrix} \tag{5.8}$$

$$\begin{cases} C_{\rho\rho}(\underline{x},\underline{x}') = \sigma_\rho^2 \exp\left(-\tfrac{1}{2}\dfrac{\|\underline{x}-\underline{x}'\|^2}{L^2}\right) \\ C_{\lambda\lambda}(\underline{x},\underline{x}') = \sigma_\lambda^2 \exp\left(-\tfrac{1}{2}\dfrac{\|\underline{x}-\underline{x}'\|^2}{L^2}\right) \\ C_{\mu\mu}(\underline{x},\underline{x}') = \sigma_\mu^2 \exp\left(-\tfrac{1}{2}\dfrac{\|\underline{x}-\underline{x}'\|^2}{L^2}\right) \end{cases} \tag{5.9}$$

and

$$\sigma_\rho = 0.4 \text{ g cm}^{-3}$$

$$\sigma_\lambda = 0.6 \text{ MPa}$$

$$\sigma_\mu = 0.3 \text{ MPa} ,$$

which means that we estimate ρ at any point to be of the order of (2.4 ± 0.4) g cm^{-3} , λ to be of the order of (0.8 ± 0.6) MPa , (0.8 ± 0.6) MPa , and μ to be of the order of (0.6 ± 0.3) MPa , we assume the functions $\rho(\underline{x})$, $\lambda(\underline{x})$, and $\mu(\underline{x})$ to be smooth functions with smoothness length of the order of L , and we assume that errors in our a priori model $\underline{m}_{PRIOR}$ are uncorrelated between $\underline{\rho}$, $\underline{\lambda}$, and $\underline{\mu}$. Of course, more elaborated choices for $\underline{m}_{PRIOR}$ and $\underline{C}_M$ can be used.

As suggested in Section (2.4), the final solution of the inverse problem will strongly depend on the particular choice for $\underline{m}_{PRIOR}$ and $\underline{C}_M$ only in those regions of the space poorly resolved by the data set.

5.3. - Resolution of the inverse problem. Once the data set $\underline{v}_{OBS}$ and the covariance operator $\underline{C}_V$ have made been precise, the a priori model $\underline{m}_{PRIOR}$ and the a priori covariance operator $\underline{C}_M$ have been chosen, and the forward problem $(\underline{v} = \underline{f}(\underline{m}))$ has been fully understood, the inverse problem can be set up (Section 2.4).

To simplify the discussions, we choose here the simpliest of the algorithms of resolution, namely, the (unpreconditioned) steepest descent algorithm. Let me recall here equations (2.61):

$$\text{(2.61 again)}\quad \begin{cases} \underline{v}_n = \underline{f}(\underline{m}_n) \\ \underline{\gamma}_n = \underline{C}_M \underline{F}_n^T \underline{C}_v^{-1} (\underline{v}_n - \underline{v}_{OBS}) + (\underline{m}_n - \underline{m}_{PRIOR}) \\ \underline{b}_n = \underline{F}_n \underline{\gamma}_n \\ \alpha_n = 1/(1 + \underline{b}_n^T \underline{C}_v^{-1} \underline{b}_n / \underline{\gamma}_n^T \underline{C}_M^{-1} \underline{\gamma}_n) \\ \underline{m}_{n+1} = \underline{m}_n - \alpha_n \underline{\gamma}_n \ . \end{cases}$$

Using

$$\underline{m} = \begin{bmatrix} \underline{\rho} \\ \underline{\lambda} \\ \underline{\mu} \end{bmatrix}$$

$$\underline{F}_n = \left(\frac{\partial \underline{v}}{\partial \underline{m}}\right)_n = \left[\left(\frac{\partial \underline{v}}{\partial \underline{\rho}}\right)_n \ \left(\frac{\partial \underline{v}}{\partial \underline{\lambda}}\right)_n \ \left(\frac{\partial \underline{v}}{\partial \underline{\mu}}\right)_n \right] ,$$

assuming that a priori errors in ρ , λ , and μ are mutually uncorrelated (equation 5.8), and assumed uncorrelated errors in the data set (equation 5.6), we obtain the following algorithm (5.10):

$$\underline{v}_n = \underline{f}(\underline{\rho}_n, \underline{\lambda}_n, \underline{\mu}_n)$$

$$\delta\hat{\underline{v}}_n = (\underline{v}_n - \underline{v}_{OBS})/\sigma_v^2$$

$$\begin{cases} \delta\hat{\underline{\rho}}_n = \left(\dfrac{\partial \underline{v}}{\partial \underline{\rho}}\right)_n^T \delta\hat{\underline{v}}_n \\ \delta\hat{\underline{\lambda}}_n = \left(\dfrac{\partial \underline{v}}{\partial \underline{\lambda}}\right)_n^T \delta\hat{\underline{v}}_n \\ \delta\hat{\underline{\mu}}_n = \left(\dfrac{\partial \underline{v}}{\partial \underline{\mu}}\right)_n^T \delta\hat{\underline{v}}_n \end{cases}$$

$$\begin{cases} \Delta\underline{\rho}_n = \underline{C}_{\rho\rho}\delta\hat{\underline{\rho}}_n + (\underline{\rho}_n - \underline{\rho}_{PRIOR}) \\ \Delta\underline{\lambda}_n = \underline{C}_{\lambda\lambda}\delta\hat{\underline{\lambda}}_n + (\underline{\lambda}_n - \underline{\lambda}_{PRIOR}) \\ \Delta\underline{\mu}_n = \underline{C}_{\mu\mu}\delta\hat{\underline{\mu}}_n + (\underline{\mu}_n - \underline{\mu}_{nPRIOR}) \end{cases}$$

$$A_n = \Delta\underline{\rho}_n^T\underline{C}_{\rho\rho}^{-1}\Delta\underline{\rho}_n + \Delta\underline{\lambda}_n^T\underline{C}_{\lambda\lambda}^{-1}\Delta\underline{\lambda}_n + \Delta\underline{\mu}_n^T\underline{C}_{\mu\mu}^{-1}\Delta\underline{\mu}_n$$

$$\underline{b}_n = \left(\frac{\partial \underline{v}}{\partial \underline{\rho}}\right)_n \Delta\underline{\rho}_n + \left(\frac{\partial \underline{v}}{\partial \underline{\lambda}}\right)_n \Delta\underline{\lambda}_n + \left(\frac{\partial \underline{v}}{\partial \underline{\mu}}\right)_n \Delta\underline{\mu}_n$$

$$B_n = \underline{b}_n^T \underline{b}_n/\sigma_v^2$$

$$\alpha_n = 1/(1 + B_n/A_n)$$

$$\left\{ \underline{\rho}_{n+1} = \underline{\rho}_n - \alpha_n \Delta\underline{\rho}_n \right.$$

$$\begin{cases} \underline{\lambda}_{n+1} = \underline{\lambda}_n - \alpha_n \, \Delta\underline{\lambda}_n \\ \underline{\mu}_{n+1} = \underline{\mu}_n - \alpha_n \, \Delta\underline{\mu}_n \end{cases}$$

To simplify further this example, we assume that we have no a priori information on density, and Lame's parameters:

$$\underline{C}_{\rho\rho} = c \, \sigma_\rho^2 \, \underline{I}$$

$$\underline{C}_{\lambda\lambda} = c \, \sigma_\lambda^2 \, \underline{I}$$

$$\underline{C}_{\mu\mu} = c \, \sigma_\mu^2 \, \underline{I}$$

where we take the limit

$$c \to \infty \, .$$

The reader will easily verify that equations (5.10) simplify to

$$\underline{v}_n = \underline{f}(\underline{\rho}_n, \underline{\lambda}_n, \underline{\mu}_n)$$

$$\delta\underline{v}_n = \underline{v}_{OBS} - \underline{v}_n$$

$$\begin{cases} \delta\underline{\rho}_n = \sigma_\rho^2 \left(\frac{\partial \underline{v}}{\partial \underline{\rho}}\right)_n^T \delta\underline{v}_n \\ \delta\underline{\lambda}_n = \sigma_\lambda^2 \left(\frac{\partial \underline{v}}{\partial \underline{\lambda}}\right)_n^T \delta\underline{v}_n \\ \delta\underline{\mu}_n = \sigma_\mu^2 \left(\frac{\partial \underline{v}}{\partial \underline{\mu}}\right)_n^T \delta\underline{v}_n \end{cases}$$

$$A_n = \delta\underline{\rho}_n^T \ \delta\underline{\rho}_n/\sigma_\rho^2 + \delta\underline{\lambda}_n^T \ \delta\underline{\lambda}_n/\sigma_\lambda^2 + \delta\underline{\mu}_n^T \ \delta\underline{\mu}_n/\sigma_\mu^2$$

$$\underline{b}_n = \left(\frac{\partial \underline{v}}{\partial \underline{\rho}}\right)_n \delta\underline{\rho}_n + \left(\frac{\partial \underline{v}}{\partial \underline{\lambda}}\right)_n \delta\underline{\lambda}_n + \left(\frac{\partial \underline{v}}{\partial \underline{\mu}}\right)_n \delta\underline{\mu}_n$$

$$B_n = \underline{b}_n^T \ \underline{b}_n$$

$$\alpha_n = A_n/B_n$$

$$\begin{cases} \underline{\rho}_{n+1} = \underline{\rho}_n + \alpha_n \ \delta\underline{\rho}_n \\ \underline{\lambda}_{n+1} = \underline{\lambda}_n + \alpha_n \ \delta\underline{\lambda}_n \\ \underline{\mu}_{n+1} = \underline{\mu}_n + \alpha_n \ \delta\underline{\mu}_n \ , \end{cases}$$

where some of the variables have been redefined.

Let $v^i(R,t;S)_n$ and $G^i_j(\underline{x},t;\underline{x}',o)_n$ respectively denote the computed values of seismograms and the Green's functions for the current model $\rho(\underline{x})_n$, $\lambda(\underline{x})_n$, and $\mu(\underline{x})_n$. Using the results of Chapter 4 and Section (5.1), equations (5.11) can be explicitly written

(5.12a) $$\delta v^i(R,t;S)_n = v^i(R,t;S)_{OBS} - v^i(R,t;S)_n$$

(5.12b) $$\delta w^i(R,t;S)_n = \delta v^i(R,t;S)_n * R(-t)$$

(5.12c) $$\psi^i(\underline{x},t:S)_n = \sum_R \sum_\beta G^i_j(\underline{x},o;\underline{x}^\beta_R,t) * \delta w^i(R,t;S)$$

(5.12d) $$u^i(\underline{x},t;S)_n = \sum_\alpha G^i_j(\underline{x},t;\underline{x}^\alpha_S,o) * S^j(t)$$

$$(5.12e)\begin{cases} \delta\rho(\underline{x})_n = \sigma_\rho^2 \sum_S \int dt\ \dot{\psi}^i(\underline{x},t;S)_n\ \dot{u}_i(\underline{x},t;S)_n \\ \delta\lambda(\underline{x})_n = \sigma_\lambda^2 \sum_S \int dt\ \psi_i^i(\underline{x},t;S)_n\ u_j^j(\underline{x},t;S)_n \\ \delta\mu(\underline{x})_n = \sigma_\mu^2 \sum_S \int dt\ \psi_j^i(\underline{x},t;X)_n\ u_i^j(\underline{x},t;S)_n \end{cases}$$

$$(5.12f)\quad A_n = \frac{1}{\sigma_\rho^2}\int_\nu dV(\underline{x})\,\delta\rho(\underline{x})_n^2 + \frac{1}{\sigma_\lambda^2}\int_\nu dV(\underline{x})\,\delta\lambda(\underline{x})_n^2 + \\ + \frac{1}{\sigma_\mu^2}\int dV(\underline{x})\,\delta\mu(\underline{x})_n^2$$

$$(5.12g)\quad \Delta f^i(\underline{x},t;S)_n = -\delta\rho(\underline{x})_n\ddot{u}^i(\underline{x},t;S)_n \\ + \frac{D}{Dx_i}[\delta\lambda(\underline{x})_n u_i^i(\underline{x},t;S)_n] + 2\,\frac{D}{Dx_j}[\delta\mu(\underline{x})_n u_j^i(\underline{x},t;S)_n]$$

$$(5.12h)\quad \Delta T^i(\underline{x},t;S)_n = -\delta\lambda(\underline{x})_n\ n^i(\underline{x})\ u_j^j(\underline{x},t;S)_n \\ - 2\ \delta\mu(\underline{x})_n\ n^j(\underline{x})\ u_j^i(\underline{x},t;S)_n$$

$$(5.12i)\quad b^i(R,t;S)_n = \sum_\beta \{\int_\nu dV(\underline{x})G_j^i(\underline{x}_R^\beta,t;\underline{x},0)_n * \Delta f^j(\underline{x},t;S)_n \\ + \int_S dS(\underline{x})G_j^i(\underline{x}_R^\beta,t;\underline{x},0)_n * \Delta T^j(\underline{x},t;S)_n\}\ .$$

$$(5.12j)\quad B_n = \sum_S \int dt \sum_R b^i(R,t;S)_n\ b_i(R,t;S)_n$$

(5.12k) $\alpha_n = A_n / B_n$

(5.21ℓ) $$\begin{cases} \rho(\underline{x})_{n+1} = \rho(\underline{x})_n - \alpha_n \, \delta\rho(\underline{x})_n \\ \lambda(\underline{x})_{n+1} = \lambda(\underline{x})_n - \alpha_n \, \delta\lambda(\underline{x})_n \\ \mu(\underline{x})_{n+1} = \mu(\underline{x})_n - \alpha_n \, \delta\mu(\underline{x})_n \, . \end{cases}$$

In these equations we have defined

$$\psi_{ij}(\underline{x},t;S) = \frac{1}{2} \left(\frac{D\psi_i}{Dx^j}(\underline{x},t;S) + \frac{D\psi_j}{Dx^i}(\underline{x},t;S) \right) .$$

Let me discuss in some detail the physical significance of the main steps of the algorithm:

1) The starting point of the iterative loop consists in the values $\rho(\underline{x})_n$, $\lambda(\underline{x})_n$, and $\mu(\underline{x})_n$ obtained from the previous iteration (or with some arbitrary initial values $\rho(\underline{x})_o$, $\lambda(\underline{x})_o$, and $\mu(\underline{x})_o$ for the first iteration, as for instance uniform values).

In the following, and until step 6, let us assume that we have chosen a particular source point S .

2) (Equation 5.12a): As, for given S , $v^i(R,t;S)_n$ represents the solution of the forward problem (synthetic seismograms), $\delta v^i(R,t;S)_n$ clearly corresponds to the <u>data residuals</u>.

3) (Equation 5.12b): The cross-correlation of $\delta v^i(R,t;S)_n$ with the impulse response of the instrument, $R(t)$, gives $\delta w^i(R,t;S)_n$, which can be interpreted as the residuals <u>deconvolved</u> of the instrument effect.

4) (Equation 5.12c): For given R and S , $\psi^i(R,t;S)_n$ is obtained by propagation of the residuals $\delta w^i(R,t;S)_n$ <u>backwards in time</u>, all the points $\underline{x}_R^\beta$ ($\beta = 1,2,\ldots$) (i.e., all the geophone locations of a single array) radiating simultaneously the same time function (the residual of the corresponding seismic trace). This means that, for given S , $\psi^i(\underline{x},t;S)_n$ is the field which

propagated forward in time, gives, at the receiver locations, the residuals. As the residuals represent that part of the data set which is not accounted for by the current model, $\psi^i(\underline{x},t;S)_n$ has to be interpreted as the <u>missing field</u> for the current model.

5) (Equation 5.12d): The field $u^i(\underline{x},t;S)_n$ simply corresponds to the predicted displacement fields in the current model.

6) (Equation 5.12e): This is the most important of all equations. The values $\delta\rho(\underline{x})_n$, $\delta\lambda(\underline{x})_n$, and $\delta\mu(\underline{x})_n$ here computed will be the corrections to be added to the current model $(\rho(\underline{x})_n$, $\lambda(\underline{x})_n$, $\mu(\underline{x})_n)$ for obtaining the updated model $(\rho(\underline{x})_{n+1}$, $\lambda(\underline{x})_{n+1}$, $\mu(\underline{x})_{n+1})$ as seen in equation (5.12ℓ). Equations (5.12e) show that the current model has to be corrected at points $\underline{x}$ of the space <u>where the missing field $\psi^i(\underline{x},t;S)_n$ is correlated with the field $u^i(\underline{x},t;S)_n$ predicted by the current model</u>. More precisely, if for a given source S, at a given point $\underline{x}$ of the space the velocity of the missing field, $\dot{\psi}^i(\underline{x},t;S)_n$ is time correlated with the velocity of the current field, $\dot{u}^i(\underline{x},t;S)_n$ the density at that point has to be corrected by an amount proportional to the value of the correlation coefficient, i.e., proportional to $\int dt\ \dot{\psi}(\underline{x},t:S)_n\ \dot{u}^i(\underline{x},t;S)_n$. If the dilatation $\psi^i_i(\underline{x},t;S)_n$ of the missing field is time correlated with the dilatation $u^i_i(\underline{x},t:S)_n$ of the current field, the value of the first Lame's parameter has to be corrected by an amount proportional to $\int dt \psi^i_i(\underline{x},t;S)_n u^j_j(\underline{x},t;S)_n$. If the deformation $\psi^i_j(\underline{x},t;S)_n$ is correlated with the deformation $u^i_j(\underline{x},t;S)_n$ then, the second Lame's parameter has to be corrected proportionally to $\int dt\ \psi^i_j(\underline{x},t;S)_n\ u^j_i(\underline{x},t;S)_n$. Let us consider the first iteration and assume that the starting model is very smooth. As the computed values of data, $v^i(R,t;R)_0$, only give the direct wave, the residuals $\delta v^i(R,t;S)_n = v^i(R,t;S)_{OBS} - v^i(R,t;S)_0$ corresponds to the reflected waves. The field $\psi^i(\underline{x},t;S)_0$ clearly corresponds to what Claerbout (1971) named the "upgoing field." $u^i(\underline{x},t;S)_0$ is then the "downgoing field" and the formulas (5.12e) correspond then to the migration equation of Claerbout (1971, Eq. 5). Here instead of defining a "reflectivity" we can compute physical properties of rocks. We also see that a <u>migration is not the final solution but corresponds to the first step of the iterative solution of the inverse problem.</u> Of course, if the starting model is accurate enough, one iteraction will suffice, but in general, some iterations will be needed. The second and subsequent iterations clearly correspond to a migration of the corresponding residuals.

7) (Equation 5.12f-5.12k): All these equations are not very important because they only serve to compute the scalar α_n which, in any case, has only to be known approximately. We can remark that $b^i(R,t;S)_n$ corresponds to the data generated by the perturbations $\delta\rho(\underline{x})_n$, $\delta\lambda(\underline{x})_n$, $\delta\mu(\underline{x})_n$ when considered as Born's secondary sources. As α_n has only to be known approximatley, the propagation of these secondary sources can be performed in a uniform model, thus replacing the Green's functions $G^i_j(\underline{x},t;\underline{x}',0)_n$ in (5.12i) by $G^i_j(\underline{x},t;\underline{x}',0)_0$ which are analytical if the starting model is homogeneous (see for instance, Aki and Richards, 1980).

8) (Equation 5.12ℓ): This equation finally gives the updated model $\rho(\underline{x})_{n+1}$, $\lambda(\underline{x})_{n+1}$, and $\mu(\underline{x})_{n+1}$, to be introduced int a new iterative loop until convergence.

Equation (5.12a)-(5.12e) can also be written as follows

$$\delta v^i(R,t;S)_n = v^i(R,t;S)_{OBS} - v^i(R,t;S)_n \tag{5.13a}$$

$$\delta w^i(R,t;S)_n = \delta v^i(R,t;S)_n * R(-t) \tag{5.13b}$$

$$\delta z^{ij}(R,t;S)_n = \delta w^i(R,t;S)_n * S^j(-t) \tag{5.13c}$$

$$\delta\rho(\underline{x})_n = \sigma_\rho^2 \sum_S \sum_R \int dt \Big(\sum_\alpha \dot{G}_{kj}(\underline{x},t;\underline{x}_S^\alpha,0)_n\Big) * \Big(\sum_\beta \dot{G}_i^k(\underline{x},t;\underline{x}_R^\beta,0)_n\Big)\ \delta z^{ij}(R,t;S)_n$$

$$\delta\lambda(\underline{x})_n = \sigma_\lambda^2 \sum_S \sum_R \int dt \Big(\sum_\alpha G_{kj}^k(\underline{x},t;\underline{x}_S^\alpha,0)_n\Big) * \Big(\sum_\beta G_{\ell i}^\ell(\underline{x},t;\underline{x}_R^\beta,0)_n\Big)\ \delta z^{ij}(R,t;S)_n \tag{5.13d}$$

$$\delta\mu(\underline{x})_n = 2\,\sigma_\mu^2 \sum_S \sum_R \int dt \Big(\sum_\alpha G_{\ell j}^k(\underline{x},t;\underline{x}_S^\alpha,0)_n\Big) * \Big(\sum_\beta G_{ki}^\ell(\underline{x},t;\underline{x}_R^\beta,0)_n\Big)\ \delta z^{ij}(R,t;S)_n$$

Prior to the interpretation of these formulas, let me make two assumptions which correspond to usual practice in seismic exploration:

i) the vibrator source is assumed to have only a vertical component

$$\underline{S}(t) = \begin{bmatrix} 0 \\ 0 \\ S(t) \end{bmatrix} ;$$

ii) only the vertical component of the particle displacement, $v(R,t;S)$, is monitored.

Equations (5.13) can then be rewritten as follows:

(5.14a) $$\delta v(R,t;S)_n = v(R,t;S)_{OBS} - v(R,t;S)_n$$

(5.14b) $$\delta w(R,t;S)_n = \delta(R,t;S)_n * R(-t)$$

(5.14c) $$\delta z(R,t;S)_n = \delta w(R,t;S)_n * S(-t)$$

(5.14d) $$\begin{cases} \delta\rho(\underline{x})_n = \sigma_\rho^2 \sum_S \sum_R \int dt \; (\sum_\alpha \dot{G}_{i3}(\underline{x},t;\underline{x}_S^\alpha,o)_n) \\ \quad * (\sum_\beta \dot{G}_3^i(\underline{x},t;\underline{x}_R^\beta,o)_n) \; \delta z(R,t;S)_n \\ \delta\lambda(\underline{x})_n = \sigma_\lambda^2 \sum_S \sum_R \int dt \; (\sum_\alpha G_{i3}^i(\underline{x},t;\underline{x}_S^\alpha,o)_n) \\ \quad * (\sum_\beta G_{j3}^i(\underline{x},t;\underline{x}_R^\beta,o)_n) \; \delta z(R,t;S)_n \\ \delta\mu(\underline{x})_n = 2\sigma_\mu^2 \sum_S \sum_R \int dt \; (\sum_\alpha G_{j3}^i(\underline{x},t;\underline{x}_S^\alpha,o)_n) \\ \quad * (\sum_\beta G_{i3}^j(\underline{x},t;\underline{x}_R^\beta,o)_n) \; \delta z(R,t;S)_n \end{cases}$$

Equation (5.14c) shows that after cross-correlation with the impulse response of the instrument, $R(t)$, the residuals have to be cross-correlated with the vibrator source time function $S(t)$. This corresponds to the usual practice when using vibrators sources. Equations (5.14d) have to be examined with some care.

Assume that we are using a rough ray theory which neglect multiples and other complications. For given S, the Green's functions $\sum_\alpha G^i_3(\underline{x},t;\underline{x}^\alpha_S,0)$ represent the response at $(\underline{x},t)$ to the array vibrators. This response can be approximated by considering a single virtual source, located at the barycenter of the array, and radiating with a conveniently defined "radiation pattern". Similarly, for given R, $\sum_\beta G^i_3(\underline{x},t;\underline{x}^\beta_R,0)$ can be simulated by the introduction of the radiation pattern of the array of geophones. Each one of the factors $\sum_\alpha G^i_3(\underline{x},t;\underline{x}^\alpha_S,0)$ and $\sum_\beta G^i_3(\underline{x},t;\underline{x}^\beta_R,0)$ (in fact their derivatives) becomes simply a sum of two terms, a P-wave term and a S-wave term. Each term is essentially a directionality factor times a Dirac's function of the form $\delta(t-T)$ where T is the travel time (for P or S waves) between the points $\underline{x}_S$ and $\underline{x}$ or $\underline{x}_R$ and $\underline{x}$ (Aki and Richards, 1980). The convolution of Green's functions in (5.14d) gives then the sum of four Dirac's functions. The integral over the time can then be performed directly, giving the result that, for given R and S (i.e., for a given seismogram), the correction of the density and Lame's parameters at a point $\underline{x}$ is a sum of the amplitudes of the seismogram $\delta z(R,t;S)_n$ at four chosen times t corresponding to the travel times of waves from $\underline{x}_S$ to $\underline{x}$ and from $\underline{x}$ to $\underline{x}_R$ (PP waves, SS waves and converted PS and SP waves), each amplitude been weighted by a geometric factor depending on the relative positions of $\underline{x}_S$, $\underline{x}$, and $\underline{x}_R$. From that point of view, equations (5.14d) correspond to a Kirchhoff <u>migration</u> (in the sense defined by French, 1974, or Schneider, 1978), here generalized to the elastodynamic (as opposed to acoustic) problem.

As is usual in seismic reflection experiments the vibrator is vertical ("P-wave vibrator"), and only the vertical component of particle displacement is monitored, only the amplitude of the PP wave will probably be significant. Nevertheless, the elastodynamic migration will be different from the acoustic migration because the dependence of the diffraction (or reflection) coefficient as a function of the offset $||\underline{x}_S - \underline{x}_R||$ is explicitly taken into account.

6. Conclusion. The use of gradient methods allows for an efficient resolution of large sized inverse problems (the algorithms thus obtained do not invoke the resolution of large linear systems). If $\underline{d} = \underline{f}(\underline{m})$ represents the resolution of the forward problem, the inverse problem is solved by an iterative applications to the residuals of the operator $\underline{F}^T$, the transpose of the Frechet derivative of $\underline{f}$, and it is shown that $\underline{F}^T$ essentially represents a migration of the residuals (in the usual sense in today seismic reflection philosophy).

The quality of the results obtained with the method here developed, will strongly depend on the realism of the synthetic seismograms we are able to generate (i.e., will strongly depend on the quality of the results of the forward problem). As far as this condition is satisfied, we can hope that the values of Earth's parameters (i.e., seismic images of the Earth) obtained with the present method for high quality data can be of great help as well as for understanding deep processes in the Earth, as for petroleum trap identification.

7. Aknowledgments. Many of the results of this paper were born during helpful discussions with R.W. Clayton, P. Lailly, A. Nercessian, and B. Valette. The research has been supported by the Centre National de la Recherche Scientifique (CNRS) under contracts ATP Geophysique Applique, ATP Geothermie, and RCP Problemes Inverses.

8. Appendix. Demonstration for Formulae (4.7). Up to the first order, equation (4.6) is written

$$\delta u_i(\underline{x}_R,t) = \int_V dV(\underline{x})\; G_{ij}(\underline{x}_R,t;\underline{x},0) * \Big\{\delta f^j(\underline{x},t) - \frac{\partial^2 u^j}{\partial t^2}(\underline{x},t)\,\delta\rho(\underline{x}) + \frac{D}{Dx^k}[\delta c^{jk\ell m}(\underline{x})u_{\ell m}(\underline{x},t)]\Big\}$$

$$+ \int_V dS(\underline{x})\; G_{ij}(\underline{x}_R,t;\underline{x},0) * \Big\{\delta T^j(\underline{x},t) - n_k(\underline{x})\,\delta c^{jk\ell m}(\underline{x})\,u_{\ell m}(\underline{x},t)\Big\} .$$

Using

$$G_{ij}(\underline{x}_R,t;\underline{x},0) * \frac{D}{Dx^k}[\delta c^{jk\ell m}(\underline{x})\,u_{\ell m}(\underline{x},t)] =$$

$$- \frac{DG_{ij}}{Dx^k} (\underline{x}_R, t; \underline{x}, 0) * u_{\ell m}(\underline{x}, t) \; \delta c^{jk\ell m}(\underline{x})$$

$$+ \frac{D}{Dx^k} [G_{ij}(\underline{x}_R, t; \underline{x}, 0) * u_{\ell m}(\underline{x}, t) \; \delta c^{jk\ell m}(\underline{x})] \; ,$$

using the divergence theorem

$$\int_V dV(\underline{x}) \frac{D\underline{Q}}{Dx^k} (\underline{x}) = \int_S dS(\underline{x}) \; n_k(\underline{x}) \; \underline{Q}(\underline{x}) \; ,$$

and using the property that for fields with a quiescent past

$$f(t) * \dot{g}(t) = \dot{f}(t) * g(t) \; ,$$

we arrive at

$$\delta u_i(\underline{x}_R, t) = \int_V dV(\underline{x}) \; G_{ij}(\underline{x}_R, t; \underline{x}, 0) * \delta f^j(\underline{x}, t)$$

$$- \int_V dV(\underline{x}) \; \dot{G}_{ij}(\underline{x}_R, t; \underline{x}, 0) * \dot{u}^j(\underline{x}, t) \; \delta\rho(\underline{x})$$

$$- \int_V dV(\underline{x}) \frac{DG_{ij}}{Dx^k} (\underline{x}_R, t; \underline{x}, 0) * u_{\ell m}(\underline{x}, t) \; \delta c^{jk\ell m}(\underline{x}$$

$$+ \int_S dS(\underline{x}) \; G_{ij}(\underline{x}_R, t; \underline{x}, 0) * \delta T^j(\underline{x}, t) \; .$$

Using the reciprocity theorem (3.10) and equation (4.4) we finally arrive at

$$\delta u_i(\underline{x}_R, t) = \int_V dV(\underline{x}) \; G_i^j(\underline{x}, t; \underline{x}_R, 0) * \delta f_j(\underline{x}, t)$$

$$- \int_V dV(\underline{x}) \; \dot{G}_i^j(\underline{x}, t; \underline{x}_R, 0) * \dot{u}_j(\underline{x}, t) \; \delta\rho(\underline{x})$$

$$- \int_V dV(\underline{x}) \; G_{\ell i}^{\ell}(\underline{x}, t; \underline{x}_R, 0) * u_m^m(\underline{x}, t) \; \delta\lambda(\underline{x})$$

$$- 2 \int_V dV(\underline{x}) \; G^{\ell}_{mi}(\underline{x},t;\underline{x}_R,o) * u^m_{\ell}(\underline{x},t) \; \delta\mu(\underline{x})$$

$$+ \int_S dS(\underline{x}) \; G^j_i(\underline{x},t;\underline{x}_R,o) * \delta T_j(\underline{x},t) \; .$$

where $G^i_{jk}(\underline{x},t;\underline{x}',o)$ has been defined in equation (3.8).

9. References

[1] K. AKI and P. RICHARDS, Quantitative seismology, San Francisco, W.H. Freeman and Co., 1980.

[2] A. BEN-MENAHEM and S.J. SINGH, Seismic waves and sources, New York, Springer-Verlag, 1981.

[3] G. BACKUS and F. GILBERT, The resolving power of gross earth data, Geophys. J.R. Astr. Soc., 16 (1968), pp. 169-205.

[4] G. BACKUS and F. GILBERT, Uniqueness in the inversion of inaccurate gross earth data, Philos. Trans. R. Soc. London, 266 (1970), pp. 13-192.

[5] J.F. CLAERBOUT, Toward a unified theory of reflector mapping, Geophysics, V. 36 (1971), pp. 467-481.

[6] J.F. CLAERBOUT and F. MUIR, Robust modelling with erratic data, Geophysics, 38, 5 (1973), pp. 826-844.

[7] A.J. DEVANEY, A computer simulation study of diffraction tomography, IEEE Transactions on Biomedical Engineering, Vol. BME-30, No. 7, (1983) pp. 377-386.

[8] R. FLETCHER and C.M. REEVES, Function minimization by conjugate gradients, The Computer Journal, 7, (1964) pp. 149-154.

[9] J.N. FRANKLIN, Well-posed stochastic extensions of ill-posed linear problems, J. Math. Analy. Appl., 31, (1970) pp. 682-716.

[10] W.S. FRENCH, Two-dimensional and three-dimensional migration of model-experiment reflection profiles, Geophysics, 39, (1974) pp. 265-277.

[11] J.A. HUDSON and J.R. HERITAGE, The use of Born approximation in seismic scattering problems, Geophys. J.R. Astr. Soc., 66, (1981) pp. 221-240.

[12] D.D. JACKSON, The use of a priori data to resolve non-uniqueness in linear inversion, Geophys. J.R. Astr. Soc., 57, (1979) pp. 137-157.

[13] V.I. KEILIS-BOROK and YANOVSKAYA, Inverse problems in seismology, Geophys. J.R. Astr. Soc., 13, (1967) pp. 223-234.

[14] A. LICHNEROWICZ, Elements de calcul tensoriel, Armand Collin, Paris, 1960.

[15] F. PRESS, Earth models obtained by Monte-Carlo inversion, J. Geophys. Res., Vol. 73, No. 16, (1968) pp. 5223-5234.

[16] V.S. PUGACHEV, Theory of Random Functions, Pergamon, New York, 1965.

[17] F. RIESZ, Les systemes d'equations lineaires a une infinite d'inconnues, Gauthier-Villars, Paris, 1913.

[18] W.S. SCHNEIDER, Integral formulation for migration in two and three dimensions, Geophysics, 43, (1978) pp. 49-76.

[19] A. TARANTOLA and B. VALETTE, Inverse problems: Quest for information, J. Geophys., 50, (1982a), pp. 159-170.

[20] A. TARANTOLA and B. VALETTE, Generalized nonlinear inverse problems solved using the least-squares criterion, Reviews of Geophys. and Space Physics, 20, No. 2, (1982b) pp. 219-232.

[21] A. TARANTOLA and A. NERCESSIAN, Three-dimensional inversion without blocks, Geophys. J.R. Astr. Soc., 76, (1984) pp. 000-000.

[22] A. TARANTOLA, Inversion of seismic reflection data in the acoustic approximation, Geophysics, (1984) in press.

[23] A.E. TAYLOR and D.C. LAY, Introduction to functional analysis, John Wiley & Sons, New York, 1980.

[24] G.R. WALSH, Methods of optimization, John Wiley & Sons, Chichester, 1975.

[25] G.A. WATSON, Approximation theory and numerical methods, John Wiley & sons, Chicherster, 1980.

[26] K.A. WATERS, Reflection seismology, John Wiley & Sons, New York, 1981.

MIGRATION METHODS: PARTIAL BUT EFFICIENT SOLUTIONS TO THE SEISMIC INVERSE PROBLEM

PATRICK LAILLY*

Abstract. The aim of this paper is to introduce migration methods and to explain the simplifications made in them allow partial solutions to the seismic (acoustic) inverse problem. After having formulated the inverse problem as a least-squares problem, we study successively the approximations used in a migration method. We consider briefly the "before stack migration" problem (which is treated in detail in Berkhout's lecture) and treat more deeply the "migration of zero offset data". To describe numerical methods, we consider only finite difference methods which are very popular. This presentation gives insights concerning the advantages and drawbacks of migration.

1. Introduction. An exploration seismology experiment is described on Figure 1. A seismic source is located at a point S_1, near the surface of the earth and is fired at time $t = 0$. The seismic waves propagate from the source into the earth and are reflected, transmitted, diffracted, etc. when they meet the interfaces between geologic layers. The upgoing wave field (for instance the pressure in offshore exploration) is measured at different point $G_1, \ldots G_N$ (G stands for geophone) on the time interval $[0,T]$ and we obtain the data $P^{obs}_{S_1}(\vec{r}_G, t)$** for $t \in [0,T]$ and $\vec{r}_G$ (vector defining the location of one receiver) $\in R_G(S_1)$ which is defined as the set of receiver locations for a source in S_1. This was shot number 1.

* Institut Français du Pétrole, 1-4, avenue de Bois Préau - 92506 Rueil-Malmaison, France.

**The subscript S1 indicates that these data have been obtained for a source located in S1 .

Then the source and the receivers are moved to other places and shot number 2 is run the same way: it gives the data $p_{S_2}^{obs}(\vec{r}_G,t)$ for $t \varepsilon [0,T]^*$ and $\vec{r}_G \epsilon RG(S2)$ (S2 defines the source location for shot #2). An so on for the following shots.

Usually (for a 2D survey), the source and the receivers are moved along a line which determines a seismic profile.

Let us give some ideas about the dimension of the seismic data set. Let us consider a small seismic profile and suppose that the maximum investigated depth is 4 km. Typical dimensions of the data set are:

- duration of observation: 4 s,
- frequency bandwith: 10-50 Hz,
- time sampling rate: 2 ms,
- number of observation points for one shot: 100 distributed every 20 m from 100 m to 2000 m from the shot. The distance between the source and one receiver is called the offset,
- number of shot points: 1000 (every 20 m).

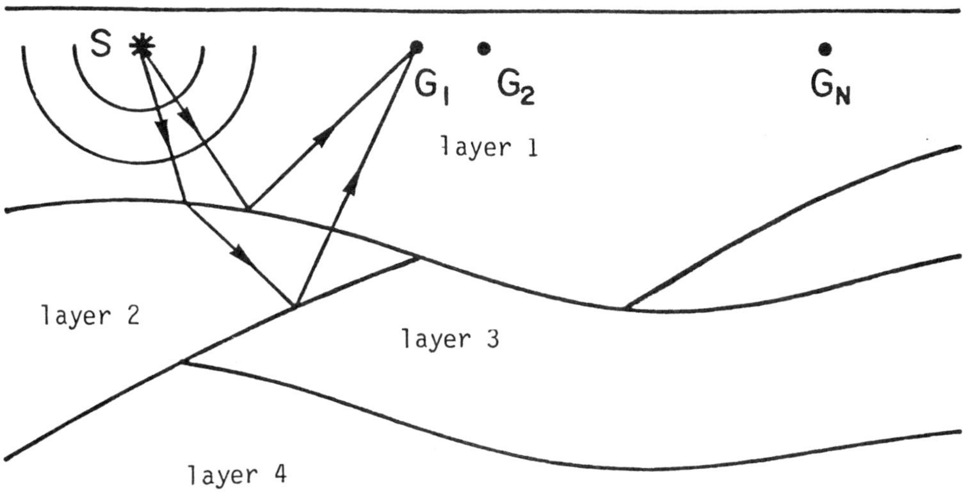

Figure 1: An exploration seismology experiment.

* For each shot, the time t=0 is the time when the source is fired.

Then the number of samples recorded along the profile is 2×10^8. This is the amount of information that must be processed by the computer.

When one deals with 3D surveys (several seismic profiles near each other or geophones and sources moving along a surface) the amount of recorded information increases considerably!

The exploration geophysicist uses the data to study the geologic structure of the area under survey. The problem to be solved can be formulated as an inverse problem as follows: to determine distributed parameters in a wave equation from different observations of the wavefield. The inversion of seismic profiles using the elastic wave equation has been formulated by Tarantola (1984b). Here we deal with the acoustic wave equation: the simplicity of this mathematical model makes it very popular among geophysicists. So we assume that the geologic layers are constituted of liquid media (shear modulus = 0) and furthermore that the density is constant (for simplicity). Let us introduce the following notations:

$\vec{r}$	vector characterizing one point of coordinates (x,y,z) in the space $\mathbb{R}^3$
z	depth
x	direction of the seismic profile
y	coordinate perpendicular to x and z
Ω	half space $z > 0$ *
$c(\vec{r})$	propagation velocity at point $\vec{r}$
S	point defining one source location
$R_G(S)$	set of receiver locations for a source located in S
R_S	$= \{S_1, S_2, \ldots, S_I\}$ set of source locations
I	number of shot points along the profile

* For sake of simplicity we assume the surface of the earth to be plane.

$f_S(\vec{r},t)$ source function for a source located in S

$p_S(\vec{r},t)$ pressure at point $\vec{r}$ and time t for a source located in S (unspecified medium)

T duration of observation

$p_S^{obs}(\vec{r}_G,t)$ pressure observed at point $\vec{r}_G \varepsilon R_G(S)$ and time $t \varepsilon [0,T]$ for a source located in S

Then the forward problem can be formulated as follows:

Given $c(x,y,z)$ and f_S, find $p_S(\vec{r}_G,t)$ for $\vec{r}_G \varepsilon R_G(S)$ and $t \varepsilon [0,T]$ where p_S is the solution of the wave equation:

(2a) $$\frac{1}{c^2}\frac{\partial^2 p_S}{\partial t^2} - \nabla^2 p_S = f_S \quad \text{in} \quad \Omega \times (0,T]$$

(2b) $$p_S\,(t=0) = \frac{\partial p_S}{\partial t}\,(t=0) = 0 \quad \text{(initial conditions)}$$

(2c) $$p_S\,(z=0) = 0 \quad \text{(free surface boundary condition)}$$

To model the data acquisition along the seismic profile we need to solve the problem (2) for each S in R_S.

A classical way (Lailly (1984a), Tarantola, 1984a)) to formulate the inversion of seismic data consists in solving the following least-squares problem:

Find $\tilde{c}(x,y,z)$ that minimizes over a set C the cost function

(3) $$J(c) = \sum_{S \varepsilon R_S} \|p_S(\vec{r}_G,t) - p_S^{obs}(\vec{r}_G,t)\|_S^2$$

where $\| \; \|_S$ in a convenient norm. For instance, if we assume that the noise which corrupts the observations is additive, Guassian and not correlated, a simple L^2 norm can be chosen:

(4)

$$\|p_S(\vec{r}_G,t)-p_S^{obs}(\vec{r}_G,t)\|_S^2 = \sum_{\vec{r}_G \varepsilon R_G(S)} \int_0^T dt\,(p_S(\vec{r}_G,t)-p_S^{obs}(\vec{r}_G,t))^2$$

The set C introduced in (3) is the set of admissible parameters (see Bamberger, et al. (1979) or Mace, et al. (1984) for some examples). One can also introduce statistical information on the solution by modifying the cost function as suggested by Tarantola, et al. (1982).

We shall deal shortly (for greater detail see Berkhout (1984) and Tarantola (1984b)) with the general inversion of a seismic profile (inversion and migration before stack). We shall be most interested in inversion and migration of zero offset data (one receiver located at the source point*). This particular case is very important because this kind of inversion is by far the most used in practice because of its simplicity and its robustness. Then we shall study the simplifications (paraxial approximations) that the migration approach allows for the numerical solution of the wave equation. After the principle of the numerical schemesis given, we shall be able to compare the computing times required to solve the different formulations of the seismic inverse problem and to show the scope but also the limits of migration methods.

2. Inversion and Migration Before Stack.

2.1. Inversion before stack. We are going to solve the optimization problem (3) by the gradient method**. Such a method constructs, starting from an initial guess $c^0(\vec{r})$ of the solution, a sequence $c^n(\vec{r})$ defined by:

$$(5) \quad c^{n+1}(\vec{r}) = c^n(\vec{r}) - \lambda^n \operatorname{grad}^n J$$

where $\operatorname{grad}^n J$ is the gradient of the cost function J at point c^n (the gradient is a function of $\vec{r}$) and λ^n is a conveniently chosen positive number.

The constructed sequence is such that

$$J(c^{n+1}) < J(c^n)$$

but its convergence to $\tilde{c}$ relies on assumed convexity properties of J.

* This configuration is actually artificial: geophysicists know how to perform approximate transformations of the original experiment into this artificial one.

**Only for simplicity. For a practical implementation, it would be better to use a more refined method such as conjugate gradient. However the following results can be generalized readily to any gradient type method.

To evaluate the gradient $\mathrm{grad}^n J$ we need to calculate, if we give a perturbation $\delta c(\vec{r})$ to the velocity distribution $c^n(\vec{r})$, the first order variations δp_S that will result on the seismograms. We define:

$$(6)\quad G_S^n:\ \delta c(\vec{r}) \to \delta p_S(\vec{r}_G,t)\ ,\quad \vec{r}_G \varepsilon R_G(S) \ \text{ and } \ t\varepsilon[0,T]$$

and δp_S is the solution of:

$$(7)\begin{cases} \left(\dfrac{1}{c^n}\right)^2 \dfrac{\partial^2 \delta p_S}{\partial t^2} - \nabla^2 \delta p_S = 2\,\dfrac{\delta c}{(c^n)^3}\,\dfrac{\partial^2 p_S^n}{\partial t^2} \quad \text{in} \quad \Omega x[0,T] \\ \delta p_S\ (t=0) = \dfrac{\partial \delta p_S}{\partial t}\ (t=0) = 0 \quad \text{(initial conditions)} \\ \delta p_S\ (z=0) = 0 \quad \text{(boundary condition)} \end{cases}$$

with p_S^n solution of (2) with $c = c^n$.

Then, for the cost function defined by (3), (4), the gradient $\mathrm{grad}^n J$ is given (see Lailly, 1984a or Tarantola, 1984a) by:

$$(8)\begin{cases} \mathrm{grad}^n J = \sum\limits_S G_S^{*n}\, P_S^n \\ \text{with } P_S^n(\vec{r}_G,t) = 2\,(p_S^n(\vec{r}_G,t) - p_S^{obs}(\vec{r}_G,t)) \ \text{ for } \vec{r}_G \varepsilon R_G(S) \\ \qquad\qquad\qquad\qquad\qquad\qquad\qquad\qquad \text{and } t\varepsilon[0,T] \\ G_S^{*n} = \text{adjoint of the operator } G_S^n \end{cases}$$

$G_S^{*n} P_S^n$ can be easily computed by using the technique of adjoint equations:

$$(9)\quad G_S^{*n} P_S^n = \frac{2}{(c^n)^3} \int_0^T \frac{\partial^2 p_S^n}{\partial t^2}(\vec{r},t)\, q_S^n(\vec{r},t)\,dt$$

where q_S^n is the solution of (adjoint equations):

$$
(10)\quad \begin{cases} \left(\dfrac{1}{c^n}\right)^2 \dfrac{\partial^2 q_S^n}{\partial t^2} - \nabla^2 q_S^n = \displaystyle\sum_{\vec{r}_G \in R_G(S)} P_S^n(\vec{r}_G, t)\, \delta(\vec{r}-\vec{r}_G) \\[2ex] q^n \ (t=T) = \dfrac{\partial q_S^n}{\partial t}(t=T) = 0 \quad \text{(final conditions)} \\[2ex] q_S^n \ (z=0) = 0 \quad \text{(boundary condition)} \end{cases}
$$

Then the gradient algorithm applied to the solution of the inverse problem (3) is:

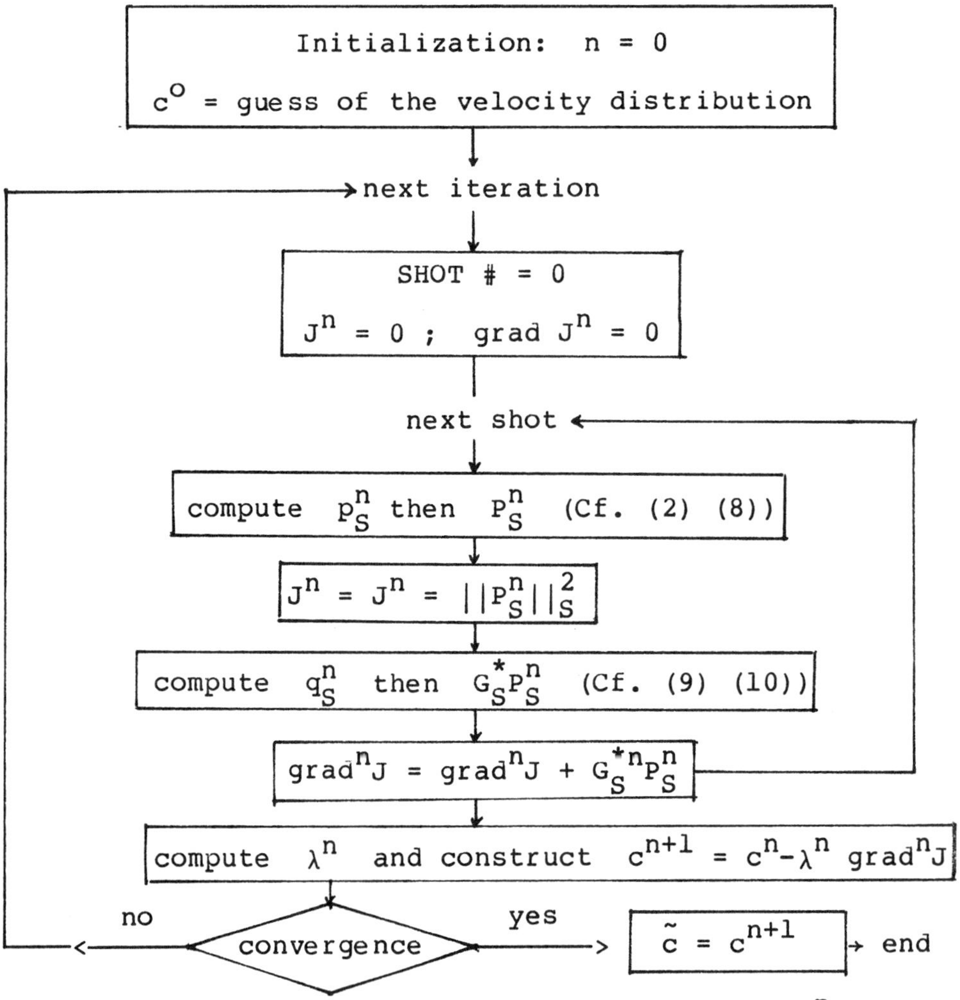

(*) different techniques may be used to compute λ^n. Basically we have to solve a (1D) optimization problem in the direction $-\,\mathrm{grad}^n J$.

2.2. Migration before stack. If we choose as initial guess c^0 a smooth function* and if we stop the algorithm after the first iteration ($\tilde{c} = c^1$) the result obtained will be the one given by a classical before stack migration.

Indeed, the only arrival in $p_S^0(\vec{r}_G,t)$ will be the direct wave (we neglect the back-scattered waves because c^0 is smooth). Then P_S^0 will only contain the contribution of reflected waves in the records and the algorithm can be compared with the before stack migration described by Berkhout (1982, §7.7): c^1-c^0 is the result given by the before stack migration algorithm in which the forward extrapolation of the source wavefield is achieved by using the second derivative of the source with respect to time (see (9)).

The physical idea which has led Claerbout (1971) to the wave equation migration is the following. Assume that an approximation c^0 of the actual velocity distribution is known. Let us try to determine the diffracting points (points where the actual velocity is different from c^0). If a diffracting point exists at some location $\vec{r}$, there will be a time coicidence at this location of the primary wave (solution of (2) with $c = c^0$ and of the scattered wave (solution of (7) with $c^n = c^0$).

This leads to the following before stack migration algorithm:

- recover for each shot, the history of the diffracted wave-field which has been recorded (downward continuation of the data). This is carried out by solving (10) with $c^n = c^0$ and $P_S^n = P_S^0$ (the final conditions in (10) mean that we look after upgoing waves).
- get a guess, for each shot, of the existence at point $\vec{r}$ of a diffracting point by evaluating the value at $t=0$ of the cross-correlation (see (9)) between the primary and the diffracted wave-fields at $\vec{r}$ (Claerbout's imaging principle).
- adding (stacking) the results obtained for each shot improves the resolution and the signal to noise ratio.

Actually, such a method works only if the reference medium is close to the actual medium. If it is not the case, the estimated diffracting points will be different from one shot to another one and the stack will decrease the

* It will be the case in practice: the geophysicist can obtain a rough (low frequency) estimate of the velocity distribution by means of velocity scans measurements in wells, etc.

resolution and the signal to noise ratio instead of improving them.

Mathematically, the result c^1 of the before stack migration is an approximation of the result $\hat{c}$ of the inversion algorithm only if:

- the linearization about c^0 is justified (c^0 close to the actual velocity distribution),

- the gradient algorithm converges very quickly (1 iteration)*,

- the noise is not too important.

For greater detail, the reader can refer to Berkhout (1984).

As a conclusion, before stack migration allows a substantial reduction (1 iteration instead of a large number) of the computations which are necessary to solve the inverse problem (3). Anyway, before stack migration still requires an important computation time to be described further and the result appears to be very sensitive to the chosen reference model. These two drawbacks have seriously limited the interest of before stack migration and geophysicists prefer, in most cases, to perform migration after stack (or zero offset).

3. Inversion and Migration of Zero Offset Data. We assume here that, for each shot, there is only one receiver which is located at the point source. It would be possible to solve the inverse problem as explained in the previous section but the computing time would be of the same order as with many receivers and the signal to noise ratio after inversion would be worse because we have less data. We are going to simplify the forward problem (and consequently the inverse problem) by taking into account the specificity of this seismic experiment: this is the aim of the "exploding reflector model".

3.1. The exploding reflector model. The exploding reflector model has been introduced by Loewenthal, et al. (1976) and different justifications of this simplification have been attempted (Stolt (1978); Cohen, et al.; (1979);

* This surprising property results from the focusing character of the operator $\sum_S G_S^* G_S$ when the source function is high frequency. This is the case in practice: the wave lengths are short with respect to the scale of variation of the geologic structure that is being investigated.

Berkhout (1982); Tarantola (1984c)). The schematic idea is the following (see Figure 2): the reflectors are replaced by sources which will explode at t=0 and the velocity is halved. The interest is that the modeling of a zero offset seismic section requires only one numerical solution of the wave equation. The drawback is that the simplification is not always justified; in particular, it relies on the assumption that the downgoing raypath is the same as the upgoing one.

To give a precise formulation of the exploding reflector model we use the approach developed by Cohen, et al. (1979) and Tarantola (1984c).

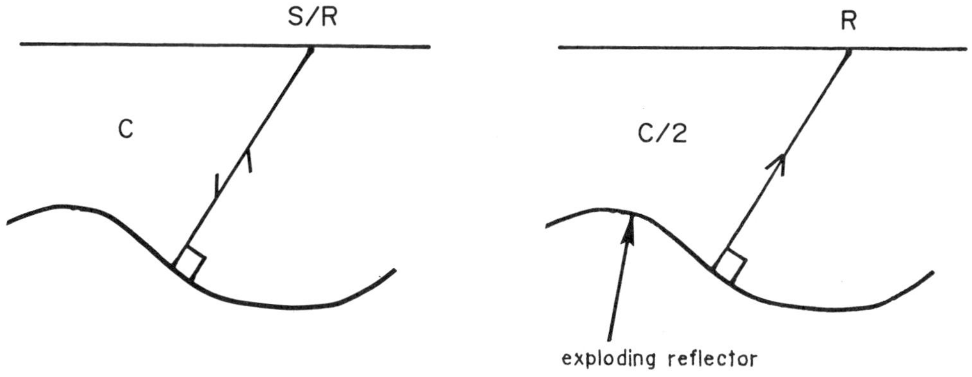

Figure 2: The exploding reflector model. Left: The wave leaves S at t=0 , is reflected at normal incidence and comes back to R=S . Right: the wave leaves the reflector at t=0 , propagates (perpendicularly to the wavefront) at the velocity c/2 and arrive at R at the same time as in the left experiment.

Let us assume:

- The seismic source f_S is a Dirac delta function at $\vec{r}=S$ multiplied by a function of time $F(t)$.

- The actual medium is not too far from a homogeneous medium. More precisely it is possible to compute with a sufficient accuracy, the synthetic seismograms associated to the actual medium by the linearization (7) of the forward problem about a homogeneous unbounded* reference medium (constant velocity c^o).

Then the problem consists in computing, with c^o and δc as data, the diffracted wavefield:

$$\delta p_S(S,t) = G_S^o \delta c$$

i.e., δp_S is the solution of (7) with $c^n = c^o$.

As the reference medium has been chosen homogeneous we can calculate analytically the operator G_S^o . We obtain (see Tarantola (1984c), formula (11)):

$$\text{(11)} \quad \delta p_S(\vec{r}_G=S,t) = \int_0^T \frac{d\tau}{(c^o)^2 \tau} \, \Theta(\vec{r}=S,\tau) \, \ddot{F}(t-\tau) \, \frac{1}{4\pi}$$

where $\ddot{F}$ is the second derivative of F and Θ the solution of

$$(\frac{2}{c^o})^2 \frac{\partial^2 \Theta}{\partial t^2} - \nabla^2 \Theta = 0 \qquad \text{in } \mathbb{R}^3 \times [0,T]$$

$$\text{(12)} \quad \Theta(t=0) = 0$$

$$\frac{\partial \Theta}{\partial t}(t=0) = \delta c(\vec{r})$$

* Consequently we must remove the effects of the free surface. This can hardly be justified from a theoretical point of view. In practice the geophysicist considers that he measures the upgoing waves and modifies conveniently the function $F(t)$ in order to take account of the images of the source and of the receiver.

When δc does not depend on the coordinate y the wave equation (12) reduces to a 2D problem: we then obtain formulae (6) in Cohen, et al. (1979) or (13) in Tarantola (1984c).

Let us insist on the fundamental interest of the exploding reflector model: modeling a zero offset seismic section requires one computation of the wave equation (12) instead of I computations of (2) in the exact approach. Unfortunately, the exact justification of this technique can be done only with the above mentioned assumptions and the second one can hardly be checked in practice*.

By use of some approximations it is possible to justify (Appendix 1) the exploding reflector model by a linearization about a smooth reference medium. Let us introduce the following exploding reflector model:

$$(13)\quad \delta p_S(\vec{r}_G=S,t) = \int_0^T d\tau\; a(\tau)\Theta(\vec{r}=S,\tau)\ddot{F}(t-\tau)$$

where Θ is the solution of

$$\left(\frac{2}{c^0(\vec{r})}\right)^2 \frac{\partial^2\Theta}{\partial t^2} - \nabla^2\Theta = 0 \quad \text{in } \mathbb{R}^3 x\,]0,T[\quad {}^{**}$$

$$(14)\quad \Theta(t=0) = 0$$

$$\frac{\partial\Theta}{\partial t}(t=0) = \frac{1}{2}\,\delta c(\vec{r})b(z)/(c^0(\vec{r}))^3$$

The functions $a(t)$ and $b(z)$ in (13) and (14) are amplitude correctors defined in Appendix 1.

Again when δc and c^0 do not depend on the coordinate y, the wave equation (14) becomes 2D.

<u>3.2. Inversion of zero offset data</u>. We come now to the inverse problem which is:

Given $c^0(\underset{\sim}{\vec{r}})$ and $\delta p_S^{obs}(\vec{r}_G=S,t)$ for $S\varepsilon R_S$ and $t\varepsilon]0,T[$ to find $\delta\tilde{c}(\vec{r})$ that minimizes the cost function:

* This is a major trouble for inversion methods using this assumption as noticed by Gray, et al. (1983).

**We have to define $c^0(x,y,z)$ also for $z<0$: one can, for instance, define $c^0(x,y,z) = c^0(x,y,0)$ for $z<0$.

$$(15)\quad J(\delta c) = \sum_{S\varepsilon R_S} \int_0^T [(G\delta c - \delta p_S^{obs})(\vec{r}_G = S, t)]^2 dt$$

with $G:\ \delta c \to \delta p(\vec{r}_G, t)$ defined by (13), (14).

This is a linear inverse problem. The gradients $\text{grad}^n J$ of J at point δc^n is given by:

$$(16)\quad \begin{cases} \text{grad}^n J = G^* P^n \\ \text{with } P^n = 2\ (G\delta c^n - \delta p_S^{obs})(S,t) \quad S\varepsilon R_S \text{ and } t\varepsilon(0,T) \end{cases}$$

The computation of $G^* P^n$ can be achieved by the technique of adjoint equations (see Appendix 2).

We define:

(17)

$$u_S^n(t) = 2a(t)\int_0^T d\tau (G\delta c^n - \delta p_S^{obs})(S,\tau)\ddot{F}(\tau - t) \text{ for } S\varepsilon R_S \text{ and } t\varepsilon(0,T)$$

and solve backwards in time the (adjoint) wave equation:

$$(18)\quad \begin{cases} \left(\frac{2}{c^o}\right)^2 \frac{\partial^2 q^n}{\partial t^2} - \nabla^2 q^n = \sum_{S\varepsilon R_S} u_S^n(t)\,\delta(\vec{r} - S) \\ q^n(t=T) = \frac{\partial q^n}{\partial t}(t=T) = 0 \end{cases}$$

Then:

$$(19)\quad G^* P^n = \frac{2}{(c^o)^3}\, bq(t=0)$$

3.3. Migration of zero offset data. The migration algorithm consists in stopping the gradient algorithm after the first iteration starting from $\delta c^0 = 0$* . Then

$$(20) \quad \delta c^1 = 2\lambda^1 G^* \delta p^{obs}$$

Thus migration is an approximate inversion which consists in applying to the records the operator G^* instead of G^{-1} (see Berkhout (1982) §7.2). As the operator G is usually not invertible, the solution of the least squares problem would be:

$$(21) \quad \tilde{\delta c} = (G^* G)^{-1} G^* \delta p^{obs} \quad (**)$$

To compare δc^1 and $\tilde{\delta c}$, we write:

$$\delta p^{obs} = G\delta c + n \quad \text{(n stands for noise)}$$

and we obtain:

$$\delta c^1 = G^* G \delta c + G^* n$$

$$\tilde{\delta c} = \delta c + (G^* G)^{-1} G^* n \quad (**)$$

Then δc^1 will be close to $\tilde{\delta c}$ only if

i) the noise level is low

ii) $G^* G$ is close to a diagonal operator (***)

* In practice, because the geophysicist is not much confident in the recorded amplitudes, the amplitude correctors $a(t)$ and $b(z)/(c^0(\vec{r}))^3$ are ignored or simplified in most cases. Furthermore the seismic source $F(t)$ being badly known, a "spiking" deconvolution is performed usually instead of the cross-correlation (17).

** The operator $G^* G$ may have small or zero eigenvalues; we should introduce a regularization term in the cost function which would give sense to these expressions.

***Again we do not look atamplitudes and are interested only in the location of the diffracting points.

The operator G^*G can be made explicit when the reference medium is homogeneous (see Appendix 3). From this result, it can be easily understood that the drawback of migration versus complete inversion (results δc given by (21)) is a limitation of the resolution. One advantage of zero-offset migration versus inversion or migration before stack is its computing time: one needs only one solution of the wave equation. Furthermore zero offset migration appears to be less sensitive to the inaccuracies in reference model as before stack migration: if, instead of applying the operator G^* to the records, we had applied an operator $\tilde{G}^*$ associated to a bad reference medium, the result would have been mainly corruption by a spatial shift which would not have been disastrous for the geologic interpretation*.

As a conclusion, in order to investigate most geologic structures, zero-offset migration appears to be a very efficient tool because it is inexpensive and robust.

4. Paraxial Approximations of the Wave Equation. Because of the smooth character of the reference medium, the wavefields we have to compute to use the migration algorithm are only constituted of downgoing waves (eq. (2)) and of upgoing waves (eq. (10)). If we assume that the records are slowly varying in x (low dip of the reflectors) and if we are interested only in the result of the migration at depths greater than z_{min}, which is supposed to be large compared to the maximum offset, we realize (see Figure 3) that we need modeling wave propagation accurately only in the directions that are not too far from the vertical. The same conclusion can be obtained for the migration of zero-offset data (solution of (18)).

The paraxial approximations of the wave equation use this specificity of our problem to simplify the computation. These approximations have been used in different scientific areas (Leontovitch, et al. (1946), Tappert (1977)). Their applications to migration problems have been developed by Claerbout (1976) and his group.

We present briefly the basic ideas in paraxial approximations. For detailed explanations the reader can refer to Bamberger, et al. (1984), Claerbout (1976) and (1984), Joly (1984), Hatton, et al. (1981), Robinson (1983) and Tappert (1977). For simplicity we shall consider only paraxial approximations for downgoing waves in 2D.

*In practice, the geophysicist often refines the reference model by using the result of migration in the shallowest layers.

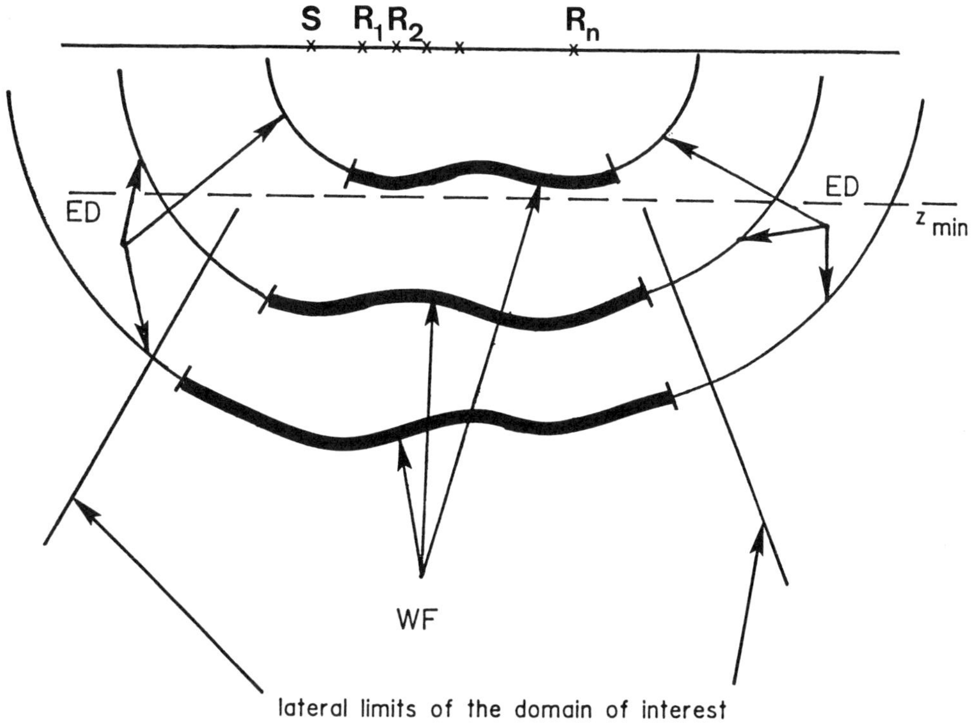

Figure 3: Location at some time t of the wave-fronts WF obtained by downward continuation of the data. To compute (9) we can neglect the contribution in q_s of the extreme parts of the edge diffractions ED . If we restrict the domain of interest to depths larger than z_{min} solving equation (2) and (10) that are not too far from the vertical.

4.1. Paraxial approximations in a homogeneous medium. We consider the solution p_s of equation (2) with $f_s = (\delta(\vec{x})F(t)$. It is well known that, far from the source $(\omega r/c >> 1)$, we have the following approximation:

$$\hat{p}_s(x,z,\omega) \simeq \frac{\hat{f}(\omega)}{2\sqrt{2\pi\omega r/c}} \exp - i \left(\frac{\omega r}{c} + \frac{\pi}{4}\right)$$

where $\hat{p}_s$ and $\hat{f}$ are the Fourier transform of p_s and f and $r = \sqrt{x^2 + z^2}$.

Even far from the source, $\hat{p}_s$ is a very irregular function: because the wavelength is small with respect to the size of the domain, a lot of samples are nesessary to represent this function correctly. The paraxial approximation consists in approximating p_s in a narrow (angle θ small) cone of vertical axis and for $\omega r/c >> 1$ in the following way:

$$
(22)\quad \begin{cases}
\hat{p}_s(x,z,\omega) = \hat{u}(x,z,\omega) + O(tg^2\theta) + O(kz\ tg^4\theta) \\
\text{with:} \\
u(x,z,\omega) = \exp(-\frac{i\omega z}{c})\,\hat{\psi}(x,z,\omega) \\
\hat{\psi}(x,z,\omega) = \hat{f}(\omega)\sqrt{\frac{c}{8\pi\omega z}}\,\exp(-i\omega x^2/2zc) \\
tg\theta = \frac{x}{z}
\end{cases}
$$

This approximation is valid in the domain shown on Figure 4. In this domain $\hat{\psi}$ is much more regular than p_s , so it will be simpler to compute $\hat{\psi}$ and p_s . It can be easily checked that $\hat{\psi}$ and $\hat{u}$ are solutions of the following equations:

$$
i\omega \frac{\partial\hat{\psi}}{\partial z} - \frac{c}{2}\frac{\partial^2\hat{\psi}}{\partial x^2} = 0
$$

$$
i\omega \frac{\partial\hat{u}}{\partial z} - \frac{\omega^2}{c^2}\hat{u} - \frac{c}{2}\frac{\partial^2\hat{u}}{\partial x^2} = 0
$$

Coming back to the time domain, it is possible to prove (see Jolly - 1984) that the solution u of

$$
(23a)\quad \frac{1}{c^2}\frac{\partial^2 u}{\partial t^2} + \frac{1}{c}\frac{\partial^2 u}{\partial z\partial t} - \frac{1}{2}\frac{\partial^2 u}{\partial x^2} = \sqrt{2}\,\delta(\vec{x})F(t)
$$

$$
(23b)\quad u(t=0) = \frac{\partial u}{\partial t}(t=0) = 0 \qquad \text{(initial conditions)}
$$

approximates p_s in the domain shown on Figure 4 and to obtain an error estimate.

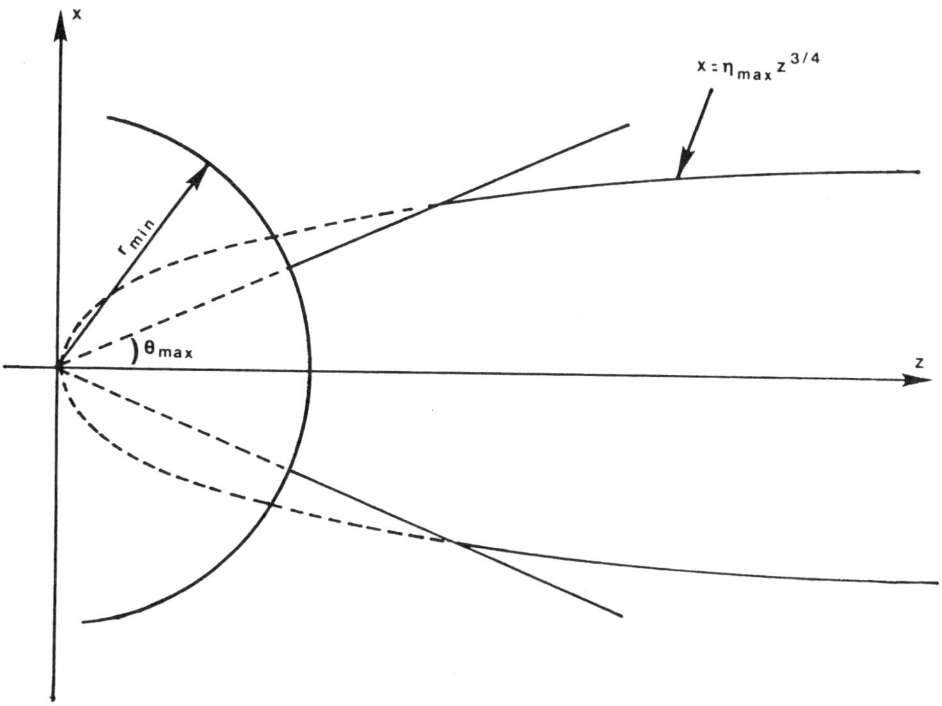

Figure 4: Domain of validity of the paraxial approximation (22) r_{min} is much larger than the (maximum) wavelength and η_{max} is much smaller than the fourth root if the (minimum) wavelength.

An other way to understand the paraxial approximation consists in comparing the dispersion relations:

$$(24) \quad \left(\frac{ck_x}{\omega}\right)^2 + \left(\frac{ck_z}{\omega}\right)^2 = 1$$

$$(25) \quad \frac{ck_z}{\omega} = 1 - \frac{1}{2}\left(\frac{ck_x}{\omega}\right)^2$$

associated to the equations (2a) and (23a) (ω, k_x and k_z are the frequencies in t, x and z respectively).

If we consider 2 harmonic waves, solutions of (2a) and (23a) respectively and propagating in the direction characterized by the angle θ with the z axis, we can see that the phase velocities are different (see Figure 5), the length ℓ being interpreted as the ratio between these phase velocities. We can conclude the following:

- the approximation is good for downgoing waves propagating in nearly vertical directions,

- the paraxial approximation (23a) has an anisotropic behavior.

It is possible to extend the paraxial approximation to propagation directions which are further from the vertical by constructing higher order paraxial approximations (see Engquist (1980), Bamberger, et al. (1984), Claerbout (1984) for instance).

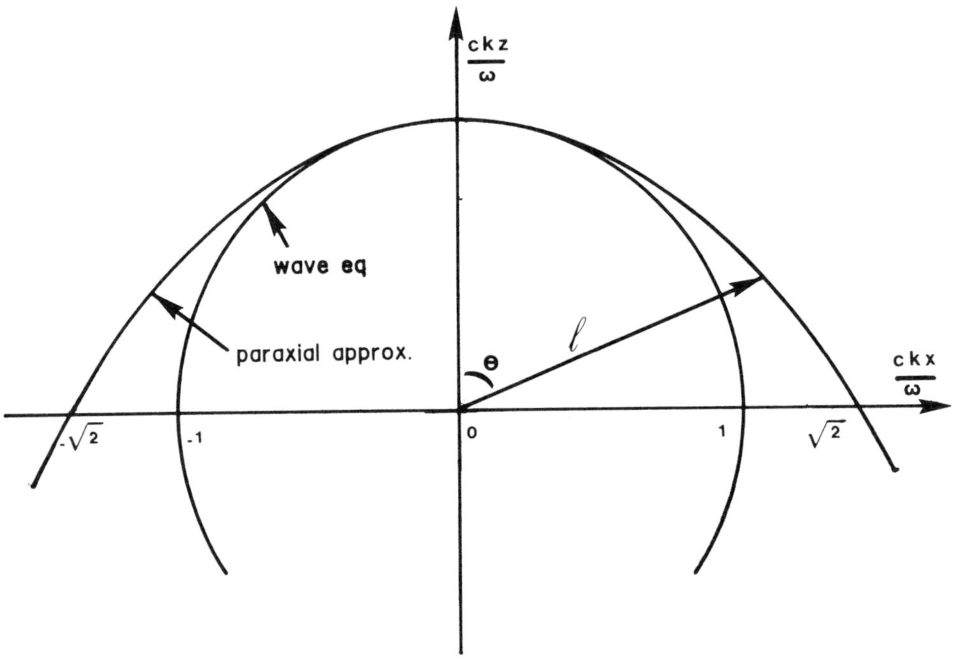

Figure 5: Comparison of the dispersion relations associated to the wave equation and the paraxial approximation (23a).

4.2. Paraxial approximation in a heterogeneous medium. When the medium is close to a homogeneous medium, immediate generalization of equation (23a) are quite satisfactory and have allowed very interesting applications to wave propagation problems in the ocean (see Tappert - 1977).

When one tries to generalize the approximation to slowly varying media but with a ratio of order 2 or 3 between the minimum and maximum velocities (which is a usual case for migration problems), some difficulties appear. Different generalizations of (23a) have been proposed for such media (see Bamberger, et al. (1984), Claerbout (1984), Joly (1984), Hatton, et al. (1981, Tappert (1977), for instance). Most approximations meet difficulties concerning the stability of the result. Moreover, even for sophisticated approximations such as the one proposed by Tappert (1977, equation 3.38) it is impossible to obtain error estimates between the exact and paraxial wavefields (see Appendix 4). These difficulties have led different authors (Hemon (1978), Whitmore (1983)) to perform migration with the exact wave equation in spite of the advantages of the paraxial approximation with respect to the computation effort.

4.3. Numerical solution of the paraxial approximation. As indicated in Section 4.1, it will be easier to discretize the equation giving ψ rather than the equation giving u. In a homogeneous medium, this can be done by the change of variable (Claerbout (1976)

$$(26)\quad \begin{cases} x' = x \\ z' = z \\ t' = t - \dfrac{z}{c} \end{cases}$$

The new function:

$$(27)\quad \psi(x', z', t') = u(x, z, t)$$

is then solution of:

$$(28)\quad \frac{\partial^2 \psi}{\partial z' \partial t'} = \frac{c}{2} \frac{\partial^2 \psi}{\partial x'^2}$$

This equation is usually discretized by a centered (Cranck-Nicholson) scheme using finite differences for the variables z and t and P1 finite element for the variable x* .

The scheme is implicit: the computation of one point in the (z,t) plane requires the solution of a linear tridiagonal system (Gauss method) linking the values of ψ at the different points in x . For greater detail concerning the numerical analysis and computational problems of paraxial approximations, see Berkhout (1982), Brysk (1983), Dubrulle (1983), Hood (1978), Jacobs (1982), Lailly (1984b), Loewenthal, et al. (1976), Stolt (1978).

As a result, the numerical solution of equation (28) can be achieved with a good accuracy by introducing only 6 samples per period in time, and respectively 5 and 3 samples per wavelength in x and z** . For the numerical solution of the exact wave equation 10 samples per wavelength in x and z and 15 samples per period in time are required (see Alford, et al. (1974)). So we can estimate the gain of computation time obtained by replacing the wave equation by a paraxial approximation to a factor greater than 10.

5. Conclusion. The table hereafter compares the computing time estimated for the different problems we have presented. The unit is the time required to process the migration of zero offset data using the paraxial approximation. N is the number of iterations required in the optimization algorithm. Typically we have I (number of shot points) $>$ 100 . N is difficult to evaluate but because the large number of unknowns, N will very likely be much greater than 100 even for sophisticated gradient type methods.

* It may be useful (see Hood (1978), for instance) to off-center slightly the scheme in order to reduce the dispersion effect.

**Actually the number of samples depends strongly on the difficulty (maximum dip) of the example.

	Using the paraxial approximation	Using the exact wave equation
Zero Offset Migration	1	10
Before Stack Migration	2 I	20 I
Before Stack Inversion		20 IN

We can see the interest of migration method and especially of zero offset migration. In spite of the limitations we have emphasized, they allow for the geological interpretation of complicated structures (see Figure 6). For the moment, they are the only ones that can be applied on realistic examples with competitive computing costs. They have also allowed applications to 3D problems (Jacubowitz, et al. (1983)).

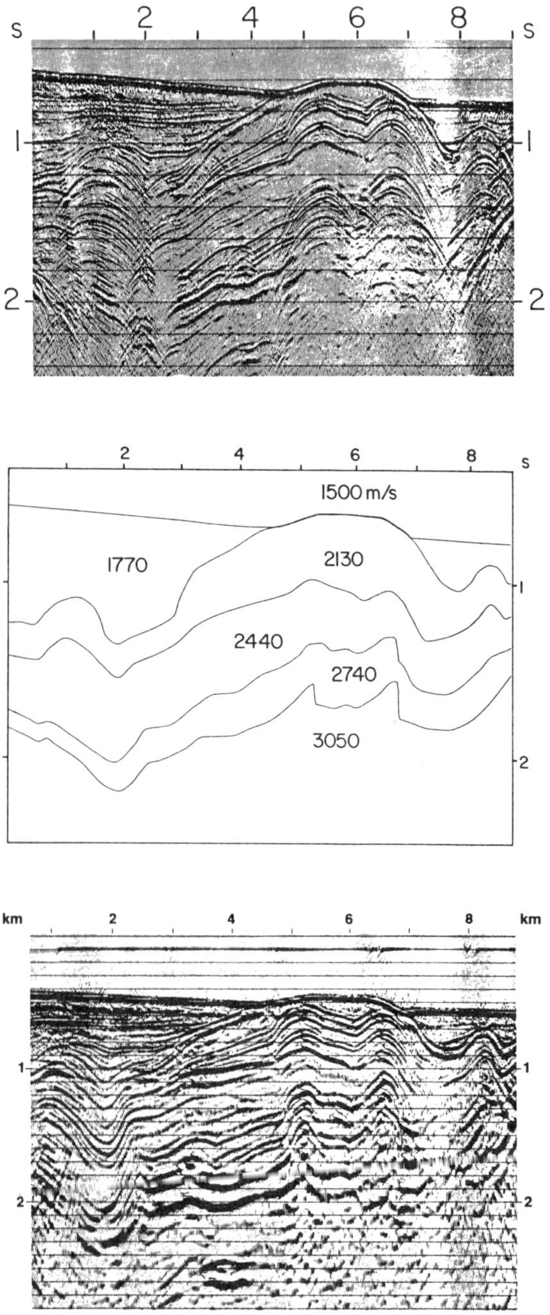

Figure 6: Migrated section (bottom) obtained from the zero offset section (top) and the velocity model displayed on the middle. Steepest dips are approximately 25 degrees. (From Hatton, et al. (1981)).

Appendix 1. The Exploding Reflector Model for a Smooth Reference Medium. We suppose the source located at the origin and the velocity $c^o(x,y,z)$ of the reference medium slowly varying in order to allow the geometrical optics (see Cerveny (1977)) approximation of the primary wave:

$$\frac{\partial^2 p_S^o}{\partial t^2}(\vec{r},t) \simeq A(\vec{r})\,\delta(t-\tau(\vec{r})) * \ddot{F}(t)$$

where $\tau(\vec{r})$ is the travel time from the source to $\vec{r}$ and $A(\vec{r})$ characterizes the amplitude.

We consider a perturbation $\delta c(\vec{r}) = \delta(\vec{r}-\vec{r}_o)$.

Then the diffracted wave, solution of (7) in which the boundary condition has been cancelled, can be approximated at the origin:

$$\delta p_S(0,t) \simeq \frac{2}{(c^o(\vec{r}_o))^3} A^2(\vec{r}_o)\,\delta(t-2\tau(\vec{r}_o)) * \ddot{F}(t)$$

By noticing that $4A(\vec{r}_o)\delta(t-2\tau(\vec{r}_o))/(c^o)^2$ is the value at origin of the geometrical optics approximation of the solution $G_{\vec{r}_o}$ of the wave equation

$$\begin{cases} \left(\frac{2}{c^o(\vec{r})}\right)^2 \frac{\partial^2 G_{\vec{r}_o}}{\partial t^2} - \nabla^2 G_{\vec{r}_o} = 0 \quad \text{in } \mathbb{R}^3 \times [0,T] \\ G_{\vec{r}_o}(\vec{r},t=0) = 0 \\ \frac{\partial G_{\vec{r}_o}}{\partial t}(\vec{r},t=0) = \delta(\vec{r}-\vec{r}_o) \end{cases}$$

we obtain:

$$\delta p(0,t) \simeq \frac{1}{2c^o(\vec{r}_o)} A(\vec{r}_o) G_{\vec{r}_o}(0,t) * \ddot{F}(t)$$

If we assume that $A(\vec{r}_o)$ can be approximated in the following way:

(A1) $A(\vec{r}_o) \simeq b(z_o)a(2\tau(\vec{r}_o))$ with z_o=depth of the point $\vec{r}_o$

(an example will be given at the end of this section) then δp can be written:

$$\delta p(0,t) \simeq a(t)G_{\vec{r}_o}(0,t) * \ddot{F}(t)$$

where $G_{\vec{r}_o}$ is now the solution of:

$$\left(\frac{2}{c^o(\vec{r})}\right)^2 \frac{\partial^2 G_{\vec{r}_o}}{\partial t^2} - \nabla^2 G_{\vec{r}_o} = 0 \quad \text{in} \quad \mathbb{R}^3 \times [0,T]$$

$$G_{\vec{r}_o}(\vec{r},t=0) = 0$$

$$\frac{\partial G_{\vec{r}_o}}{\partial t}(r,t=0) = \frac{1}{2c^o(\vec{r}_o)}\, b(z_o)\,\delta(\vec{r}-\vec{r}_o)$$

By superposition we obtain the exploding reflector model (13)-(14).

<u>An example of approximation (A1)</u>. Let us consider a ray and two points $\vec{r}$ and $\vec{r}_o$ whose position on the ray is characterized by the lengths s and s_o of the ray path starting from the source.

The amplitudes at $\vec{r}$ and $\vec{r}_o$ are related by (see Cerveny (1977)):

$$A(\vec{r}) = A(\vec{r}_o)\left[\left(\frac{J(s_o)c^o(\vec{r}_o)}{J(s)c^o(\vec{r})}\right)\right]^{1/2}$$

where $J(s)$ is the spreading of rays at point $\vec{r}$.

If we approximate (smooth variations of the curvature of the wave front):

$$\frac{J(s_o)}{J(s)} \simeq \frac{s_o^2}{s^2}$$

we obtain:

$$\frac{A(\vec{r})}{A(\vec{r})_o} \simeq \frac{s_o}{s} \left[\frac{c^o(\vec{r}_o)}{c^o(\vec{r})}\right]^{1/2}$$

Let $\tilde{c}(s)$ be the harmonic mean of the velocity along the ray from the source to $\vec{r}$. Then we have:

$$s = \tilde{c}(s)\,\tau(s) \quad \text{and} \quad s_o = \tilde{c}(s_o)\,\tau(s_o)$$

Now let us consider two different rays and the points $\vec{r}$ and $\vec{r}'$ on these rays located at depth z. We suppose that the lateral variations of the velocity in the reference medium are not too important so that the previous harmonic means of the velocity $\tilde{c}(\vec{r})$ and $\tilde{c}(\vec{r}')$ are close to each other. In other words we may write:

$$s \simeq \bar{c}(z)\,\tau(x)$$

(define for instance $\bar{c}(z)$ = arithmetic means of $\tilde{c}(\vec{r})$) at depth z.

Again, because of the previous assumption of weak lateral inhomogeneity we can approximate:

$$c^o(\vec{r}) \simeq \bar{c}(z) \quad \text{(arithmetic mean of the velocities at depth z)}$$

Then, considering two arbitrary points $\vec{r}$ and $\vec{r}_o$ (not necessarily on the same ray), we obtain an example of approximation A1:

$$\frac{A(\vec{r})}{A(\vec{r}_o)} \simeq \left[\frac{\hat{c}(z_o)}{\hat{c}(z)}\right]^{1/2} \frac{\bar{c}(z_o)}{\bar{c}(z)}\,\frac{\tau(\vec{r}_o)}{\tau(\vec{r})}$$

Appendix 2. Calculation of G^*P (proof of (19)). Let P be an arbitrary function $S \in R_S$ and $t \in [0,T]$ and $\delta c(\vec{r})$ be an arbitrary perturbation of the velocity, We set $\delta p(S,t) = G\delta c$ and evaluate:

$$Q = \sum_{S \in R_S} \int_0^T dt\ P(S,t)\,\delta p(S,t)$$

By using (13) we obtain:

$$Q = \sum_{S \in R_S} \int_0^T dt \int_0^T d\tau\ a(\tau)\,\theta(\vec{r}=S,\tau)\,\ddot{F}(t-\tau)\,P(S,t)$$

$$= \sum_{S \in R_S} \int_0^T dt\ \eta_S(t)\,\theta(\vec{r}=S,t)$$

with:

$$\eta_S(t) = a(t) \int_0^T d\tau\ P(S,\tau)\,\ddot{F}(\tau-t)$$

By using the definition (see (18)) of the adjoint equation:

$$\begin{cases} \left(\dfrac{2}{c^o}\right)^2 \dfrac{\partial^2 q}{\partial t^2} - \nabla^2 q = \displaystyle\sum_{S \in R_S} \eta_S(t)\,\delta(\vec{r}-S) \\ q(t=T) = \dfrac{\partial q}{\partial t}(t=T) = 0 \end{cases}$$

we obtain:

$$Q = \int_{\mathbb{R}^3} d\vec{r} \int_0^T dt\ \left(\left(\frac{2}{c^o}\right)^2 \frac{\partial^2 q}{\partial t^2} - \nabla^2 q\right)\theta$$

Integrations by part give:

$$Q = \int_{\mathbb{R}^3} d\vec{r}\ \left(\frac{2}{c^o}\right)^2 \frac{\partial \theta}{\partial t}(t=0)\,q(t=0)$$

$$= \int_{\mathbb{R}^3} d\vec{r}\ \frac{2}{(c^o)^3}\,\delta c\ b\ q(t=0) = \int_{\mathbb{R}^3} d\vec{r}\ \delta c\ G^*p$$

Then:

$$G^*p = \frac{2}{(c^o)^3}\, b\, q\ (t=0)$$

Appendix 3. Calculation of G^*G in 2D for a homogeneous reference medium. We consider a local perturbation of the velocity:

$$\delta c(\vec{r}) = \delta(\vec{r} - \vec{r}_o)$$

Then we obtain from (11) (12):

$$\text{(A2)} \quad [G\delta c](\vec{r},t) = \frac{1}{\pi (c^o)^4} \int_0^T \frac{d\tau}{\tau}\, G(\|\vec{r}-\vec{r}_o\|,\tau)\ddot{F}(t-\tau)$$

for $\vec{r} \in R_S$ and $t \in [0,T]$ with

$$\text{(A3)} \quad G(r,t) = \begin{cases} \dfrac{c^o}{4\pi} \dfrac{1}{\sqrt{(\frac{c^o t}{2})^2 - r^2}} & \text{if} \quad ct \geq 2r \\ 0 & \text{otherwise} \end{cases}$$

Following (17) (18) we introduce:

$$\text{(A4)} \quad u(\vec{r},t) = \frac{1}{2\pi c^o t} \int_0^T d\xi\, [G\delta c](\vec{r},\xi)\ddot{F}(\xi,t)$$

for $\vec{r} \in R_S$ and $t \in 0,T]$

and the solution q of (18) is:

$$q(\vec{r},T-t) = \sum_{\vec{r} \in R_S} \int_0^T d\xi\, u(\vec{r},\xi) G(\|\vec{r}\,'-\vec{r}\|, t-T+\xi)$$

Using (19) we obtain:

$$[G^*G\,\delta c](\vec{r}\,') = \frac{2}{(c^o)^3} \sum_{\vec{r}\in R_S} \int_0^T d\xi\, u(\vec{r},\xi) G(\|\vec{r}\,' - \vec{r}\|,\xi)$$

Now we replace the quantities defined in (A2) and (A4):

$$[G^*G\,\delta c](\vec{r}\,') = \frac{2}{\pi^2 (c^o)^3} \sum_{\vec{r}\in R_S} f(\vec{r}\,',\vec{r}_o,\vec{r})$$

with

$$\text{(A6)}\quad f(\vec{r}\,',\vec{r}_o,\vec{r}) = \int_0^T d\xi \int_0^T d\tau \frac{1}{\xi}\frac{1}{\tau} G(\vec{r}-\vec{r}_o,\xi) G(\vec{r}\,'-\vec{r},\tau) H(\xi,\tau)$$

$$H(\xi,\tau) = \int_0^T dt \ddot{F}(t-\xi)\ddot{F}(t-\tau)$$

From (A6) we can deduce the following properties of the operator G^*G :

- $\vec{r}\,'$ and $\vec{r}_o$ have symmetric roles
- if the observation points spread from $-\infty$ to $+\infty$ and if we translate $\vec{r}_o$ horizontally, $G^*G\delta c$ will be translated by the same amount.
- if the observation points spread from $-\infty$ to $+\infty$ and if $\vec{r}_1$ and $\vec{r}_2$ are symmetric with respect to the vertial of $\vec{r}_o$ then

$$[G^*G\delta c](\vec{r}_1) = [G^*G\delta c](\vec{r}_2)$$

- if the support of $H(\xi,\tau)$ can be assumed to be localized near $\xi=\tau$, the support of $f(\vec{r}\,',\vec{r}_o,\vec{r})$ will be localized near the circle whose center is $\vec{r}$ and whose radius is $\|\vec{r}-\vec{r}_o\|$. The summation over $\vec{r}$ in R_S will focus the support of $G^*G\delta c$ on $\vec{r}_o$ if the lines joining $\vec{r}_o$ to the different $\vec{r}$ in R_S spread on a sufficiently wide range of angles. Of course, the higher the frequencies, the greater the focusing on $\vec{r}_o$.

Appendix 4. <u>A kind of reciprocity principle for the paraxial approximation</u>. Let us consider two points A and B .

We consider the solution d of the approximation of downgoing waves proposed by Tappert (1977, equation 3.38):

$$
\text{(A7)} \quad \begin{cases} \dfrac{1}{c^2}\dfrac{\partial^2 d}{\partial t^2} + \dfrac{1}{c}\dfrac{\partial^2 d}{\partial z \partial t} - \dfrac{1}{2c}\dfrac{\partial}{\partial x}\left(c\,\dfrac{\partial d}{\partial x}\right) + \dfrac{1}{4c}\dfrac{\partial c}{\partial x}\, d = \sqrt{2}\,\delta(\vec{x}-A,t) \\[2ex] d(t=0) = \dfrac{\partial d}{\partial t}(t=0) = 0 \end{cases}
$$

and the solution u of the analogous approximation of upgoing waves:

$$
\text{(A8)} \quad \begin{cases} \dfrac{1}{c^2}\dfrac{\partial^2 u}{\partial t^2} - \dfrac{1}{c}\dfrac{\partial^2 u}{\partial z \partial t} - \dfrac{1}{2c}\dfrac{\partial}{\partial x}\left(c\,\dfrac{\partial u}{\partial x}\right) + \dfrac{1}{4c}\dfrac{\partial c}{\partial x}\, u = \sqrt{2}\,\delta(\vec{x}-B,t) \\[2ex] u(t=0) = \dfrac{\partial u}{\partial t}(t=0) = 0 \end{cases}
$$

We have the following reciprocity principle:

$$
\text{(A9)} \quad c(A)u(A,t) = c(B)d(B,t)
$$

From this result, we can see the limits of paraxial approximations concerning the reliability of amplitudes. We consider two points A and B satisfying the conditions required for the paraxial approximation:

- A above B
- line AB close to the vertical.

We can see that, when $c(A)$ differs from $c(B)$, the solution d of (A7) and u of (A8) cannot approximate the corresponding solutions of the wave equation.

<u>Proof of (A9)</u>. τ being a given time, we introduce

$$
v(x,z,t) = d(x,z,\tau-t)
$$

v is the solution of:

$$
\text{(A10)}\quad
\begin{cases}
\dfrac{1}{c^2}\dfrac{\partial^2 v}{\partial t^2} - \dfrac{1}{c}\dfrac{\partial^2 v}{\partial z \partial t} - \dfrac{1}{2c}\dfrac{\partial}{\partial x}\left(c\dfrac{\partial v}{\partial x}\right) = \sqrt{2}\,\delta(\vec{x}-A,\tau-t) \\[2ex]
v(t=\tau) = \dfrac{\partial v}{\partial t}(t=\tau) = 0
\end{cases}
$$

We multiply (A8) by cv and (A10) by cu , subtract and integrate over space and on the interval $[0,\tau]$ in time. Integrations by part lead to:

$$
\text{(A11)}\quad
\begin{aligned}
&\sqrt{2}\,(c(B)v(B,0) - c(A)u(A,\tau)) = \\
&\int dt \int dz \int dx \left(u\frac{\partial^2 v}{\partial z \partial t} - v\frac{\partial^2 u}{\partial z \partial t}\right)
\end{aligned}
$$

An integration by part of the right hand side of (A11) shows that it is zero because of the bounded velocity of propagation in paraxial approximations (see Joly, 1984) and the result follows.

References

[1] R.M. ALFORD, K.R. KELLY, and D.M. BOORE, Accuracy of finite difference modeling of the acoustic wave equation, Geophysics, 39, (1974) pp. 834-842.

[2] A. BAMBERGER, G. CHAVENT, and P. LAILLY, About the stability of the inverse problem in 1-D wave equations. Application to the interpretation of seismic profiles, Appl. Math. Optim., 5, (1979) pp. 1-47.

[3] A. BAMBERGER, B. ENGQUIST, L. HALPERN, and P. JOLY, Paraxial approximation in heterogeneous media: some new results, Research report, Ecole Polytechnique (to appear), Palaiseau, 1984.

[4] A.J. BERKHOUT, Seismic Migration, Elsevier, Amsterdam, 1982.

[5] A.J. BERKHOUT, Relationship between inverse scattering and seismic migration, these proceedings.

[6] BRYSK, Numerical analysis of the 45 degree finite difference equation for migration, Geophysics, 48, (1983) pp. 532-542.

[7] V. CERVENY, I.A. MOLOTKOV, and I. PSENCIK, Ray Methods in Seismology, Univerzita Karlova, Praha, 1977.

[8] J.F. CLAERBOUT, Toward a unified theory of reflection mapping, Geophysics, 36, (1971) pp. 467-581.

[9] J.F. CLAERBOUT, Fundamentals of Geophysical Data Processing, Mc Graw-Hill, New York, 1976.

[10] J.F. CLAERBOUT, Imaging the Earth's Interior, book to appear, 1984.

[11] J.K. COHEN and N. BLEISTEIN, Velocity inversion procedure for acoustic waves, Geophysics, 44, (1979) pp. 1077-1086.

[12] A.A. DUBRULLE, On numerical methods for migration in layered media, Geophysical Prospecting, 31, (1983) pp. 237-264.

[13] B. ENGQUIST, Inverse imaging methods in exploration seismology, published in, Computing Methods in Applied Sciences and Engineering, pp. 547-552, North-Holland, Amsterdam, 1980.

[14] L. HATTON, K. LARNER, and B. GIBSON, Migration of seismic data from inhomogeneous media, Geophysics, 46, (1981) pp. 751-767.

[15] C. HEMON, Equation d'ondes et modeles, Geophysical Prospecting, 26, (1978) pp. 790-821.

[16] P. HOOD, Finite difference and wave number migration, Geophysical Prospecting, 26, (1978) pp. 773-789.

[17] H. JACUBOWITZ and S. LEVIN, A simple exact method of 3D migration-theory, Geophysical Prospecting, 31, (1978) pp. 34-56.

[18] P. JOLY, Etude mathematique de l'approximation paxaxiale 15 degres en milieu stratifie, INRIA research report (to appear), Le Chesnay, 1984.

[19] P. LAILLY (a), The seismic inverse problem as a sequence of before stack migrations, Proceedings of the Conference on inverse scattering, theory and application, SIAM, Philadelphia, 1984.

[20] P. LAILLY (b), Etude de la dispersion des ondes dans des schemas de migration par differences finies, IFP report (to appear), Rueil Malmaison, 1984.

[21] M. LEONTOVICH and V. FOCK, Journ. Phys. USSR, 10, (1946) pp. 13-24.

[22] D. LOEWENTHAL, L. LU, R. ROBERSON, and J.W.C. SHERWOOD, The wave equation applied to migration, Geophysical Prospecting, 24, (1976) pp. 380-399.

[23] D. MACE and P. LAILLY, A solution of an inverse problem in the 1D wave equation, application to the inversion of vertical seismic profiles, Proceedings of the 6th International Conference on Analysis and Optimization of Systems, Springer Verlag, Berlin, 1984.

[24] E.A. ROBINSON, Migration of Geophysical Data, International Human Resources Development Corporation, Boston, 1983.

[25] R.H. STOLT, Migration by Fourier transform, Geophysics, 43, (1978) pp. 23-48.

[26] F.D. TAPPERT, The parabolic equation method, published in Wave Propagation and Underwater Acoustics, Springer Verlag, Berlin, 1977.

[27] A. TARANTOLA and B. VALETTE, Generalized non-linear inverse problems solved using the least-squares criterion, Reviews of Geophysics and Space Physics, 20, pp. 219-232, 1982.

[28] A. TARANTOLA (b), The seismic reflection inverse problem, these proceedings, 1984.

[29] A. TARANTOLA (c), Iterative migration = inversion (linearized problem for an homogeneous reference model), Geophysical Prospecting (to appear), 1984.

[30] A.B. WEGLEIN and S.H. GRAY, The sensitivity of Born inversion to the choice of the reference velocity: a simple example, Geophysics, 48, (1983) pp. 36-38.

[31] N.D. WHITMORE, Iterative depth migration by backward time propagation, paper presented at the 53rd SEG meeting, Las Vegas, 1983.

RELATIONSHIP BETWEEN LINEARIZED INVERSE SCATTERING AND SEISMIC MIGRATION

A. J. BERKHOUT*

Abstract. A discrete formulation is given of the forward model that is used in the linearized inverse scattering theory. From this model an inversion scheme is derived. This inversion scheme is compared with the procedure in pre-stack seismic migration.

It is shown that if the backscatter response of the reference medium may be neglected then seismic migration and linearized inversion define identical processes.

1. Introduction. Seismic migration techniques are based on the acoustic wave equation

$$\rho \nabla \cdot \left(\frac{1}{\rho} \nabla P\right) + k^2 P = 0 , \tag{1}$$

where P represents the Fourier transformed pressure of the acoustic wave field, ρ equals the density of the medium, c equals the acoustic velocity of the medium and $k = \omega/c$.

Based on a solution of (1) in integral form (Schneider, 1978) or in terms of wave numbers (Stolt, 1978; Gazdag, 1978) or in finite-difference form (Claerbout, 1976), inversion schemes have been proposed which all assume independent wave propagation of a downward travelling source wave field (P^+) and upward travelling reflected wave fields (P^-). A further simplification is always made by using the exploding reflector model (Loewenthal et al., 1974) for post-stack data. In the near future we may expect that post-stack migration techniques, based on the exploding reflector model, will be gradually replaced by more accurately formulated pre-stack migration techniques.

*Delft University of Technology, Laboratory of Seismics and Acoustics, P.O. Box 5046, 2600 GA Delft, The Netherlands.

Multi-dimensional linearized inversion techniques are based on the acoustic wave equation

$$(2) \quad \bar{\rho}\nabla \cdot \left(\frac{1}{\bar{\rho}} \nabla P\right) + \bar{k}^2 P = -(\gamma\bar{k}^2 + \nabla \ln \beta \cdot \nabla) P \ ,$$

where

$\beta = \bar{\rho}/\rho$

$\bar{\rho} = \bar{\rho}(x,y,z)$ represents the density of a reference medium for $P = o$

$\rho = \rho(x,y,z)$ represents the density of the actual medium for $P = o$

$\gamma = (\bar{c}/c)^2 - 1$

$\bar{c} = \bar{c}(x,y,z)$ represents the propagation velocity of a reference medium

$c = c(x,y,z)$ represents the propagation velocity of the actual medium

$\bar{k} = \omega/\bar{c}$.

A solution of (2) can be given in integral form, which contains the wave field $\bar{P}$ in the reference medium and a deviation term $\Delta P = P - \bar{P}$ which depends on the deviation parameters β and γ . Inversion schemes have been proposed (a.o. Clayton and Stolt, 1981; Cohen and Bleistein, 1979; Prosser, 1980; Raz, 1981; Weglein, 1982) which compute β and γ from a given ΔP at a surface bounding the medium under investigation. So far, practical schemes always make use of the simplification that the incident wave field for the inhomogeneities (as defined as β and γ) equals the reference wave field. Hence linear expressions in β and γ are assumed (Born approximation). This approximation is only valid if the reference medium is sufficiently close to the actual medium.

Inversion methods based on the theory of seismic migration and linearized inversion appear rather different, but in this paper it will be shown that both methods are closely related. We will see that the 'one-way decomposition' assumption in seismic migration plays a similar role as the 'Born approximation' in linearized inversion.

2. Matrix Formulation of the Linearized Forward Model. The linearized solution of (2) can be formulated in terms of (see, e.g., Bath and Berkhout, 1984, Chapter 15):

$$(3)\quad \Delta P(\vec{r}_o) = \frac{1}{4\pi}\int_V \bar{G}'(\vec{r}_o,\vec{r})[\gamma\bar{k}^2 + \nabla \ln\beta\cdot\nabla]\bar{P}(\vec{r})dV ,$$

where

$$\bar{G}'(\vec{r}_o,\vec{r}) = \bar{\rho}(\vec{r}_o)\bar{G}(\vec{r}_o,\vec{r})\bar{\rho}^{-1}(\vec{r})$$

and $\bar{G}(\vec{r}_o,\vec{r})$ being the pressure response in $\vec{r}_o$ due to a monopole in the reference medium at $\vec{r}$:

$$\bar{\rho}\nabla\cdot(\frac{1}{\bar{\rho}}\nabla\bar{G}) + \bar{k}^2\bar{G} = -4\pi\delta(\vec{r}_o - \vec{r}) .$$

First let us consider the linearized contribution due to the velocity inhomogeneities:

$$(4a)\quad \Delta P_1 = \frac{1}{4\pi}\int_V \bar{G}'\ [\gamma\bar{k}^2]\bar{P}dV$$

or, if we discretize along the z-direction,

$$(4b)\quad \Delta P_1 = \frac{\Delta z}{4\pi}\sum_{m=1}^{M}\int_x\int_y \bar{G}'_{m\Delta z}[\gamma\bar{k}^2]_{m\Delta z}\bar{P}_{m\Delta z}dxdy ,$$

where $\Delta z < \frac{1}{4}\lambda_{min}$, λ_{min} being the smallest wave length of the integrand and $M\Delta z$ the maximum depth of interest. Now, if we discretize along x and y as well, we will make use of the following matrix notation

$\vec{\bar{P}}(z_m)$ = illumination vector which represents the discretized version of $\bar{P}$ at depth level $z_m = m\Delta z$

$\bar{\mathbf{G}}'(z_o,z_m)$ = propagation matrix related to source depth level z_m and receiver depth level z_o , the columns of which represent the discretized version of $(\Delta x\Delta y/2\pi)\bar{G}'$

$\Gamma(z_m)$ = diagonal matrix, the diagonal elements representing the discretized version of $\frac{1}{2}\Delta z\gamma\bar{k}^2$ at depth level $z_m = m\Delta z$

$\Delta\vec{P}_1(z_o)$ = deviation vector which represents the discretized linear version of $P - \bar{P}$ at the surface $z_o = 0$ due to velocity inhomogeneities.

Then the matrix version of (4b) is given by

$$\Delta\vec{P}_1(z_o) = \sum_m \bar{\mathbf{G}}'(z_o,z_m)\boldsymbol{\Gamma}(z_m)\vec{\bar{P}}(z_m) \quad . \tag{5}$$

Similarly, for the density contribution we may write

$$\Delta P_2 = \Delta P_{21} + \Delta P_{22} + \Delta P_{23}$$

with

$$\Delta P_{21} = \frac{1}{4\pi}\int_V \bar{G}' \frac{\partial \ln\beta}{\partial x}\frac{\partial\bar{P}}{\partial x}\, dV \tag{6a}$$

or, after discretization,

$$\Delta\vec{P}_{21}(z_o) = \sum_m \bar{\mathbf{G}}'(z_o,z_m)\mathbf{B}_x(z_m)\,\partial x\vec{\bar{P}}(z_m) \tag{6b}$$

and

$$\Delta P_{22} = \frac{1}{4\pi}\int_V \bar{G}' \frac{\partial \ln\beta}{\partial y}\frac{\partial\bar{P}}{\partial y}\, dV \tag{7a}$$

or, after discretization,

$$\Delta\vec{P}_{22}(z_o) = \sum_m \bar{\mathbf{G}}'(z_o,z_m)\mathbf{B}_y(z_m)\,\partial_y\vec{\bar{P}}(z_m) \tag{7b}$$

and

$$(8a)\quad \Delta\vec{P}_{23} = \frac{1}{4\pi}\int_V \bar{G}' \frac{\partial \ln\beta}{\partial z}\frac{\partial \bar{P}}{\partial z}\,dV$$

or, after discretization,

$$(8b)\quad \Delta\vec{P}_{23}(z_o) = \sum_m \bar{\mathbf{G}}'(z_o,z_m)\mathbf{B}_z(z_m)\partial_z\vec{\bar{P}}(z_m) \ .$$

In the above expressions $\mathbf{B}_x$, $\mathbf{B}_y$ and $\mathbf{B}_z$ are diagonal matrices which contain the discretized values of $\partial_x \ln\beta$, $\partial_y \ln\beta$ and $\partial_z \ln\beta$ respectively.

The relationship between the pressure values at depth level z_m and the pressure values at the surface is given by

$$(9a)\quad \begin{bmatrix} \vec{\bar{P}}(z_m) \\ \partial_z\vec{\bar{P}}(z_m) \end{bmatrix} = \begin{bmatrix} \bar{\mathbf{X}}_1(z_m,z_o) & \bar{\mathbf{X}}_2'(z_m,z_o) \\ \bar{\mathbf{X}}_3'(z_m,z_o) & \bar{\mathbf{X}}_4'(z_m,z_o) \end{bmatrix} \begin{bmatrix} \vec{\bar{P}}(z_o) \\ \partial_z\vec{\bar{P}}(z_o) \end{bmatrix}$$

and

$$(9b)\quad \begin{aligned} \partial_x\vec{\bar{P}}(z_m) &= \mathbf{D}_x\vec{\bar{P}}(z_m) \\ \partial_y\vec{\bar{P}}(z_m) &= \mathbf{D}_y\vec{\bar{P}}(z_m) \ , \end{aligned}$$

where the rows of $\mathbf{D}_x$ and $\mathbf{D}_y$ contain a discrete version of ∂_x and ∂_y respectively and the operators $\mathbf{X}$ are defined by the two-way wave equation (see, e.g., Bath and Berkhout, 1984, Chapter 15).

Equations (6) - (9) can be combined into one matrix equation

$$(10)\quad \Delta\vec{P}(z_o) = \sum_m \bar{\mathbf{G}}'(z_o,z_m)\Delta\mathbf{Y}(z_m)\vec{\bar{Q}}(z_m) \ ,$$

where

$$\Delta\vec{P}(z_o) = \Delta\vec{P}_1(z_o) + \Delta\vec{P}_2(z_o) \ ,$$

$$\Delta\mathbf{Y}(z_m) = (\mathbf{\Gamma}(z_m)\mathbf{B}_x(z_m)\mathbf{B}_y(z_m)\mathbf{B}_z(z_m)) \ ,$$

$$\vec{\bar{Q}}(z_m) = \begin{pmatrix} \vec{\bar{P}}(z_m) \\ \partial_x\vec{\bar{P}}(z_m) \\ \partial_y\vec{\bar{P}}(z_m) \\ \partial_z\vec{\bar{P}}(z_m) \end{pmatrix} \ .$$

For the special situation that we may take $\partial_x \ln\beta = 0$ and $\partial_y \ln\beta = 0$ then $\Delta\mathbf{Y}$ simplifies to

$$\Delta\mathbf{Y}(z_m) = (\mathbf{\Gamma}(z_m)\mathbf{B}_z(z_m))$$

and $\bar{Q}$ simplifies to

$$\vec{\bar{Q}}(z_m) = \begin{matrix} \vec{\bar{P}}(z_m) \\ \partial_z\vec{\bar{P}}(z_m) \end{matrix} \ .$$

If the elements of vector $\vec{\bar{Q}}$ are ordered such that the quantities $\vec{\bar{P}}$, $\partial_x\vec{\bar{P}}$, $\partial_y\vec{\bar{P}}$, $\partial_z\vec{\bar{P}}$ are grouped together for one subsurface grid point then $\Delta\mathbf{Y}$ represents a matrix (Fig. 1) with four non-zero elements at the position (n,4n-3) - (n,4n) for n ranging from 1 to N , N being the length of data vector $\Delta\vec{P}(z_o)$.

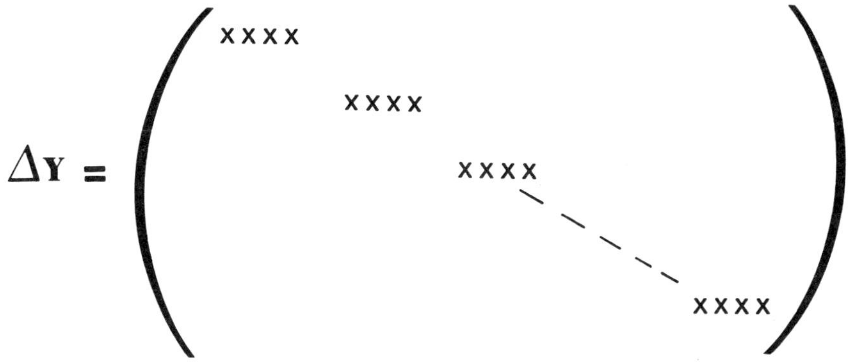

Figure 1: Structure of admittance matrix $\Delta\mathbf{Y}$.

If we include the properties of the detection system, (10) can be extended to

$$\text{(11)} \quad \Delta\vec{P}(z_o) = \mathbf{D}(z_o) \sum_m \bar{\mathbf{G}}'(z_o,z_m)\Delta\mathbf{Y}(z_m)\vec{\bar{Q}}(z_m) \ ,$$

where $\mathbf{D}(z_o)$ represents the detector matrix (Berkhout, 1982, Chapter 6).

Finally, expression (11) formulates the difference response due to one seismic source (or source array). If $\vec{\bar{Q}}_i(z_m)$ represents the wave field in the reference medium at depth level z_m due to the ith source (array), and we define the matrix

$$\bar{\mathbf{Q}}(z_m) = (\vec{\bar{Q}}_1(z_m)\ \vec{\bar{Q}}(z_m)\ldots\vec{\bar{Q}}_i(z_m)\ldots)$$

then the multi-record version of (11) may be formulated as

$$\text{(12a)} \quad \Delta\mathbf{P}(z_o) = \mathbf{D}(z_o) \sum_m \bar{\mathbf{G}}'(z_o,z_m)\Delta\mathbf{Y}(z_m)\bar{\mathbf{Q}}(z_m) \ ,$$

where the ith column of $\Delta\mathbf{P}(z_0)$ represents the difference response at the surface due to the ith source (array).

Similarly,

$$\text{(12b)} \quad \partial_z\{\Delta\mathbf{P}(z_0)\} = \mathbf{D}(z_0) \sum_m \bar{\mathbf{G}}'_z(z=z_0,z_m)\Delta\mathbf{Y}(z_m)\bar{\mathbf{Q}}(z_m) \ ,$$

where

$$\bar{\mathbf{G}}'_z \ (z=z_0,z_m) = \mathbf{D}_\rho(z_0)[\partial_z\bar{\mathbf{G}}(z=z_0,z_m)]\mathbf{D}_\rho^{-1}(z_m) \ .$$

Remarks:

1. For a homogeneous reference medium each column of $\bar{\mathbf{G}}'$ is given by the classical Rayleigh-I operator

$$\frac{e^{-j\bar{k}\Delta r}}{2\pi\Delta r}$$

2. For a homogeneous reference medium (9a) simplifies to

$$\vec{\bar{P}}(z_m) = \bar{\mathbf{G}}'_z(z_m,z=z_0)\vec{\bar{P}}(z_0)$$

$$\partial_z\bar{P}(z=z_m) = -j\bar{\mathbf{H}}_1\vec{\bar{P}}(z_m) \ ,$$

where each column of $\bar{\mathbf{G}}'_z$ is given by the classical Rayleigh-II operator

$$[\frac{1+j\bar{k}\Delta r}{\Delta r}\cos\phi]\ \frac{e^{-j\bar{k}\Delta r}}{2\pi\Delta r}$$

and each column of $\bar{\mathbf{H}}_1$ is given by (Berkhout, 1982, Appendix E)

$$\frac{\bar{k}}{2\pi}\ \frac{j_1(\bar{k}\Delta r)}{\Delta r} \ ,$$

j_1 respesenting the first-order spherical Bessel function.

3. Inversion According to the Linearized Inverse Scattering Approach. Based on the acoustic model we have derived, the following linearized inversion scheme can be presented:

a. Specify the reference medium and compute the responses at the surface

$$\bar{\mathbf{P}}^-(z_o)\ ,\quad \partial_z\bar{\mathbf{P}}^-(z_o)$$

due to the source wave fields

$$\bar{\mathbf{P}}^+(z_o)\ ,\quad \partial_z\bar{\mathbf{P}}^+(z_o)\ .$$

b. Calculate the deviation pressures at the surface

$$\Delta\vec{P}_i(z_o) = \vec{P}_i(z_o) - \vec{\bar{P}}_i(z_o) \qquad i = 1,2,\ldots$$

c. Downward continue the total reference wave field as given by

$$\bar{\mathbf{P}}(z_o)\ ,\quad \partial_z\bar{\mathbf{P}}(z_o)$$

from the surface z_o to depth level z_m according to (9a) and, by using also (9b), compute $\mathbf{Q}(z_m)$.

d. Compensate for the monopole responses $\mathbf{D}(z_o)\bar{\mathbf{G}}'(z_o,z_m)$ by matrix inversion

(13a) $$\Delta[\partial_z\mathbf{P}(z_m)] = \bar{\mathbf{F}}(z_m,z_o)\Delta\mathbf{P}(z_o)\ ,$$

where

(13b) $$\bar{\mathbf{F}}(z_m,z_o) = [\mathbf{D}(z_o)\bar{\mathbf{G}}'(z_o,z_m)]^{-1}$$

in some stable sense.

e. Compute for each grid point at depth level z_m the density and velocity information by matrix inversion

(14a) $$\Delta \mathbf{Y}(z_m) = \frac{\Delta\omega}{2\pi} \sum_{\omega} \Delta[\partial_z \mathbf{P}(z_m)]\bar{\mathbf{F}}(z_m) \ ,$$

where

(14b) $$\bar{\mathbf{F}}(z_m) = \bar{Q}^{-1}(z_m)$$

in some stable sense.

As only very few elements of matrix $\Delta\mathbf{Y}$ need to be computed it is advantageous to solve (14a) per subsurface grid point.

For K subsurface grid points at depth level z_m we have to solve for $4N$ unknowns. This means that for this grid point range at least 4 different source positions should be used to illuminate the subsurface and N different detector positions must be used to register the response.

Inversion according to (13) and (14) must be carried out in a spatially band-limited way as in practical situations inversion for the evanescent part of the wave field should be avoided for noise enhancement reasons (Berkhout, 1982, Chapter 7).

Finally, the _spatial_ band width of both the data and the two inversion operators determine the _lateral_ resolution of the inversion result; the _temporal_ band width of the data (given by the number of independent frequency components above the noise level) defines the _vertical_ resolution of the inversion result (Berkhout, 1984).

4. Inversion According to the Seismic Migration Approach. For the seismic migration application we will choose a subsurface model such that:

(15a) 1. $P = P^+ + P^-$ everywhere in V, where P^+ and P^- fulfill the one-way wave equation (hence multiple scattering is neglected).

(15b) 2. P^+ and P^- are coupled at each layer boundary by the elastic boundary conditions.

Using conditions (15), we may write

a. For the downward travelling source wave fields at depth level z_m

(16a) $$\mathbf{P}^+(z_m) = \mathbf{W}^+(z_m, z_o)\mathbf{P}^+(z_o) .$$

b. For the upward travelling reflected wave fields due to the layer boundaries at depth level z_m

(16b) $$\mathbf{P}^-(z_m) = \mathbf{R}(z_m)\mathbf{P}^+(z_m) .$$

c. For the upward travelling reflected wave fields at the surface z_o

(16c) $$\mathbf{P}^-(z_o) = \sum_m \mathbf{W}^-(z_o, z_m)\mathbf{P}^-(z_m) .$$

Combining expressions (16) for all depth levels, and including the detector matrix $D(z_o)$, yields

(17a) $$\mathbf{P}^-(z_o) = \mathbf{D}(z_o)[\sum_m \mathbf{W}^-(z_o, z_m)\mathbf{R}(z_m)\mathbf{W}^+(z_m, z_o)]\mathbf{P}^+(z_o) .$$

Similarly, using $\mathbf{W}^+ = -j\mathbf{V}^+\mathbf{H}_1^+$,

(17b) $$\mathbf{P}^-(z_o) = \mathbf{D}(z_o)[\sum_m \mathbf{W}^-(z_o, z_m)\mathbf{R}(z_m)\mathbf{V}^+(z_m, z_o)]\partial_z\mathbf{P}^+(z_o)$$

or using $\mathbf{W}^- = j\mathbf{V}^-\mathbf{H}_1^-$,

(17c) $$\mathbf{P}^-(z_o) = \mathbf{D}(z_o)[\sum_m \mathbf{V}^-(z_o, z_m)\mathbf{Y}(z_m)\mathbf{W}^+(z_m, z_o)]\mathbf{P}^+(z_o)$$

or, using both,

(17d) $$\mathbf{P}^-(z_0) = \mathbf{D}(z_0)[\sum_m \mathbf{V}^-(z_0,z_m)\mathbf{Y}(z_m)\mathbf{V}^+(z_m,z_0)]\partial_z\mathbf{P}^+(z_0) \ .$$

Note that impedance matrix $\mathbf{Y}$ and reflectivity matrix $\mathbf{R}$ are related according to

(18) $$\mathbf{Y}(z_m) = j\mathbf{H}_1^-(z_m)\mathbf{R}(z_m) \ .$$

Expression (17a) defines the forward model used by Berkhout (1982) for the description of pre-stack data.

From (17a) it follows that in the pre-stack migration procedure inversion should be carried out according to

$$\mathbf{X}^-(z_m) = \mathbf{F}^+(z_m,z_0)\mathbf{P}^-(z_0)\mathbf{F}^-(z_0,z_m)$$

with

$$\mathbf{F}^-(z_0,z_m) = [\mathbf{W}^+(z_m,z_0)\mathbf{P}^+(z_0)]^{-1}$$

and

$$\mathbf{F}^+(z_m,z_0) = [\mathbf{D}(z_0)\mathbf{W}^-(z_0,z_m)]^{-1} \quad \text{in some stable sense.}$$

Using the imaging principle (Claerbout, 1976),

$$\vec{R}(z_m) = \frac{\Delta\omega}{2\pi}\sum_\omega [\text{diagonal elements of } \mathbf{R}(z_m)]$$

$$= \frac{\Delta\omega}{2\pi}\sum_\omega [\text{diagonal elements of } \mathbf{X}^-(z_m)] \ ,$$

where $\vec{R}(z_m)$ defines the zero-offset reflectivity distribution at depth level z_m averaged over all available temporal frequencies.

So far we have dealt with the pre-stack situation only.

For zero-offset data expression (17a) can be rewritten as

$$\text{(19a)} \quad \vec{P}^-(z_o) = \mathbf{D}(z_o) \sum_m \mathbf{W}^{(2)}(z_o, z_m)\vec{R}_s(z_m) \ ,$$

where $\mathbf{R}(z_m)$ is assumed to be a diagonal matrix containing the zero-offset reflection coefficients and

$$\text{(19b)} \quad \vec{R}_s(z_m) = \mathbf{R}(z_m)\vec{P}^+(z_o) \ .$$

Each element of $\mathbf{W}^{(2)}$ equals the squared version of the related element of $\mathbf{W}$. For simple models $\mathbf{W}^{(2)}$ is generally replaced by $\mathbf{W}$ taking half the medium velocity (exploding reflector model).

Note that for zero-offset migration one matrix inversion for each depth level need be applied only:

$$\vec{P}^-(z_m) = \mathbf{F}^{(2)}(z_m, z_o)\vec{P}^-(z_o) \ ,$$

where

$$\mathbf{F}^{(2)}(z_m, z_o) = [\mathbf{D}(z_o)\mathbf{W}^{(2)}(z_o, z_m)]^{-1} \quad \text{in some stable sense.}$$

In fore mentioned practical applications two types of simplifying assumptions are often being made:

1. Hyperbolic assumption, <u>nonrecursive</u> approach, implying (Fig. 2a)

$$\text{(20a)} \quad W(\vec{r}_o, \vec{r}_m) = \frac{z_m - z_o}{2\pi} \frac{1 + jk\Delta r}{\Delta r^3} e^{-jk\Delta r} \ ,$$

where $\Delta r = |\vec{r}_o - \vec{r}_m|$ and $k = \omega / v_{migr}$.

Hence, using assumption (20a), the reference medium is considered as a pseudo-homogeneous fluid where the dipole response $\partial_z G$ from each subsurface point (x_l, y_j, z_m) can be described by a spherical wave front (or hyperbolic diffraction curve) at the surface, the curvature being defined by velocity parameter $v_{migr}(x_i,y_j,z_m)$.

For restricted dip angles inversion based on (20a) may have surprisingly good 'focussing properties', even in complicated situations, but the inversion result may contain serious amplitude- and spatial distortions (Robinson, 1982).

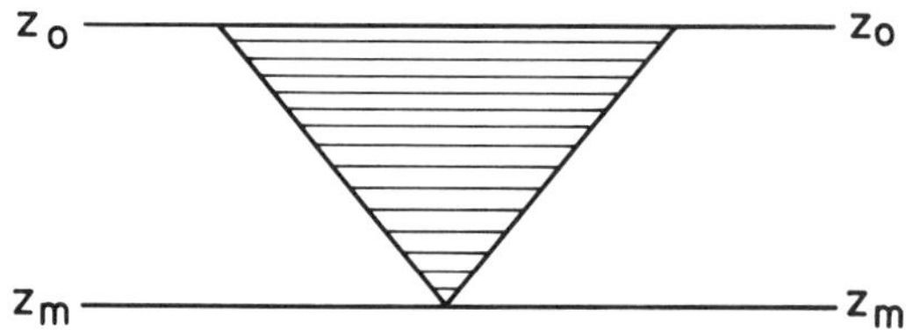

Figure 2a: Nonrecursive approach.

2. Hyperbolic assumption, recursive approach, implying (Fig. 2b)

$$\mathbf{W}(z_o,z_m) = \mathbf{W}(z_o,z_1)\mathbf{W}(z_1,z_2)\ldots\mathbf{W}(z_{m-1},z_m) \ ,$$

where

$$\text{(20b)} \quad W(\vec{r}_{i-1},\vec{r}_i) = \frac{z_i-z_{i-1}}{2\pi} \frac{1+jk\Delta r}{\Delta r^3} e^{-jk\Delta r}$$

with $\Delta r_i = |\vec{r}_{i-1}-\vec{r}_i|$ and $k = \omega/c$.

Hence, using assumption (20b), the medium is considered to be locally homogeneous, the density and velocity being taken constant within each sub-aperture volume. This means that for each sub-aperture the hyperbolic assumption applies. Of course, this property may not be valid at all for the full aperture.

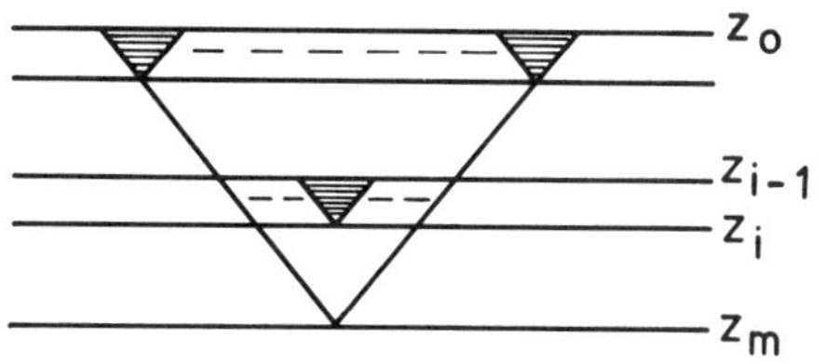

Figure 2b: Recursive approach.

Simplified assumptions (20a) and (20b) also mean that transmission losses are not considered.

5. Comparison Between Linearized Inversion and Seismic Migration. Let us assume that we have chosen for our linearized inversion problem a reference medium with the following properties:

a. The total wave field $\bar{P}_{tot}$ can be decomposed in an upward and downward travelling wave field

$$\bar{P}_{tot} = \bar{P}^+ + \bar{P}^-$$

$\bar{P}^+$ and $\bar{P}^-$ fulfilling the one-way wave equations.

b. The layer boundaries consist of gradual transition zones in density and velocity such that

$$|\bar{P}^-| << |\bar{P}^+| \quad \text{and} \quad |P^-| << \sqrt{|P^-|^2 + N^2} \ ,$$

where N^2 represents the noise power spectrum.

For this particular choice of the reference medium it follows from the foregoing that the following properties hold:

1. The illumination pressure is defined by a downward travelling wave field

$$\bar{P}(z_m) = \bar{P}^+(z_m) = \bar{W}^+(z_m, z_o)\bar{P}^+(z_o) \ .$$

2. The deviation pressure is defined by an upward travelling wave field

$$\Delta \mathbf{P}(z_m) = \mathbf{P}^-(z_m) .$$

3. $\bar{\mathbf{Q}}(z_m) = \bar{\mathbf{Q}}^+(z_m)$ can be computed from $\bar{\mathbf{P}}^+(z_m)$:

$$\bar{\mathbf{Q}}(z_m) = \begin{bmatrix} \bar{\mathbf{P}}^+(z_m) \\ \mathbf{D}_x\bar{\mathbf{P}}^+(z_m) \\ \mathbf{D}_y\bar{\mathbf{P}}^+(z_m) \\ -j\bar{\mathbf{H}}_1^+\bar{\mathbf{P}}^+(z_m) \end{bmatrix}$$

and downward continuation of $\bar{\mathbf{Q}}(z_0)$ can be realized by forward extrapolation of $\bar{\mathbf{P}}^+(z_0)$ only.

4. Using property 3, and assuming that linearization is justified, **R** can be expressed in terms of elastic parameters β and γ :

(22a) $$\mathbf{R}(z_m) = [+j\bar{\mathbf{H}}_1^-(z_m)]^{-1}\mathbf{Y}_L(z_m)$$

with

(22b) $$\mathbf{Y}_L(z_m) = \Gamma(z_m) + \mathbf{B}_x(z_m)\mathbf{D}_x + \mathbf{B}_y(z_m)\mathbf{D}_y - j\mathbf{B}_z(z_m)\bar{\mathbf{H}}_1^+(z_m) .$$

5. Upward propagation is defined by the one-way operators

$$\bar{\mathbf{V}}^-(z_0,z_m) \quad \text{or} \quad \bar{\mathbf{W}}^-(z_0,z_m) .$$

As a consequence, if the migration velocity distribution at each depth level is derived from a reference medium with the above properties and, in addition, if $\bar{\mathbf{Q}}(z_m)$ is expressed in terms of $\bar{\mathbf{P}}(z_m)$ then migration and linearized inversion define identical inversion processes:

$$\chi(z_m) = \mathbf{F}^+(z_m, z_o)\mathbf{P}^-(z_o)\mathbf{F}^-(z_o, z_m) \ .$$

Figure 3 summarizes the difference between the two approaches of linearized inversion. Note that

P^+ and $\bar{Q}$ (illumination)

P^- and ΔP (response)

play comparable roles.

Note that in seismic migration linearization (22b) is not assumed!

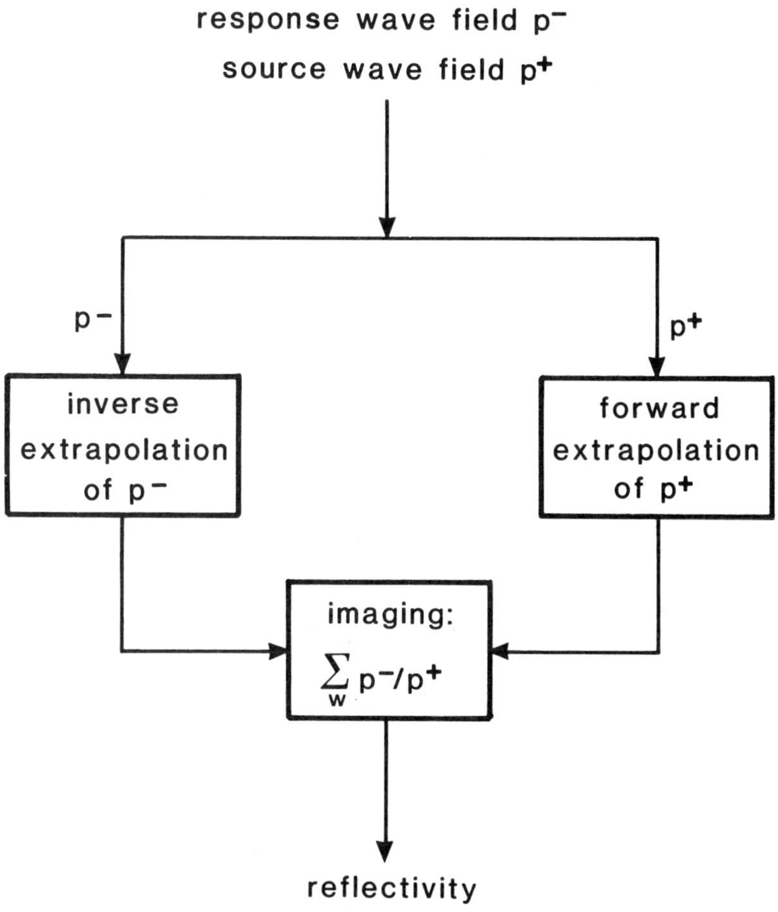

Figure 3a: Principle of linearized inversion according to the seismic migration approach.

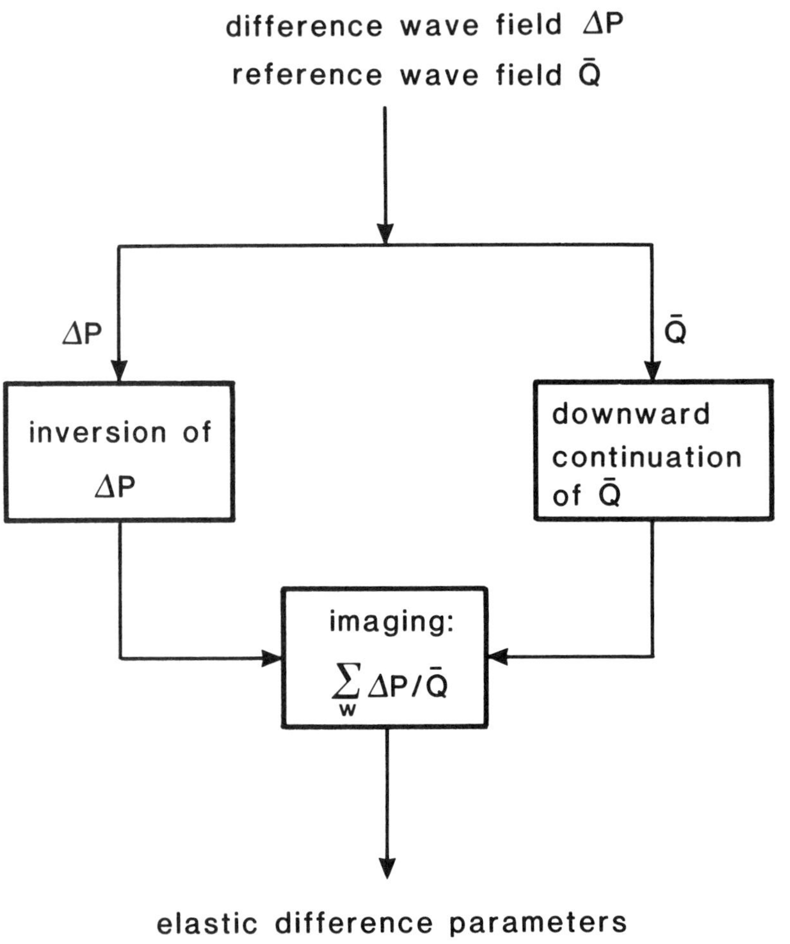

Figure 3b: Principle of linearized inversion according to the inverse scattering approach.

6. Conclusions.

1a. In the seismic migration approach to inversion, use is made of the forward model

$$P^-(z_0) = D(z_0)[\sum_m W^-(z_0,z_m)R(z_m)P^+(z_m)] \quad , \tag{23}$$

where $R(z_m)$ transforms the downward-travelling illuminating wave field $P^+(z_m)$ into the upward-travelling reflected wave field $P^-(z_m)$ and $W^-(z_0, z_m)$ transfers the reflected wave field to the surface.

b. In seismic migration (pre-stack formulation) band-limited inversion is carried out for the illuminating wave field $P^+(z_m)$ and the upward transfer function $D(z_0)W^-(z_0, z_m)$.

2a. In the linearized inverse scattering approach, use is made of the forward model

$$\Delta P(z_0) = D(z_0)\,[\sum_m \bar{G}'(z_0, z_m)\,\Delta Y(z_m)\bar{Q}(z_m)]$$

(24) with

$$\bar{Q}(z_m) = [\bar{P}(z_m),\ \partial_x\bar{P}(z_m),\ \partial_y\bar{P}(z_m),\ \partial_z\bar{P}(z_m)]^T ,$$

where $\Delta Y(z_m)$ transforms the illuminating wave field $\bar{Q}(z_m)$ into the scattered wave field $\Delta P(z_m)$ and $\bar{G}'(z_0, z_m)$ transfers the scattered wave field to the surface.

b. In the linearized inverse scattering problem band-limited inversion is carried out for the illuminating wave field $\bar{Q}(z_m)$ and the transfer function $\bar{G}'(z_0, z_m)$.

3a. If the total wave field $\bar{P}$ in the reference medium can be decomposed in one-way wave fields $\bar{P}^+$ and $\bar{P}^-$ such that

$$\text{a.}\quad |\bar{P}^-| \ll |\bar{P}^+|$$

(25)

$$\text{b.}\quad |\bar{P}^-| \ll \sqrt{|P^-|^2 + N^2} ,$$

N^2 representing the noise power spectrum, then forward model (23) and (24) are identical.

b. In addition

$$\text{(26)}\quad \mathbf{R} = (j\bar{\mathbf{H}}_1^-)^{-1}\mathbf{Y}_L$$

with

$$\text{(27a)}\quad \mathbf{Y}_L = [\bar{\mathbf{G}}^{-1}\mathbf{G}]\mathbf{Y}[\mathbf{W}^+(\bar{\mathbf{W}}^+)^{-1}]$$

$$\text{(27b)}\quad = \Gamma + \mathbf{B}_x\mathbf{D}_x + \mathbf{B}_y\mathbf{D}_y - j\mathbf{B}_z\bar{\mathbf{H}}_1^+ \,.$$

Note that linearization (27b) is only valid if the difference between $\mathbf{G}$ and $\bar{\mathbf{G}}$ is small.

In seismic migration linearization (27b) is not used.

VII. Reference.

[1] A.J. BERKHOUT, Seismic Migration (2nd edition), Elsevier, Amsterdam, 1982.

[2] A.J. BERKHOUT, Seismic Resolution, Geophysical Press, Amsterdam, 1984.

[3] M. BATH and A.J. BERKHOUT, Mathematical Aspects of Seismology, Geophysical Press, Amsterdam, 1984.

[4] J.F. CLAERBOUT, Fundamentals of Geophysical Data Processing, McGraw-Hill, New York, 1976.

[5] R.W. CLAYTON and R.H. STOLT, A Born WKBJ inversion method for acoustic reflection data, Geophysics, vol. 46, no. 11, 1981.

[6] J.K. COHEN and N. BLEISTEIN, Velocity inversion for acoustic waves, Geophysics, Vol. 44, No. 6, 1979.

[7] J. GAZDAG, Wave equation migration with the phase shift method, Geophysics, Vol. 23, No. 7, 1978.

[8] D. LOEWENTHAL, L. LU, R. ROBERSON, and J. SHERWOOD, The wave equation applied to migration, Geophysical Prospecting, Vol. 24, No. 2, 1974.

[9] R.T. PROSSER, Formal solutions of inverse scattering problems, J. Math. Phys. 21, 1980.

[10] S. RAZ, Three-dimensional velocity profile inversion from finite offset scattering data, Geophysics, Vol. 45, No. 6, 1981.

[11] E.A. ROBINSON, Migration of geophysical data, IHRDC Publishers, Boston, 1984.

[12] W.A. SCHNEIDER, Integral formulation for migration in two and three dimensions, Geophysics, Vol. 23, No. 1, 1978.

[13] R.H. STOLT, Migration by Fourier transform, Geophysics, Vol. 23, No. 1, 1978.

[14] A.B. WEGLEIN, Nearfield inverse scattering formalism for the three-dimensional wave equation: The inclusion of a priori velocity information, J. Acoust. Soc. Am., Vol. 71, No. 5, 1982.

PROJECT REVIEW ON GEOPHYSICAL AND OCEAN SOUND SPEED PROFILE INVERSION

NORMAN BLEISTEIN*, JACK K. COHEN*, JOHN A. DESANTO* AND FRANK G. HAGIN*

Abstract. This report summarizes the status of the research projects on inverse problems at the Colorado School of Mines. We are engaged in two major projects in inverse problems: (i) inverse methods for reflector imaging and (ii) inverse methods for profile inversion in the ocean. In the first project we use "high frequency" methods to invert backscatter (monostatic) observations. Our primary applications are to seabed mapping and seismic exploration, although we have also implemented our method to image flaws in solids. Computer implementation of the current algorithm is discussed, and some indication of directions for extending the method is given. A unique feature of the second project is that the direction of variation of the sound speed profile (depth) is orthogonal to the direction of propagation (range). Again, high frequency methods are used and computer implementation is briefly discussed.

* Center for Wave Phenomena, Department of Mathematics, Colorado School of Mines, Golden, Colorado 80401, prepared under support of the Selected Research Opportunities Program of the Office of Naval Research.

1. Introduction. The purpose of this article is to report on the present status of the research project on inverse problems at the Center for Wave Phenomena, Colorado School of Mines, partially supported by the Selected Research Program of the Office of Naval Research. The support of ONR has provided a strong impetus to an ongoing research program which, at the time of this writing, includes four faculty members and nine graduate students.

We are engaged in two major research projects in inverse problems. They are (i) inverse methods for reflector imaging and (ii) inverse methods for profile inversion in the ocean.

In the first project we use "high frequency" methods to invert backscatter (monostatic) observations. Our primary applications are to seabed mapping and seismic exploration, although we have also implemented our method to image flaws in solids.

A unique feature of the second project is that the direction of variation of the sound speed profile (depth) is orthogonal to the direction of propagation (range). Again, high frequency methods are used.

The term high frequency, means that the length scales of the problem are "many" wavelengths long — at least three — for the frequencies of the sources in the experiments used to probe the medium of interest. For the backscatter experiments, the relevent parameter is

$$\lambda = 4\pi fL/c \tag{1}$$

where f is the frequency of interest, c is an average sound speed and L is a typical length scale. There is an extra factor of 2 in this expression compared to the more traditional dimensionless parameter, due to the two-way traveltime of the backscatter experiment.

For seismic exploration, typical ranges of the frequency and velocity are

$$4\text{Hz} \lesssim f \lesssim 40\text{ Hz}, \quad 1500\text{m/s} \lesssim c \lesssim 6000\text{m/s} \quad . \tag{2}$$

The length scales of interest are the range to the reflector or the Gaussian curvature of the reflector or the separation between reflections. For $L \geq 400\text{m}$, $\lambda \geq \pi$, our high frequency methods are applicable. It is reasonable to expect the first two length scales to satisfy this constraint. However, layer spacing may be much less. It is often not possible to separate nearby layers from backscatter experiments <u>whether or not we use high frequency asymptotic analysis</u> to simplify our solution.

For near surface seabed mapping, typical values for frequency and velocity are

$$100 \text{ Hz} \lesssim f \lesssim 400 \text{ Hz} \quad , \quad 1500 \text{ m/s} = c \quad . \tag{3}$$

In this case, λ is "large" for L on the order of 4m.

In non-destructive testing applications, one might have parameters such as

$$f \approx 4 \times 10^{6} \text{Hz} \quad , \quad c \approx 6000 \text{m/s} \quad , \quad L \gtrsim 440 \times 10^{-6} \text{m} \quad , \tag{4}$$

for which

$$\lambda > \pi \quad . \tag{5}$$

2. **Reflector Imaging for the Seismic Inverse Problem.** The model for the reflector imaging problem is as follows. An acoustic source is set off at each point on the surface of the earth ($z = 0$) and the backscattered signal, due to scattering by the inhomogeneities in the earth, is observed. The earth is assumed to be an acoustic medium with only the sound speed varying. As will be seen below, our solution formulas take account of bandlimiting in the Fourier domain and temporal and spatial discretization of the data in real experiments. However, we begin our problem formulation as if we have a continuum of data in space and time. If $U(\underline{x},t;\underline{\xi})$, $\underline{x} = (x,y,z)$, $\underline{\xi} = (\xi,\eta,0)$, (z positive "downward") is the impulse response in a single backscatter experiment at $\underline{\xi}$, then this function satisfies the equation

$$\nabla^2 U - \frac{1}{v^2}\frac{\partial^2}{\partial t^2} U = - \delta(\underline{x} - \underline{\xi})\, \delta(t) \quad , \quad t > 0 \quad . \tag{6}$$

The objective is to find the sound speed, $v = v(x,y,z)$, from observations of the ensemble of backscatter responses, $U_S(\underline{\xi},t;\underline{\xi})$, to experiments carried out for all $\underline{\xi} = (\xi,\eta,0)$.

We use a perturbation method to solve this problem as described in Cohen and Bleistein [1] and Cohen and Bleistein [2]. Thus, we introduce a reference velocity, c, and a perturbation, α, as follows:

$$\frac{1}{v^2} = \frac{1}{c^2}(1 + \alpha) \quad , \quad \alpha \equiv 0 \text{ for } z < 0 \quad . \tag{7}$$

We then define U_I to be the response to the impulsive source in the unperturbed medium:

$$(8)\qquad \nabla^2 U_I - \frac{1}{c^2}\frac{\partial^2 U_I}{\partial t^2} = -\,\delta(\underline{x} - \underline{\xi})\,\delta(t)$$

and set

$$(9)\qquad U = U_I + U_S \quad .$$

From (6–9) we conclude that the scattered field U_S must satisfy the equation

$$(10)\qquad \nabla^2 U_S - \frac{1}{c^2}\frac{\partial^2 U_S}{\partial t^2} = \frac{\alpha}{c^2}\frac{\partial^2}{\partial t^2}\left[U_I + U_S\right]$$

Under the assumption of small α, we are led to discard the product $\alpha\,\partial^2 U_S/\partial t^2$ as being quadratic in α while the term $\alpha\,\partial^2 U_I/\partial t^2$ is linear in α. Rewriting the linearized equation in frequency domain, we have

$$(11)\qquad \nabla^2 u_S(\underline{x},\omega;\underline{\xi}) + \frac{\omega^2}{c^2} u_S(\underline{x},\omega;\underline{\xi}) = -\frac{\omega^2\alpha}{c^2} u_I(\underline{x},\omega;\underline{\xi}) \quad ;$$

$$(12)\qquad u_S(\underline{x},\omega;\underline{\xi}) = \int_0^{x} u_S(\underline{x},t,\underline{\xi})\exp\{i\omega t\}\,dt \quad .$$

We remark that $u_S\ (\underline{x},\omega;\underline{\xi})$ is just the Green's function for (11). Thus, we have the integral representation of the solution to (11):

$$(13)\qquad u_S(\underline{\xi},\omega,\underline{\xi}) = \omega^2 \int \frac{\alpha}{c^2} u_I^2(\underline{x},\omega;\underline{\xi})\,dV \quad .$$

This is a Fredholm integral equation of the first kind for $\alpha(\underline{x})$ with the left side being the "known" observations on the upper surface. For constant background speed, c, the function u_I is just the free space Green's function:

$$(14)\qquad u_I = \frac{\exp\{i\omega|\underline{x}-\underline{\xi}|\}}{4\pi|\underline{x}-\underline{\xi}|} \quad .$$

We remark that for this case the squared absolute value of the kernel, $|u_I|^4/c^4$ is not integrable in all of its variables, $(\underline{x},\underline{\xi},\omega)$. Thus, the kernel is not compact and the integral equation does not suffer the ill-conditional properties of compact operators. In fact the kernel in (13) is "Fourier-like", hence (13) inverts nicely if care is taken.

Parenthetically, let us consider the case in which the support of α is finite and far from the surface of observations, $\underline{\xi}$. Revising our coordinate system for a moment, with the origin "near" the scattering domain, we can characterize this case by

$$(15) \qquad |\underline{x}|/|\underline{\xi}| << 1 ,$$

and use the far field approximations

$$(16) \qquad \begin{aligned} |\underline{x}-\underline{\xi}| &= |\underline{\xi}| - \underline{x}\cdot\tilde{\xi} + O(|\underline{x}|/|\underline{\xi}|) \\ &= |\underline{\xi}| \, [1 + O(|\underline{x}|/|\underline{\xi}|)] , \quad \hat{\xi} = \underline{\xi}/|\underline{\xi}| , \end{aligned}$$

in the phase and amplitude of (14), respectively. We use these results in (13) to obtain

$$(17) \qquad u_S(\underline{\xi},\omega,\underline{\xi}) = \frac{\omega^2 \exp\{2i\omega|\underline{\xi}|/c\}}{[4\pi c(\underline{\xi})]^2} \int \alpha(\underline{x}) \exp\{2i\omega\underline{x}\cdot\hat{\xi}/c\} \, dV .$$

The integral here can be recognized as the Fourier transform $\tilde{\alpha}(\underline{k})$ of $\alpha(\underline{x})$ evaluated at

$$(18) \qquad \underline{k} = \frac{2\omega}{c} \hat{\xi} .$$

Thus an ensemble of experiments provides an aperture limited Fourier transform of $\tilde{\alpha}(\underline{k})$, the direction of $\underline{k}$ ranging over $\hat{\xi}$ sgn ω and the magnitude of $\underline{k}$ ranging over the bandwidth of the experiments. Coincidentally, this result was presented by us at another meeting at Cornell [3], as a basis for imaging flaws in solids. Subsequently, we found that an equivalent result could be derived by using the Kirchhoff approximation, rather than the Born approximation. This has the advantage that the reflection strength at the boundary of the flaw need not be small. That method was successfully used to obtain aperture limited images of flaws in solids [4].

For seabed mapping and seismic exploration, we cannot apply the farfield approximation. For the real seismic experiment, finite in space and time, both the lateral extent and depth of the experiment are comparable to or greater than the range of the support of α.

We return, therefore, to the coordinate system as originally stated with $\underline{\xi} = (\xi,\eta,0)$ and $\alpha(x,y,z)$ nonzero for $z>0$. We consider first, the case of constant background. Thus, we use (14) in (13). As reported in Cohen and Bleistein, [2], this equation admits a closed form solution which we write as

$$\tilde{\alpha}(k_1,k_2,k_3) = \text{const.} \frac{\partial}{\partial\omega} \left[\frac{\tilde{u}_S(k_1,k_2,\omega)}{\omega^2} \right] ;$$

(19)

$$k_3 = \operatorname{sgn}\omega \sqrt{\frac{\omega^2}{c^2} - k_1^2 - k_2^2} , \quad \frac{\omega^2}{c^2} \geq k_1^2 + k_2^2 .$$

In this equation,

$$\tilde{u}_S(k_1,k_2;\omega) = \int d\xi d\eta \; u_S(\underline{\xi},\omega;\underline{\xi}) \exp\{-2i[k_1\xi - k_2\eta]\} . \tag{20}$$

In (19), we have only stated the "dispersion relation" for k_3, real. The function $\alpha(\underline{x})$ can be reconstructed from its Fourier transform over only real values of k_1, k_2, k_3. Thus, there is no need to evaluate $\tilde{\alpha}$ for imaginary k_3.

3. **The Singular Function and the Reflectivity Function.** We must now address the problem of the limited aperture of the data for $\tilde{\alpha}$ in the $\underline{k}$ domain. In particular, for high frequency data, $|\underline{k}| = |\omega|/c$ must be large, as well. Thus, we must obtain information about $\alpha(\underline{x})$ from high frequency aperture limited information about its Fourier transform. This is a problem in Fourier analysis, not peculiar to the particular application considered here.

It is well known that trend information about a function is contained in the low frequency part of the spectrum and rapid variation or discontinuity information is contained in the high frequency portion of the spectrum. Thus, lacking low frequency information, we abandon any attempt to obtain trend information about $\alpha(\underline{x})$ and seek only information about the discontinuities of $\alpha(\underline{x})$.

Let us suppose, therefore, that the interior of the earth is layered. That is, it consists of regions of constant velocity whose boundaries are arbitrary — non-planar — surfaces, the reflectors of acoustic energy. In this case, from (7), $\alpha(\underline{x})$ is a piecewise constant functions, as well, with the same reflectors as the boundaries of the regions of constant α.

In a series of papers culminating in Cohen and Bleistein, [5] we have developed a theory for imaging the boundaries of piecewise constant functions such as $\alpha(\underline{x})$ from high frequency aperture limited Fourier data for $\tilde{\alpha}$. In order to understand this theory, we first introduce the singular function $\gamma(\underline{x})$ of a surface S. We take this function to be a Dirac delta function of normal distance to S. Alternatively, we define $\gamma(\underline{x})$ by its action on test functions, namely,

$$\int f(\underline{x})\gamma(\underline{x})dV = \int_S f(\underline{x})dS \quad . \tag{20}$$

In this integral, the left hand side is an integral over all space. In particular,

$$\tilde{\gamma}(\underline{k}) = \int_S e^{i\underline{k}\cdot\underline{x}}\, dS \quad . \tag{21}$$

We remark that mathematical imaging of a surface is equivalent to determining its singular function. Let us suppose, now, that we have high frequency aperture limited Fourier data for $\tilde{\gamma}(\underline{k})$. In Bleistein and Cohen, [3], we have shown that the Fourier transform of the aperture limited Fourier data produces a bandlimited Dirac delta function of normal distance to S along every normal for which the direction of the normal vector is in the aperture of the Fourier data. Since bandlimited high frequency data will adequately depict Dirac delta functions, a section of S — the part for which it's normal is in the aperture — can be imaged from the aperture limited Fourier data [6].

We have also developed a theory in [3] and [7] to relate the high frequency data of a piecewise constant function, such as α, to the Fourier data of the array of singular functions of boundaries of constant α. The result of that theory is that

$$\tilde{\beta}(\underline{k}) = \sum R_j\tilde{\gamma}_j(\underline{k}) \sim \frac{i\omega}{2c}\,\tilde{\alpha}(\underline{k}) \quad . \tag{22}$$

In this equation, $\gamma_j(\underline{x})$ is the singular function of the jth reflector, R_j is the normal reflection strength and we call $\beta(\underline{x})$, the sum of all of these weighted singular functions, the reflectivity function. Thus, multiplication by $i\omega/2c$ in this first line of (19) produces a high frequency approximation of the Fourier transform of the reflectivity function of the interior in terms of the observed backscattered data at the surface. Fourier inversion of this data yields a partial image of the interior.

4. Further Reduction of the Solution Formula. A number of simplifications of this Fourier inversion are possible. First, since we are focusing our attention on the leading order high frequency solution, we need retain only the leading order term on the right side of (19). Secondly, it is usually the case in seismic experiments that only a line of experimental data is collected rather than a planer array of data as we have assumed. In this case, we cannot hope to produce a three dimensional image of the subsurface. To accommodate such a data set, we assume that α is independent of the variable orthogonal to the line of data. In our solution formula, we assume that the observed data is independent of y. The Fourier transform in y implicit in (19) can now be calculated explicitly, as can the inverse transform in k_2 in the inversion of (22). The resulting formula for β produces a cylindrical or two dimensional reflectivity function in response to a wave which propagates in three dimensions. We refer to this case — of great interest in processing real seismic data — as the <u>two-and-one-half dimensional</u> case.

A further simplification can be achieved by carrying out the Fourier transform in $\underline{k}_1$ in the inversion of (22) by the method of stationary phase. The result of carrying out all of these simplifications is a leading order asymptotic inversion formula for reflectivity function in the two-and-one-half dimensional case. The result (see [7]) is:

$$\beta(x,z) = \frac{4z}{\sqrt{\pi c_0}} \int \frac{d\xi}{\rho^{3/2}} \int \sqrt{|\omega|}\, d\omega \exp\{-2i\omega\rho/c - i\pi/4 \operatorname{sgn} \omega\}$$

$$\cdot \int t\, U_S(\underline{\xi},t;\underline{\xi}) \exp\{i\omega t\} \quad , \tag{23}$$

$$\rho = \sqrt{(x-\xi)^2 + z^2} \quad , \quad \underline{\xi} = (\xi,0,0) \quad .$$

It is this multifold integral which must be carried out over the ensemble of backscattered data.

5. **<u>Computational</u> <u>Considerations</u>.** The second line in (23) represents a Fourier time transform of each data trace (fixed ξ). This step can be carried out using a Fast Fourier transform (FFT) algorithm. That result, for each trace, must be multiplied by the indicated filter in the frequency domain. Any additional filtering to "taper" the data in the Fourier domain should be carried out at this step as well. The inverse Fourier transform is thus calculated, again by employing the FFT. The result of these two operations is to produce a function on a discrete time mesh, t_j, $j = 1\cdots N_t$, for each trace, ξ_m, $m = 1,\cdots, N_\xi$. Let us call this function $V(\xi_m,t_j)$. To carry out the ξ integration (summation) we require the function $V(\xi_m, 2\rho/c)$. We calculate this function by interpolation in the table $V(\xi_m,t_j)$.

The integration in time is carried out over the entire time record of the trace. The integral over ω must include the bandwidth of the data. The integration over ξ is constrained in three ways. The first constraint is the length of the available data set. The second constraint is the length of the available data set. The second constraint is that $2\rho/c$ cannot exceed the maximum time on the data trace. The third constraint arises from the discreteness of the integral in the spatial domain. It is necessary that the transverse component of the wave number $2k_1$ in (19) be bounded by the Nyquist limit $\pi/\Delta\xi$, where $\Delta\xi$ is the spacing between disrried out by the method of stationary phase. The condition of stationarity relates to k_1 to the spatial variables and the frequency, leading to the condition,

$$\frac{|\underline{x} - \underline{\xi}|}{\rho} \le \frac{4\pi f_{max}}{c} , \tag{24}$$

where f_{max} is the maximum frequency of the data in Hertz. (See [7] for details).

It is the processing of that last integral which dominates the computer time of this algorithm. If N_x and N_z are the number of output points in x and z, respectively, where we seek β, and $\tilde{N}_\xi$ is an average number of integration points in the ξ integral, then the processing time is $O(\tilde{N}_\xi N_x N_z)$. We note that $\tilde{N}_\xi$ is often significantly smaller than N_ξ. Moreover, sampling considerations typically dictate that N_z be substantially smaller than N_t, thus further reducing the operation count.

In tests by Bleistein and Gray [8] this algorithm has been demonstrated to be comparable to or faster than a K-F migration algorithm. In Cohen and Bleistein [2] output for synthetically generated data is depicted. In Bleistein and Cohen [9] both synthetic and real data examples are presented. As an example of computer processing time, we quote the result for one data set presented in that paper. For a set of 475 traces with 1024 data points per trace, approximately 240,000 output points were generated on a CDC Cyber 76 in under 10 minutes.

6. Recent Extension and Future Research. Recently [8], this method has been extended to a depth dependent background velocity. In this extensive, the two-way traveltime $2\rho/c$ in (23) is replaced by 2τ, the two-way traveltime on the geometrical optics ray connecting the output point at depth (x,z) with the integration point on the surface, $(\xi,0)$. This algorithm allows for "ray bending" in the propagation from the reflector to the surface. For steeply dipping reflectors in an otherwise stratified medium — such as on the flanks of a saltdome — this extension provides a significantly more accurate placement of reflectors. (See [8] for examples.) We view this extension as the first of a suite of methods dealing with progressively more complex background structures and with offset, as well. These extensions are made possible by introducing asymptotics earlier in the inversion procedure through use of a geometrical optics U_I in (13). We then invert the integral equation asymptotically, as well.

7. **Sound Speed Profile Inversion in the Ocean.** The method we have developed to invert the ocean sound speed profile is based on a typical ocean experimental situation. We have a point source located at depth z_S in the ocean waveguide and a point receiver located at the range and depth point (r,z). In general $z \neq z_S$. A time series of the data is recorded and Fourier transformed to k-space (wide-band data). An additional transform on this wide-band data, described below, yields the sound speed profile correction from an assumed profile guess. In the ocean the guess must be a function of depth in order to correctly represent, even in lowest order, the refraction paths on which the sound propagates [10], [11].

The two-dimensional acoustic velocity potential $\phi(r,z,z_S)$ is the solution of the Helmholtz equation (in range and depth) with a point source. Since the sound speed c(z) or index of refraction $n(z) = c_0/c(z)$ is only a function of depth, ϕ can be expressed using the Fourier-Bessel representation (see [12]).

$$\phi(r,z,z_S) = \frac{k^2}{4\pi} \int_C H_0^{(1)}(kr\beta) F(z,z_S,k,\beta)\beta d\beta \tag{25}$$

Here β is the square-root of the separation parameter, $H_0^{(1)}$ is the outgoing wave Hankel function, the contour C is along the upper branch of the Hankel function and closes in the upper half of the complex β-plane, and F is the solution of

$$\frac{d^2F}{dz^2} + k^2\left[n^2(z) - \beta^2\right] F = -\delta(z - z_S) \tag{26}$$

In the travel length coordinate

$$\tau(z) = \int_0^z \left[n^2(z') - \beta^2 \right]^{1/2} dz' \equiv \int_0^z f(z',\beta)dz' \tag{27}$$

the differential equation for F contains a first derivative term which is treated as a perturbation. Choosing the Green's function $G(\tau,\tau_S,\beta)$ to satisfy

(28) $$\frac{d^2G}{d\tau^2} + k^2G = -\delta(\tau - \tau_s)$$

and, using Green's theorem, yields an integral equation for $F(\tau,\tau_S,\beta)$ (the details can be found e.g. in [12]),

(29) $$F(\tau,\tau_S,\beta) = \frac{G(\tau,\tau_S,\beta)}{f(z_S,\beta)} + \int_0^d G(\tau,\tau',\beta)\left[\frac{d}{dz'}\ln f(z',\beta)\right]\frac{dF}{d\tau'}\,dz'$$

where d is the waveguide bottom.

If we define the reduced data D as

(30) $$D(k,r,z,z_S) = \phi(r,z,z_s) - \frac{k^2}{4\pi}\int_C \frac{H_0^{(1)}(kr\beta)G(\tau,\tau_S,\beta)}{f(z_S,\beta)}\,\beta\,d\beta$$

and use the Born (WKB) approximation

(31) $$\frac{dF}{d\tau'} \cong \frac{1}{f(z_S,\beta)}\frac{dG}{d\tau'} ,$$

in (29), and the far-field representation of the Hankel function, we get a linear transformation between the profile correction

(32) $$\varepsilon(z') = \frac{d}{dz'}n^2(z') ,$$

and the data D given by

33) $$D(k,r,z,z_S) = A(k,r)\int_0^d K(k,r,z,z',z_S)\varepsilon(z')dz'$$

where A is given by

(34) $$A(k,r) = k^{3/2}\frac{\exp\{i\pi/4\}}{\pi^{3/2}4(2r)^{1/2}}$$

and the kernel K of the transform is

$$(35)\quad K(k,r,z,z',z_S) = \int_C \frac{e^{ikr\beta}G(\tau,\tau',\beta)\,\frac{d}{d\tau'}\,G(\tau',\tau_S,\beta)\beta^{1/2}}{f(z_S,\beta)f^2(z',\beta)}\,d\beta \ .$$

To derive (33), substitute (31) into (29), and the result in (25) using (30). Choosing G to be the free-space WKB solution

$$(36)\qquad G(\tau,\tau',\beta) = -\,(2ik)^{-1}\,\exp[ik|\tau(z)-\tau(z')|]$$

we can then asymptotically evaluate the kernel K. It's integrand contains three branch points, two fixed (at $\beta = n(z)$ and $\beta = n(z_S)$) once we choose our input profile guess and given source and receiver positions, and one moving (at $\beta = n(z')$). The steepest descent calculation of K depends on the lowest value of the index of refraction at these three branch points and this effectively divides up the ocean into different horizontal inversion regions. The inversions in each region must be combined to yield the full inverted profile.

As an example, assume $n(z')$ is the minimum branch point. Asymptotically K is given by

$$(37)\qquad K(k,r,z,z',z_S) \sim F(z',z_S)\,\exp\{ik\Phi(r,z,z',z_S)\}$$

where

$$(38)\qquad F(z,z_S') = -\frac{2\pi i}{3}\,\frac{1}{\sqrt{n(z')}\,\sqrt{n^2(z_S)-n^2(z')}}\ ,$$

and

$$(39)\qquad \Phi(r,z,z',z_S) = rn(z') + \frac{2}{3}\,\mathrm{sgn}(z-z')\,\frac{\left[\,n^2(z)-n^2(z')\,\right]^{3/2}}{dn^2(z)/dz} - \frac{2}{3}\,\mathrm{sgn}\,(z'-z_S)\,\frac{\left[\,n^2(z_S)\,-\,n^2(z')\right]^{3/2}}{dn^2(z_S)/dz_S}\ .$$

To do the inversion, we substitute this result in (30), multiplying the result by $\exp\{-ik\Phi(r,z,\bar{z}',z_S)\}$ and integrate over the bandwidth of the data in k. The phase integral in k yields a single delta function in z' (if the phase is monotonic) and this is used to evalutate the z' integral. One result is an explicit representation for the profile correction in the region where the K asymptotics are valid

$$(40) \qquad \varepsilon(z') = \frac{\left|\frac{d\Phi}{dz'}\right|}{2\pi F(z',z_S)} \int e^{-ik\Phi(r,z,z',z_S)} \frac{D(k,r,z,z_S)}{A(k,r)}\, dk \ .$$

The final transform of the data in (40) we call a phase surface transform inversion. The result is iterative and the profile correction can be updated.

Present research consists in the computational implementation of (33) and (40), the latter to include all horizontal ocean regions. If ε is known, (33) can be used to generate (synthetic) data. This is used in the inversion (40) to test the stability of the algorithm. Of particular interest is the bandwidth of the data necessary to effect a successful inversion.

REFERENCES

[1] J.K. COHEN, N. BLEISTEIN, An inverse method for determining small variations in propagation speed, SIAM J. Appl. Math. 32, (1977), p. 4.

[2] J.K. COHEN, N. BLEISTEIN, Velocity inversion procedure for acoustic waves, Geophysics, 44, (1979a), p.6.

[3] N. BLEISTEIN, J.K. COHEN, Application of physical optics inversion scattering to non-destructive testing, Denver Research Institute Report, MS-R-8007, (1979b).

[4] J.K. COHEN, N. BLEISTEIN, Progress on a mathematical technique for non-destructive evaluation, Wave Motion, 2, (1980).

[5] J.K. COHEN, N. BLEISTEIN, The singular function of a surface and physical optics inversion scattering, Wave Motion, 1, (1979).

[6] N. BLEISTEIN, Mathematical methods for wave phenomena, Academic Press, San Diego, CA, 1984.

[7] N. BLEISTEIN, J.K. COHEN, F.G. HAGIN, Computational and asymptotic aspects of velocity inversion, Center for Wave Phenomena Report CWP-004, submitted to Geophysics, (1984).

[8] N. BLEISTEIN, S.H. GRAY, An extension of the born inversion method to a depth dependent reference profile, Center for Wave Phenomena Report CWP-007, submitted to Geophysical Prospecting, (1984).

[9] N. BLEISTEIN, J.K. COHEN, The velocity inversion problem: present status, new directions, Geophysics 47, November, (1982).

[10] J.A. DeSANTO, Theoretical methods in ocean acoustics, Ocean Acoustics, ed. by J. A. DeSanto, Topics in Current Physics, Vol. 8, Springer-Verlag, Heidelberg (1979).

[11] J.A. DeSANTO, Ocean acoustics, Encyclopedia of Physics, ed. by R.M. Besancon, and in press.

[12] J.A. DeSANTO, Oceanic sound speed profile inversion, IEEE J. Ocean Eng., OE-9, (1984), pp. 12-17.

ACOUSTIC TOMOGRAPHY

A. J. DEVANEY*

Abstract

The mathematical foundations of acoustic tomography are presented for weakly inhomogeneous objects whose compressibility and density fluctuations are sufficiently weak to admit the Rytov approximation for the forward transmitted acoustic field. The Rytov approximation for such objects is reviewed and shown to lead to a linear, nonlocal mapping between a certain linear functional of the density and compressibility fluctuations of the object and the complex phase of the transmitted field observed in tomographic experiments. In the short wavelength limit the mapping is shown to reduce to a *projection* of the sum of the density and compressibility fluctuations onto the measurement line and is, thus, invertible by the *filtered backprojection algorithm* of X-ray computed tomography. The generalized theory appropriate to finite wavelengths is presented and shown to lead to a "generalized" projection-slice theorem and a generalized reconstruction algorithm called the filtered back*propagation* algorithm. The filtered backprojection and backpropagation algorithms are applied to synthetic, computer simulated examples of acoustic tomography within the Rytov approximation. The filtered backprojection algorithm is found to yield high quality reconstructions only at wavelengths significantly shorter than the overall dimension of the object, while the filtered backpropagation algorithm is found to yield high quality reconstructions, independent of the wavelength of the probing acoustic field.

1.0 Introduction

The use of X-ray computed tomography is, by now, a widely known method for obtaining cross-sectional images of the human body [1]. Less well known are the applications of X-ray tomography in non-destructive evaluation of plastic, wood, and metal objects [2]. X-ray tomography has even been employed to obtain cross-sectional images of standing trees in a forest environment [3] and a commercial scanner for that purpose is currently under development [4].

Although X-ray tomography has been very successful in a number of medical and industrial applications, the method has several inherent drawbacks and deficiences. Among these is the invasive character of X-rays to living organisms. This limits the medical applications to areas of the body where the risk from moderate to high exposure is not great or to situations where the tomographic scan

* **Schlumberger-Doll Research, P.O. Box 307, Ridgefield, CT 06877.**

is deemed essential and worth the potential, long range effects caused by excessive exposure to X-rays. The invasive character of X-rays can also be a problem in industrial applications of X-ray tomography where the operators of the scan equipment run the risk of exposure to stray radiation. A second drawback of X-ray tomographic scanners are their size, complexity and cost. A full body "CAT" (Computer Aided Tomography) scanner can cost in excess of a million dollars. The size of the scanners make their use in applications requiring portability doubtful at best.

The limitations of X-ray tomographic scanners have motivated a number of researchers to investigate the use of ultrasound as a radiation source for tomography. Perhaps the earliest studies in this area were those of James Greenleaf and co-workers at the Mayo Clinic who investigated the use of ultrasound tomography for early diagnosis of breast cancer in women [5-7]. In these early investigations the ultrasound was considered simply as a replacement for the X-rays. The ultrasound wave was assumed to interact with an object in a manner completely analogous to the interaction occurring between an X-ray beam and an object. In place of the X-ray attenuation coefficient which is recovered in an X-ray scan, the ultrasound scan system would recover the acoustic attenuation coefficient [5]. However, because it is possible to measure both the amplitude and phase of an acoustic wave, the ultrasound scanner had the added potential attraction of yielding the phase velocity of the object as well as the attenuation coefficient [6].

The results obtained in the early investigations of ultrasound tomography were very discouraging. Rather than the crystal clear high resolution images associated with X-ray scanners, ultrasound tomographic scans produced blurred, "out of focus" images. One of the principal reasons for the poor results obtained with ultrasound was the lack of appropriate processing software. As mentioned above, it was naively assumed that an ultrasound beam would behave like an X-ray beam when passing through an object. Thus, the standard reconstruction algorithms of X-ray tomography such as the filtered backprojection algorithm [8,9] were employed to process the ultrasound and reconstruct the images of the object's acoustic attenuation profile and/or acoustic velocity profile. The ultrasound was not so accommodating, however, and did not obey the simple transport law obeyed by X-rays. Diffraction and coherent scattering play an important role in the propagation of acoustic signals through even weakly inhomogeneous objects whereas such effects are negligible for X-rays. The important role played by diffraction in ultrasound tomography has led to the name *diffraction tomography* being applied to it. It is important to note, however, that "diffraction tomography" is more general than ultrasound tomography since it encompasses any tomographic technique employing diffracting wavefields such as Coherent Optical Tomography [10].

A number of researchers have contributed to our current propagation models for ultrasound tomography and the resulting reconstruction algorithms. The earliest investigations appear to be those of Mueller and co-workers [11,12], Stenger and Johnson [13] and Iwata and Nagata [14]. These early investigations all make use of the work of Wolf [15] who formulated the optical inverse scattering problem for weakly inhomogeneous objects within the first Born approximation [16]. Mueller et. al. [11,12] and Iwata and Nagata [14] formulated the tomographic reconstruction problem both within the first Born approximation and the first Rytov approximation [17] and showed how both formulations fit into the theoretical framework developed earlier by Wolf [15]. It has since been determined both from theoretical and experimental considerations that the Rytov approximation is to be preferred over the Born approximation for plane beam, transmission tomography which is the type most often employed and which will be the subject of this paper.

The essential ingredient of the formulations of ultrasound tomography discussed above is the relationship existing between the complex phase of the acoustic field, assumed known (measured) over a plane surface lying outside the confines of the object being scanned, and the scattering potential associated with the acoustic object. For objects having constant density but space varying velocity this relationship is identical to that found by Wolf [15] for the Born approximation in the optical case: namely, *the two-dimensional Fourier transform of the observed complex phase is equal to the three-dimensional Fourier transform of the scattering potential evaluated over a hemisphere in Fourier space.* This relationship is a natural generalization of the *straight line* relationship that exists in Fourier space between the measured attenuation of an X-ray beam and the attenuation profile of an object. In the X-ray case the relationship is known as the *projection-slice* theorem. For obvious reasons the relationship in the case of ultrasound tomography has become known as the *generalized* projection-slice theorem.

The generalized projection-slice theorem paved the way for the development of ultrasound tomographic reconstruction algorithms. There exist essentially two classes of reconstruction algorithms: (1) Fourier space interpolation algorithms, and (2) the filtered backpropagation algorithm. The Fourier space algorithms [18,19] were the first to be employed and work essentially like the Fourier space interpolation algorithms of X-ray tomography [20]. These algorithms also suffer the same limitations of the analogous X-ray algorithms such as computational complexity, need of large core memory, and generally inferior image quality compared to the image quality obtained via the filtered backpropagation algorithm. The filtered back*propagation* algorithm [21,22] is the generalization of the filtered back*projection* algorithm of X-ray tomography to the case of ultrasound tomography. Like the filtered backprojection algorithm it requires only a small amount of memory, is easy to implement and produces very high quality images. However, because of the added complication introduced by the diffracting nature of ultrasound, the algorithm requires much more computation time than the X-ray algorithm and, for real time operation, would require an array processor or dedicated microchip.

In the present paper we present a tutorial review of ultrasound tomography. Our goal is to review the underlying wave model upon which the reconstruction algorithms are based and to compare and contrast this model and the filtered backpropagation algorithm to the simple straight line ray model and filtered backprojection algorithm of X-ray tomography. In the process we shall find that the theory and reconstruction algorithms of ultrasound tomography reduce identically to those of X-ray tomography as the wavelength of the ultrasound tends to zero.

Although the paper is, for the most part, a review of known material, some of the material presented is not widely known. Thus, for example, the formulation of the Rytov model for the propagation of the phase of the acoustic wave through the object presented in Section 2.0 allows the object's density to be variable. A similar treatment has been published for ultrasound tomography within the Born approximation [23] but the treatment of the variable density case within the Rytov approximation is new. A similar remark applies to the treatment of the generalized projection-slice theorem and filtered backpropagation algorithm presented in Section 4.0. Prior to this paper the generalized projection-slice theorem and filtered backpropagation algorithm were established only for constant density objects or for variable density objects within the Born approximation [23]. Also, the treatment of the short wavelength limit presented in Sections 3.0, and especially the interpretation of arrival time data presented in Section 3.3, appear to be new.

2.0 The Linearized Model

We consider the situation illustrated in Fig. 1 of an acoustic object immersed in a fluid bath having constant density ρ_0 and constant compressibility κ_0. The object is assumed to be insonified by a plane wave pressure pulse

$$P_0(\underline{r},t) = \frac{1}{2\pi} \int_{-\infty}^{\infty} d\omega \, P_0(\omega) e^{i(k\underline{s}_0 \cdot \underline{r} - \omega t)} \tag{2.1}$$

where $P_0(\omega)$ is the spectrum of the pulse, $k = \omega / C_0$ the fluid wavenumber, with $C_0 = \sqrt{1/\kappa_0 \rho_0}$ being the compressional wave velocity in the fluid, and $\underline{s}_0$ a unit vector in the direction of propagation of the plane wave. We consider only the two-dimensional case where the density and compressibility of the object vary as a function of in-plane coordinates x, y and the unit propagation vector $\underline{s}_0$ lies in this plane. The results are, however, readily extended to the full three-dimensional case following the treatment presented in Ref. 21.

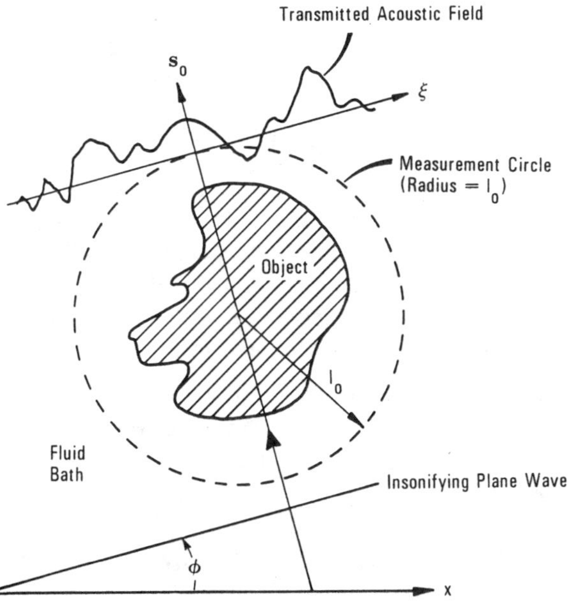

Figure 1: Experimental configuration for ultrasonic, parallel beam, transmission tomography.

The insonifying wave $P_0(\underline{r},t)$ interacts with the object to generate a total pressure field

$$P(\underline{r},t) = \frac{1}{2\pi}\int_{-\infty}^{\infty} d\omega\, P(\underline{r},\omega)\, e^{-i\omega t} \tag{2.2}$$

where the Fourier amplitude $P(\underline{r},\omega)$ of the induced pressure field $P(\underline{r},t)$ obeys the wave equation [23]

$$(\nabla^2 + k^2)\, P(\underline{r},\omega) = -k^2\gamma_\kappa(\underline{r},\omega)\, P(\underline{r},\omega) + \nabla\cdot[\gamma_\rho(\underline{r})\, \nabla\, P(\underline{r},\omega)] \tag{2.3}$$

where γ_r and γ_ρ are the normalized fluctuations in compressibility and density:

$$\gamma_\kappa \equiv \frac{\kappa(\underline{r},\omega) - \kappa_0}{\kappa_0} \tag{2.4a}$$

$$\gamma_\rho \equiv \frac{\rho(\underline{r}) - \rho_0}{\rho(\underline{r})}\ . \tag{2.4b}$$

In writing Eq. (2.3) that we have allowed for dispersion (viscoelasticity) by letting the (generally complex) compressibility $\kappa(\underline{r},\omega)$ and, hence, γ_κ be frequency dependent. Throughout the paper we will deal mainly with the Fourier amplitude $P(\underline{r},\omega)$ rather than with the time dependent field $P(\underline{r},t)$. For convenience, and when there is no possibility of confusion, we will refer to the Fourier amplitude as simply the "pressure field".

The tomographic reconstruction problem we address here can be stated as follows: A sequence of experiments is performed using insonifying plane waves having different directions of propagation $\underline{s}_0$. In each experiment, (i.e., for each valve of $\underline{s}_0$), the total pressure field $P(\underline{r},\omega)$ is measured over the line $\underline{r} = (\xi, \eta = l_0)$ which is perpendicular to $\underline{s}_0$ and tangent to a circle of radius l_0 which is centered within and completely surrounds the object. We wish to determine γ_κ and γ_ρ from this data. The experimental geometry is illustrated in Fig. 1.

The tomographic reconstruction problem as stated above is one form of the multidimensional *inverse scattering problem.* The exact formulation of this problem is essentially intractable due to the fact that the material parameters γ_κ and γ_ρ are *nonlinearly* and *nonlocally* related to the pressure field $P(\underline{r},\omega)$. If, however, the normalized fluctuations are small compared to unity, the problem can be linearized and approximate inversion schemes can be devised to "solve" this linearized problem.

There exist basically two ways in which the required linearization of the problem can be achieved: The *Born approximation* [16] and the *Rytov approximation* [17]. The first Born approximation results in a linear but nonlocal relationship between the pressure field $P(\underline{r},\omega)$ and γ_κ and γ_ρ. This approximation depends for its validity on both $|\gamma_\kappa|$ and $|\gamma_\rho|$ being small compared to unity and on the total scattering volume (volume of acoustic object) being small [25,26]. This second condition will generally not hold in tomographic applications where the object can be several hundred wavelengths in extent.

The Rytov approximation again requires that the normalized fluctuations be small compared to unity. However, for forward transmission problems employing plane wave insonification such as we have in the tomographic problem under consideration here, the condition that the scattering volume be small is removed [25,26]. Because of this, the Rytov approximation is to be preferred over the Born approximation and will be employed here. However, essentially all of our results, including the reconstruction alogrithms, can be readily modified to be applicable within the first Born approximation.

The Rytov approximation is not applied to the wave equation (2.3) but rather to the nonlinear Ricatti equation satisfied by the complex *phase* of the pressure field. In particular, if we substitute

$$P(\underline{r},\omega) \equiv P_0(\omega)\, e^{ikW(\underline{r},\omega)} \tag{2.5}$$

into Eq. (2.3) we obtain for $W(\underline{r},\omega)$ the Ricatti equation

$$(1-\gamma_\rho)\left[ik\nabla^2 W - k^2(\nabla W)^2\right] + k^2 = -k^2\gamma_\kappa + ik\nabla W \cdot \nabla\gamma_\rho \;. \tag{2.6}$$

To linearize Eq. (2.6) we replace γ_ρ and γ_κ by $\epsilon\gamma_\rho$ and $\epsilon\gamma_\kappa$, respectively, where ϵ is a strength parameter that will eventually be set equal to unity. If we now expand the phase $W(\underline{r},\omega)$ into a power series in ϵ viz.,

$$W(\underline{r},\omega) = \sum_{n=0}^{\infty} \epsilon^n\, W_n(\underline{r},\omega) \tag{2.7}$$

and substitute the expansion (2.7) into (2.6) we find that

$$ik\nabla^2 W_0 - k^2(\nabla W_0)^2 + k^2 = 0 \;, \tag{2.8a}$$

$$ik\nabla^2 W_1 - 2k^2\nabla W_0 \cdot \nabla W_1 = -k^2(\gamma_\kappa + \gamma_\rho) + ik\nabla W_0 \cdot \nabla\gamma_\rho \;. \tag{2.8b}$$

The first Rytov approximation consists of truncating (2.7) after the first two terms so that

$$W_R(\underline{r},\omega) = W_0(\underline{r},\omega) + \epsilon\, W_1(\underline{r},\omega) \tag{2.9}$$

where W_R is the Rytov approximation to W and where W_0 and W_1 are given by the solutions to Eqs. (2.8a) and (2.8b), respectively.

Eq. (2.8a) is readily shown to be the Ricatti equation satisfied by the phase of the insonifying plane wave propagating in the fluid bath. This conclusion follows immediately upon noting that Eq. (2.6) reduces to (2.8a) when γ_κ and γ_ρ are zero. Since the insonifying pressure wave is given by $P_0(\underline{r},\omega) = P_0(\omega)\exp(ik\underline{s}_0 \cdot \underline{r})$, we conclude that $W_0(\underline{r},\omega) = \underline{s}_0 \cdot \underline{r}$. On substituting $\nabla W_0 = \underline{s}_0$ into (2.8b) we then obtain for W_1 the equation

$$\frac{1}{2ik}\nabla^2 W_1 + \underline{s}_0 \cdot \nabla W_1 = \frac{1}{2}(\gamma_\kappa + \gamma_\rho) + \frac{1}{2ik}\underline{s}_0 \cdot \nabla \gamma_\rho \ . \tag{2.10}$$

To solve Eq. (2.10) we employ the Ansatz [17]

$$W_1(\underline{r},\omega) \equiv e^{-ik W_0} U(\underline{r},\omega) = e^{-ik\underline{s}_0 \cdot \underline{r}} U(\underline{r},\omega). \tag{2.11}$$

Substituting (2.11) into (2.10) and simplifying the resulting equation yields

$$(\nabla^2 + k^2) U(\underline{r},\omega) = [ik(\gamma_\kappa + \gamma_\rho) + \underline{s}_0 \cdot \nabla \gamma_\rho] e^{ik\underline{s}_0 \cdot \underline{r}} \tag{2.12}$$

from which is follows immediately that

$$W_1(\underline{r},\omega) = -\frac{i}{4}\int d^2 r' \{ik[\gamma_\kappa(\underline{r}',\omega) + \gamma_\rho(\underline{r}')] + \underline{s}_0 \cdot \nabla' \gamma_\rho(\underline{r}')\} \tag{2.13}$$

$$X\, H_0(k|\underline{r} - \underline{r}'|)\, e^{-ik\underline{s}_0 \cdot (\underline{r} - \underline{r}')}$$

where $H_0(kR)$ is the zero order Hankel function of the first kind.

Eq. (2.13) is the fundamental equation underlying the theory of acoustic tomography presented in this paper. It is seen to relate a certain linear functional of the compressibility and density fluctuations to the (measured) complex phase $W_1(\underline{r},\omega)$. In the usual treatments of acoustic tomography [11-13,18,19,21,22] the density is assumed to be constant so that γ_ρ vanishes and W_1 is then linearly related to γ_κ alone. We shall find that the presence of γ_ρ does not introduce any essential difficulties in the mathematical treatment of the inverse problem, but does affect the type of information generated by the reconstruction algorithms. For example, in the constant density case it is possible to (approximately) recover the fluctuation in acoustic velocity from a set of tomographic experiments while the same is not true in the variable density case. Rather, in this latter case one reconstructs a certain linear functional of the density and compressibility fluctuations from which the velocity fluctuation cannot, in general, be uniquely inferred.

3.0 Conventional Computed Tomography

3.1 Filtered Backprojection Algorithm

In this section we consider the case where the wavelength $\lambda = 2\pi/k$ tends to zero. This limiting case is best treated starting from Eq. (2.10). In particular, we find in this limit that Eq. (2.10) reduces to

$$\underline{s}_0 \cdot \nabla W_1^{(0)} = \frac{1}{2}(\gamma_k + \gamma_\rho) \ , \tag{3.1}$$

where $W_1^{(0)}$ stands for the short wavelength limit of W_1. If we now introduce the ξ, η coordinate system which is obtained from the x, y system via a counter clockwise rotation of ϕ degrees (see Fig. 1) we then have that $\underline{s}_0 \cdot \nabla = \partial/\partial\eta$ so that Eq. (3.1) becomes

$$\frac{\partial}{\partial\eta} W_1^{(0)}(\xi,\eta,\omega) = \frac{1}{2}[\gamma_\kappa(\xi,\eta,\omega) + \gamma_\rho(\xi,\eta)] \tag{3.2}$$

where we have expressed $W_1^{(0)}$, γ_κ, and γ_ρ in the ξ,η coordinate system. We note for future reference that the x,y and ξ, η systems are related via the set of equations

$$\xi = x\cos\phi + y\sin\phi \ , \tag{3.3a}$$

$$\eta = -x\sin\phi + y\cos\phi \ . \tag{3.3b}$$

Eq. (3.2) is readily integrated and we find that

$$W_1^{(0)}(\xi,\eta = l_0,\omega) = \frac{1}{2}\int_{-\infty}^{l_0} d\eta' [\gamma_\kappa(\xi,\eta',\omega) + \gamma_\rho(\xi,\eta',\omega)]. \tag{3.4}$$

Eq. (3.4) states that $W_1^{(0)}$ is proportional to the *projection* of the sum of the compressibility and density fluctuations onto the measurement line $\eta = l_0$. Since this measurement line rotates about the object, remaining always perpendicular to $\underline{s}_0$, it is then possible to obtain projections of this sum onto lines having different orientations simply by varying the direction of propagation $\underline{s}_0$ of the insonifying plane wave. In the short wavelength limit the acoustic tomographic reconstruction problem thus reduces to the reconstruction problem of conventional computed tomography; i.e., of reconstructing a function from an ensemble of its projections.

The data $W_1^{(0)}$ yields information only about the *sum* of the compressibilty and density fluctuations. Consequently, only this sum can be reconstructed from the available data in the short wavelength limit. We shall see that this is not necessarily true at finite wavelengths where data taken at more than one frequency can in some cases yield separate reconstructions of γ_κ and γ_ρ.

The reconstruction problem of conventional computed tomography (CT) is best treated in Fourier transform space. In particular, if we perform a one-dimensional Fourier transform with respect to the ξ coordinate of both sides of Eq. (3.4) we find that

$$\tilde{W}_1^{(0)}(p,\eta=l_0,\omega) \equiv \int_{-\infty}^{\infty} d\xi' W_1^{(0)}(\xi',\eta=l_0,\omega)e^{-ip\xi'}$$

$$= \int_{-\infty}^{\infty} d\eta' \int_{-\infty}^{\infty} d\xi' O(\xi',\eta',\omega)\, e^{-ip\xi'}. \tag{3.5}$$

In writing Eq. (3.5) we have introduced the "object profile" $O(\xi,\eta,\omega)$ defined to be

$$O(\xi,\eta,\omega) \equiv \frac{1}{2}[\gamma_\kappa(\xi,\eta,\omega) + \gamma_\rho(\xi,\eta)] \tag{3.6}$$

and have replaced the upper limit of integration in the η' integral by ∞ since $O(\xi,\eta,\omega) \equiv 0$ if $\xi^2+\eta^2 > l_0^2$.

Eq. (3.5) states that the two-dimensional Fourier transform of the object profile, viz.,

$$\tilde{O}(\underline{K},\omega) \equiv \int d^2r'\, O(\underline{r}',\omega)\, e^{-i\underline{K}\cdot\underline{r}'} \tag{3.7}$$

evaluated on the K_ξ axis (i.e., for $\underline{K}=p\hat{\underline{\xi}}$) is equal to the one-dimensional Fourier transform of $W_1^{(0)}(\xi,\eta=l_0,\omega)$. Since, according to Eq. (3.4), $W_1^{(0)}(\xi,\eta=l_0,\omega)$ is equal to a *projection* of $O(\underline{r},\omega)$ we can state this result as follows: *the one-dimensional Fourier transform of a projection of the object profile $O(\underline{r},\omega)$ is equal to a slice through the two-dimensional Fourier transform of $O(\underline{r},\omega)$*. This result is known in X-ray computed tomography as the *Projection-Slice Theorem* and forms the basis of the filtered backprojection reconstruction algorithm of conventional CT [8,9]. The theorem is illustrated in Fig. 2 where we depict a two-dimensional Fourier transform cut by a "slice" which makes the angle ϕ with the fixed K_x Coordinate axis.

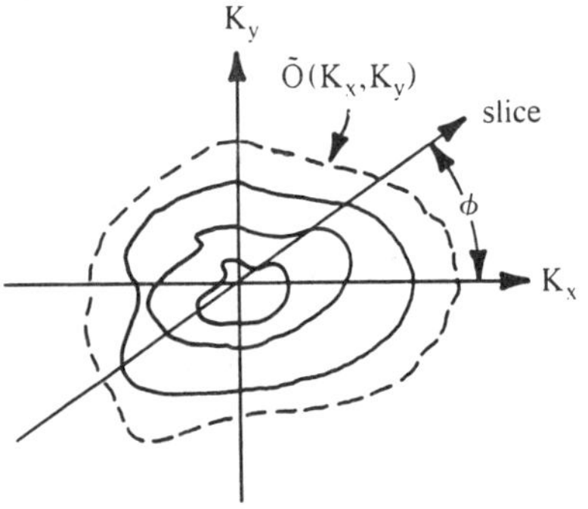

Figure 2: The Projection-Slice Theorem relates the one-dimensional Fourier transform of the data to the two-dimensional Fourier transform of the object profile evaluated over a straight line "slice" in two-dimensional transform space.

The filtered backprojection algorithm follows immediately from the *projection-slice theorem* and the Fourier integral representation of $O(\underline{r},\omega)$ expressed in circular cylindrical coordinates. In particular, we write:

$$O(\underline{r},\omega) = \frac{1}{2(2\pi)^2}\int_{-\pi}^{\pi} d\phi \int_{-\infty}^{\infty} dp|p|\tilde{O}(\underline{K},\omega)e^{i\underline{K}\cdot\underline{r}} \tag{3.8}$$

where p, ϕ are the polar coordinates of $\underline{K}$; i.e.,

$$K_x = p\cos\phi\ , \tag{3.9a}$$

$$K_y = p\sin\phi\ , \tag{3.9b}$$

and where we have extended the p range of integration to $-\infty$ for later convenience. For fixed ϕ, $\tilde{O}(\underline{K},\omega)$ in the p integrand of (3.7) is simply the "slice" of $\tilde{O}(\underline{K},\omega)$ making the angle ϕ with the positive K_x axis. By making use of the projection-slice theorem, viz a vie, Eq. (3.5) we can then express Eq. (3.8) in the form:

$$O(\underline{r},\omega) = \frac{1}{2(2\pi)^2}\int_{-\pi}^{\pi} d\phi \int_{-\infty}^{\infty} dp|p|\tilde{W}_1^{(0)}(p,\eta = l_0,\omega)\, e^{ip[x\cos\phi + y\sin\phi]}\ . \tag{3.10}$$

It is customary to decompose Eq. (3.10) into two successive operations: *convolutional filtering* of $W_1^{(0)}$ followed by *backprojection.* We define the *filtered projections* of $O(\underline{r},\omega)$ by:

$$Q_\phi(t) \equiv \frac{1}{2\pi}\int_{-\infty}^{\infty} dp|p|\tilde{W}_1^{(0)}(p,\eta = l_0,\omega)e^{ipt}$$

$$= \int_{-\infty}^{\infty} d\xi' W_1^{(0)}(\xi',\eta = l_0,\omega)h(t-\xi')\ , \tag{3.11}$$

where

$$h(t) \equiv \frac{1}{2\pi}\int_{-\infty}^{\infty} dp|p|e^{ipt}\ . \tag{3.12}$$

In terms of the filtered projections $Q_\phi(t)$, Eq. (3.10) becomes:

$$O(\underline{r},\omega) = \frac{1}{4\pi}\int_{-\pi}^{\pi} d\phi Q_\phi(x\cos\phi + y\sin\phi)\ . \tag{3.13}$$

The filtered projections $Q_\phi(t)$ are thus obtained by simply convolutionally filtering the phase $W_1^{(0)}$ with the filter $h(t)$. The filtered projections are then *backprojected* into the object space (the x,y plane). The backprojection operation consists simply of assigning all points along the *straight line* ray path $t = x\cos\phi + y\sin\phi$ the value of the filtered projection $Q_\phi(t)$. Finally, the totality of filtered backprojections are linearly superimposed to generate the final reconstruction of the object profile. The algorithm is called the *filtered* [Eq. (3.11)] *backprojection* [Eq. (3.13)] algorithm.

In practice, only a limited number of experiments can be performed so that only a limited number of projections will be available. The integral over filtered backprojections (3.13) is then replaced by a finite sum over the angles ϕ at which the projections are known [9]. In addition, these projections will not be known exactly but will be both bandlimited and corrupted by noise and measurement error. To account for noise and bandlimiting of the data the filter function h(t) defined in Eq. (3.12) is modified to become [9]:

$$\hat{h}(t) = \frac{1}{2\pi}\int_{-\infty}^{\infty} dp|p|g(p)e^{ipt} \tag{3.14}$$

where g(p) is a filter that is selected to minimize noise and Gibbs phenomena caused by bandlimiting of the data.

3.2 The Object Profile

The filtered backprojection algorithm reconstructs the object profile defined in Eq. (3.6). This quantity can be related to the acoustic index of refraction $n(\underline{r},\omega)$ of the object. In particular, on substituting for γ_κ and γ_ρ from Eqs. (2.4) we find that

$$\begin{aligned}\gamma_\kappa + \gamma_\rho &\equiv \frac{\rho_0}{\rho}[\frac{\kappa\rho}{\kappa_0\rho_0} - 1] \\ &= \frac{\rho_0}{\rho}[n^2 - 1] \ ,\end{aligned} \tag{3.15}$$

where the index of refraction is given by

$$n(\underline{r},\omega) = \sqrt{\kappa\rho/\kappa_0\rho_0} = \frac{C_0}{C(\underline{r},\omega)} \tag{3.16}$$

and where $C(\underline{r},\omega) = \sqrt{1/\kappa\rho}$ is the (possibly complex) phase velocity of the acoustic wave in the object. Since the Rytov approximation requires that κ and ρ be close to the background values κ_0 and ρ_0, we have that $n(\underline{r},\omega) \approx 1$ so that (3.15) reduces to

$$\gamma_\kappa + \gamma_\rho \approx 2\frac{\rho_0}{\rho}\delta n(\underline{r},\omega) \tag{3.17}$$

where

$$\delta n(\underline{r},\omega) = n(\underline{r},\omega) - 1 \tag{3.18}$$

is the fluctuation in the index of refraction. Substituting (3.17) into (3.6) then yields

$$O(\underline{r},\omega) \approx \frac{\rho_0}{\rho}\delta n(\underline{r},\omega) \ . \tag{3.19}$$

3.3 Arrival Time Data

The filtered backprojection algorithm requires as input data the phase $W_1^{(0)}(\xi,\eta=l_0,\omega)$. this quantity can be obtained from $P(\underline{r},\omega)$ via the equation

$$W_1^{(0)}(\xi,\eta=l_0,\omega) = \lim_{\lambda\to 0}\frac{1}{ik}\ln\left[\frac{P(\xi,\eta=l_0,\omega)}{P_0(\xi,\eta=l_0,\omega)}\right], \tag{3.20}$$

where ln(x) denotes the natural logarithm of x. Note, however that Eq. (3.20) will generate the principle value of the real part of $W_1^{(0)}$, whereas the filtered backprojection algorithm requires that $W_1^{(0)}$ be known as a continuous function of position. This means that a *phase unwrapping* algorithm [27] will be required to obtain the unwrapped phase needed in the algorithm.

In the short wavelength case there exists an alternative, and much simpler, method of determining the phase term $W_1^{(0)}$ as long as the object is essentially non-dispersive (non-viscoelastic) and non-attenuating over the frequency band of the insonifying pressure pulse. If these conditions hold we see from Eq. (3.4) that $W_1^{(0)}$ is purely real and independent of frequency. Within the Rytov approximation then

$$P(\xi,\eta=l_0,\omega) = P_0(\omega)e^{i[k\underline{s}_0\cdot\underline{r}_0 + W_1^{(0)}(\xi,\eta=l_0)]} \tag{3.21}$$

and, thus,

$$P(\xi,\eta=l_0,\ t) = P_0[\ \xi,\eta=l_0,\ t-\tau(\xi)] \tag{3.22}$$

where $P_0(\underline{r},\ t)$ is the *time dependent* insonifying pressure pulse defined in Eq. (2.1) and

$$\tau(\xi) = \frac{1}{C_0} W_1^{(0)}(\xi,\eta=l_0) \; . \qquad (3.23)$$

We conclude from the above that the phase $W_1^{(0)}(\xi,\eta=l_0)$ is directly proportional to the relative phase delay $\tau(\xi)$ experienced by the insonifying wave in propagating through the object. The phase delay is readily measured at each location ξ along the measurement line so that Eq. (3.22) provides a direct and simple method of obtaining the data required in the filtered backprojection algorithm in cases where the object's index of refraction is essentially real and non-dispersive over the frequency band of the insonifying wave.

3.4 Examples

In this Section we illustrate the use of the filtered backprojection algorithm on computer generated (synthetic) data. We use a circular version of the well-known Shepp and Logan head phantom as the object. This phantom, which is shown in Fig. 3, consists of a superposition of ten overlapping circular disks having radii which vary from a smallest of 1.15 units to a largest of 40.25 units for the outer disk. The density of the phantom is everywhere constant and equal to that of the background (bath) density so that γ_ρ is identically zero for the phantom. The phase of the transmitted field is generated within the Rytov approximation from Eq. (2.13) and can, thus, be expected to be realistic for weakly inhomogeneous objects satisfying the conditions required by the approximation. A full discussion of the computations required for generating the data is given in Ref. [22].

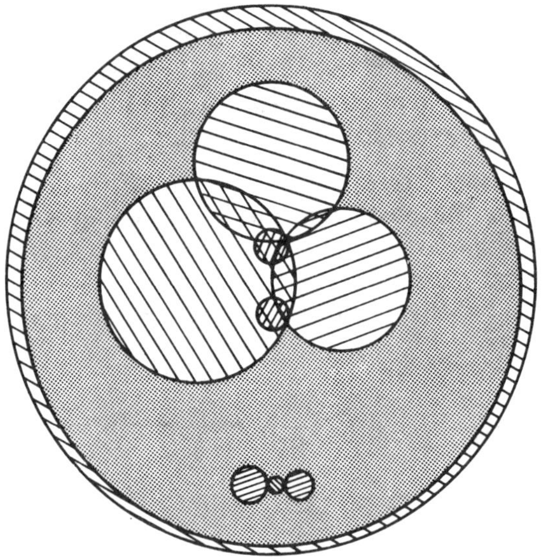

Figure 3: The circular Shepp and Logan head phantom. The outer circle has a radius of 40.25 units while the smallest circular disk (lower center) has a radius of 1.15 units. (This figure is reprinted from Ref. 22)

Shown in Fig 4 are reconstructions obtained using the filtered backprojection algorithm on *zero wavelength* data for 1,7,11, and 101 views equally spaced over 360 degrees. For this case, then, the data was generated using the short wavelength model ,Eq. (3.1), for which the filtered backprojection algorithm yields an (in principle) exact reconstruction. The high quality of the reconstructions for this zero wavelength case is evident from the figure where the only apparent image artifacts in the reconstructions are streaks caused by the use of an insufficient number of views.

We applied the filtered backprojection algorithm to the same object, but where we increased the wavelength to finite values. The results shown in Fig. 5 are for wavelengths of 0.1, 0.5, 1.0, and 3.0 units and for 101 views. The overall size of the images shown in these figures are 128X128 units, with the overall radius of the head phantom being 40.25 units and the radius of the *smallest* circular disk in the phantom being 1.15 units. Thus, the wavelengths used in the simulations are quite small compared to the overall size of the object and correspond to a/λ values (a being radius) of greater than 0.3 for all parts of the object. Even for such small wavelengths the reconstructions are seen to be considerably inferior to the 101 view, zero wavelength case shown in Fig. 4.

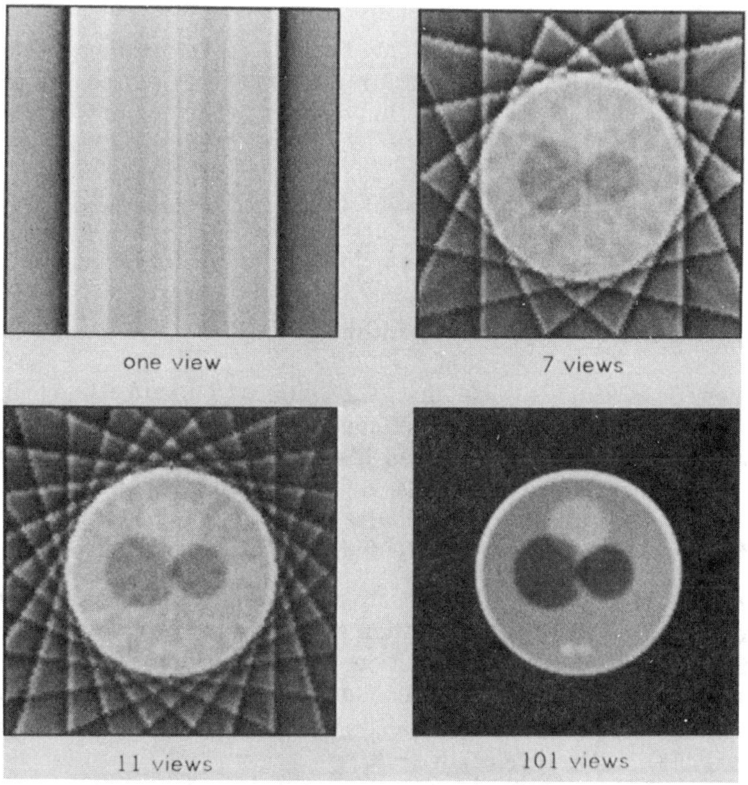

Figure 4: Reconstructions from the filtered backprojection algorithm for the zero wavelength case. (This figure is reprinted from Ref. 22).

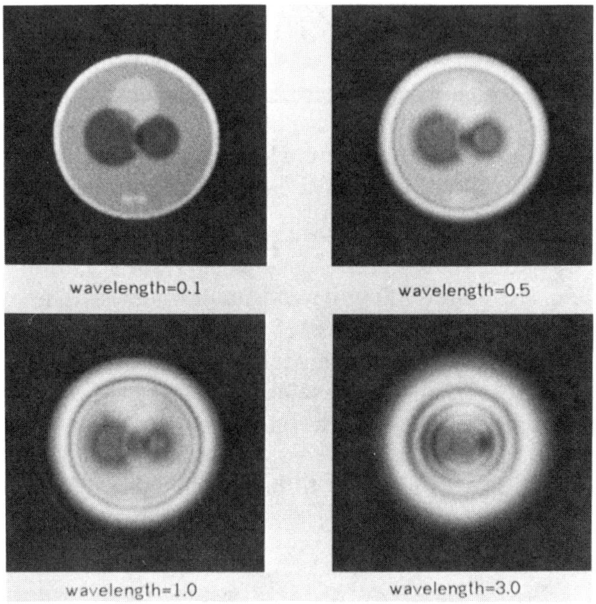

Figure 5: Reconstructions from filtered backprojection of finite wavelength data. Each reconstruction employed 101 views equally spaced over 360 degrees. (This figure is reprinted from Ref. 22).

4.0 Diffraction Tomography

4.1 Filtered Backpropagation Algorithm

We showed in Section 3.0 that the filtered backprojection algorithm of conventional computed tomography is applicable to acoustic tomography at very short wavelengths. However, we found that as the wavelength begins to approach the scale over which the object varies, this algorithm ceases to yield high quality reconstructions. This result is, of course, due to the inadequacy of the short wavelength model (3.1), upon which the filtered backprojection algorithm is based.

In order to obtain a reconstruction algorithm appropriate for finite wavelengths we must employ Eq. (2.10) as the equation governing the evolution of the phase W_1 through the acoustic object. The solution to this equation is given by Eq. (2.13) which, then replaces Eq. (3.4) as the fundamental equation relating the quantities γ_κ and γ_ρ to the observed phase $W_1(\xi,\eta=l_0,\omega)$. Eq. (3.4) will, of course, arise as the limit of Eq. (2.13) as the wavelength $\lambda \rightarrow 0$.

In treating the short wavelength case, we found it convenient to Fourier transform both sides of Eq. (3.4) with respect to the ξ coordinate. We can perform the same operation with Eq. (2.13) if we make use of the plane wave expansion of the zero order Hankel function [28]:

$$H_0(k|\underline{r}-\underline{r}'|) = \frac{1}{\pi}\int_{-\infty}^{\infty}\frac{dp}{m}\, e^{i[p(\xi-\xi')+m|\eta-\eta'|]} \tag{4.1}$$

where

$$m = \begin{cases} \sqrt{k^2-p^2} & \text{if} \quad k \geqslant p \\ i\sqrt{p^2-k^2} & \text{if} \quad k<p \ . \end{cases} \tag{4.2}$$

On making use of Eqs. (2.13) and (4.1) we find that

$$\tilde{W}_1(p,\eta=l_0,\omega) \equiv \int_{-\infty}^{\infty} d\xi \; W_1(\xi,\eta=l_0,\omega)e^{-ip\xi}$$

$$= -\frac{i}{2m}e^{i(m-k)l_0}\int d^2r'[ik(\gamma_\kappa+\gamma_\rho)+\underline{s}_0\cdot\nabla'\gamma_\rho]e^{-ik(\underline{s}-\underline{s}_0)\cdot\underline{r}'} \tag{4.3}$$

where we have introduced the unit vector $\underline{s}$ defined as

$$\underline{s} \equiv \frac{p}{k}\hat{\underline{\xi}} + \frac{m}{k}\hat{\underline{\eta}} , \tag{4.4}$$

where $\hat{\underline{\xi}}$ and $\hat{\underline{\eta}}$ are unit vectors along the ξ, η axes, respectively.

Eq. (4.3) can be simplified by integrating by parts. We find that

$$\tilde{W}_1(p,\eta = l_0,\omega) = \frac{k}{2m}e^{i(m-k)l_0}\int d^2r'[\gamma_\kappa+\underline{s}_0\cdot\underline{s}\,\gamma_\rho]e^{-ik(\underline{s}-\underline{s}_0)\cdot\underline{r}'}$$

$$=\frac{k}{2m}e^{i(m-k)l_0}\{\tilde{\gamma}_\kappa[k(\underline{s}-\underline{s}_0),\omega] + \underline{s}_0\cdot\underline{s}\;\tilde{\gamma}_\rho[(\underline{s}-\underline{s}_0)]\} , \tag{4.5}$$

where

$$\tilde{\gamma}_\kappa(\underline{K},\omega)\equiv\int d^2r'\gamma_\kappa(\underline{r}',\omega)e^{-i\underline{K}\cdot\underline{r}'} , \tag{4.6a}$$

$$\tilde{\gamma}_\rho(\underline{K})\equiv\int d^2r'\gamma_\rho(\underline{r}')e^{-i\underline{K}\cdot\underline{r}'} \tag{4.6b}$$

are the two-dimensional Fourier transforms of γ_κ and γ_ρ, respectively. Finally, by making use of the identity

$$\underline{s}\cdot\underline{s}_0 \equiv 1-\frac{1}{2}(\underline{s}-\underline{s}_0)^2 \tag{4.7}$$

we write (4.5) in the form

$$\tilde{W}_1(p,\eta=l_0,\omega) = \frac{k}{m}e^{i(m-k)l_0}\{\tilde{O}[k(\underline{s}-\underline{s}_0),\omega]$$

$$-\frac{1}{4}(\underline{s}-\underline{s}_0)^2\tilde{\gamma}_\rho[k(\underline{s}-\underline{s}_0)]\} \tag{4.8}$$

where $\tilde{O}(\underline{K},\omega)$ is the two-dimensional Fourier Transform of the "object profile" $O(\underline{r},\omega)=\frac{1}{2}(\gamma_\kappa + \gamma_\rho)$ as defined in Eq. (3.7).

Eq. (4.8) relates the one-dimensional Fourier transform of $W_1(\xi,\ \eta=l_0,\omega)$ to the two-dimensional Fourier transform of a certain linear functional of $O(\underline{r},\omega)$ and $\gamma_\rho(\underline{r})$. In particular, if we define the *generalized object profile* $F(\underline{r},\omega)$ as:

$$F(\underline{r},\omega)\equiv O(\underline{r},\omega) + \frac{1}{4k^2}\nabla^2\gamma_\rho(\underline{r}) \tag{4.9}$$

we can write Eq. (4.8) in the form:

$$\tilde{W}_1(p,\eta=l_0,\omega)=\frac{k}{m}e^{i(m-k)l_0}\tilde{F}[k(\underline{s}-\underline{s}_0),\omega] \tag{4.10}$$

where

$$\tilde{F}(\underline{K},\omega)\equiv\int d^2r\ F(\underline{r},\omega)e^{-i\underline{K}\cdot\underline{r}} \tag{4.11}$$

is the two-dimensional Fourier transform of the generalized object profile.

Eq. (4.10) is the generalization of Eq. (3.5) to the finite wavelength case. Like Eq. (3.5), the above equation has a simple geometrical interpretation in Fourier transform space. In particular, Eq. (4.10) states that the one-dimensional Fourier transform of $W_1(\xi,\eta=l_0,\omega)$ is proportional to the two-dimensional Fourier transform of the generalized object profile $F(\underline{r},\omega)$ evaluated over the set of points defined by:

$$\underline{K}\equiv k(\underline{s}-\underline{s}_0)\ , \tag{4.12}$$

where the unit vector $\underline{s}$ assumes all real values for which $\underline{s}\cdot\underline{s}_0 > 0$. This locus of points defines a semicircle in Fourier transform space centered at $\underline{k}=-k\underline{s}_0$ and having a radius equal to k. This result is illustrated in Fig. 6 where we depict a two-dimensional Fourier transform cut by a semicircular "slice" corresponding to a single unit propagation vector $\underline{s}_0$.

The relationship between W_1 and F embodied in Eq. (4.10) has become known as the *generalized projection - slice theorem.* The rational for this name derives from the relationship's geometrical interpretation in Fourier transform space as outlined above and illustrated in Fig. 6. Note that as the wavelength $\lambda \rightarrow 0$ ($k \rightarrow \infty$), the semicircular "slice" shown in Fig. 6 degenerates to a straight line slice and the figure becomes identical to Fig. 2. In this limit, the generalized object profile $F(\underline{r},\omega)$ tends to the object profile $O(\underline{r},\omega)$ and Eq. (4.10) then reduces to Eq. (3.5). The *generalized projection-slice theorem thus reduces to the conventional projection-slice theorem in the short wavelength limit.*

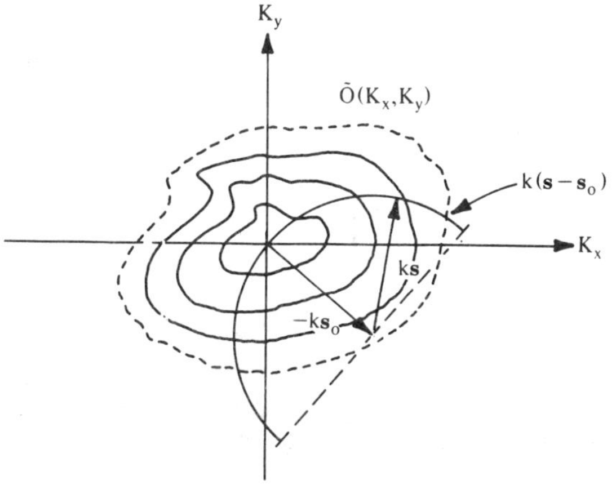

Figure 6: The Generalized Projection-Slice Theorem relates the one-dimensional Fourier transform of the data to the two-dimensional Fourier transform of the object profile evaluated over a semicircular "slice" in two-dimensional transform space.

We showed in Section 3.0 that the filtered backprojection algorithm followed directly from the Fourier integral representation of the object profile expressed in polar coordinates. The reason for this is, of course, that in the short wavelength limit the data (measured phase of the acoustic field) specifies the transform of the object profile along the radial axis in $\underline{K}$ space (projection - slice theorem). The Fourier integral representation of the object profile when expressed in polar coordinates can then be evaluated directly in terms of the data, i.e., in terms of $\tilde{W}_1^{(0)}(p,\eta=l_0,\omega)$.

A similar situation is encountered in the finite wavelength case where the data specifies the transform of the generalized object profile over semicircular arcs in $\underline{K}$ space (generalized projection-slice theorem). In place of the Fourier integral representation expressed in polar coordinates we then use the Fourier integral representation expressed in the orthogonal coordinates appropriate to these semicircular arcs. The required representation is found to given by [21]:

$$F_{LP}(\underline{r},\omega) = \frac{k}{2(2\pi)^2}\int_{-\pi}^{\pi} d\phi \int_{-k}^{k} \frac{dp}{m} |p| \tilde{F}[k(\underline{s}-\underline{s}_0),\omega] e^{i[p\xi+(m-k)\eta]} \qquad (4.13)$$

where ϕ is the angle made by the tangent to the semicircular arc with the positive x axis and $\underline{s}$ is defined in terms of p and m according to Eq. (4.4). The subscript "LP" on $F(\underline{r},\omega)$ in (4.13) means that the representation generates the *low pass filtered approximation*

$$F_{LP}(\underline{r},\omega) = \frac{1}{(2\pi)^2}\int_{|\underline{K}|\le\sqrt{2}k} d^2K\ \tilde{F}(\underline{K},\omega) e^{i\underline{K}\cdot\underline{r}} \qquad (4.14)$$

to the generalized object profile. This is, of course, a consequence of the fact that the semicircular arcs shown in Fig. 6 only sweep out the interior of a circle in Fourier space of radius $\sqrt{2}k$. If we substitute for $\tilde{F}[k(\underline{s}-\underline{s}_0),\omega]$ from Eq. (4.10) we find that Eq. (4.13) becomes

$$F_{LP}(\underline{r},\omega) = \frac{1}{2(2\pi)^2}\int_{-\pi}^{\pi} d\phi \int_{-k}^{k} dp\, |p|\, \tilde{W}_1(p,\eta=l_0,\omega) e^{i[p\xi+(m-k)(\eta-l_0)]} \ . \qquad (4.15)$$

Eq. (4.15) can be decomposed into three successive operations in analogy with what was done in the short wavelength case. The first operation is identical to that employed earlier, i.e., convolutional filtering of $W_1(\xi,\eta=l_0,\omega)$. In particular, in terms of the filtered projections $Q_\phi(t)$ defined in Eq. (3.10) with $W_1^{(0)}$ replaced by W_1 we find that Eq. (4.15) becomes

$$F_{LP}(\underline{r},\omega) = \frac{1}{4\pi}\int_{-\pi}^{\pi} d\phi \int_{-\infty}^{\infty} d\xi' Q_\phi(\xi') G(\xi-\xi',\eta-l_0) \ , \qquad (4.16)$$

where

$$G(\xi,\eta) \equiv \frac{1}{2\pi}\int_{-k}^{k} dp\ e^{i[p\xi+(m-k)\eta]} \ . \qquad (4.17)$$

The integral over $d\xi'$ in Eq. (4.16) is, for fixed η, a convolution of the filtered projections $Q_\phi(\xi')$ with the quantity $G(\xi-\xi',\eta-l_0)$. This latter quantity can be shown to be the Rytov approximation to the Green function governing the backward propagation of phase within the fluid background [21]. Thus, the integral

$$\Pi_\phi(\xi,\eta) \equiv \int_{-\infty}^{\infty} d\xi' Q_\phi(\xi') G(\xi-\xi',\eta-l_0) \qquad (4.18)$$

is interpreted as the *backpropagation* of the filtered phase $Q_\phi(\xi')$ into the region of space containing the object. The final reconstruction of F_{LP} is then obtained by summing the filtered and backpropagated phases over view angles:

$$F_{LP}(\underline{r},\omega)=\frac{1}{4\pi}\int_{-\pi}^{\pi} d\phi\ \Pi_\phi(x\cos\phi+y\sin\phi;-x\sin\phi+y\cos\phi) \tag{4.19}$$

where we have expressed ξ, η in terms of the Cartesian coordinates x, y according to Eqs. (3.3).

The reconstruction algorithm embodied in Eqs. (4.18) and (4.19) is called the *filtered backpropagation* algorithm. It differs from the filtered back*projection* algorithm discussed in the preceeding section only in the replacement of the backprojection operation by backpropagation. In this connection we note that as $\lambda \to 0$ $(k \to \infty)$, $G(\xi,\eta) \to \delta(\xi)$ and thus

$$\lim_{\lambda\to 0}\Pi_\phi(\xi,\eta) \to Q_\phi(\xi) = Q_\phi(x\cos\phi+y\sin\phi) \ . \tag{4.20}$$

The filtered backpropagation algorithm thus reduces to the filtered backprojection algorithm in the zero wavelength limit.

4.2 Examples

The filtered backpropagation algorithm was tested in computer simulations within the Rytov approximation on the modified Shepp and Logan head phantom employed in the examples presented in Section 3.4. In Fig. 7 we show reconstructions corresponding to a wavelength of unity and 1, 7, 11, and 101 views. Comparing these to the analogous reconstructions presented in Fig. 4 for the zero wavelength case, we see a rather noticeable increase in image quality in the finite wavelength reconstructions over those obtained in the zero wavelength case. This increase in image quality can be attributed to a "focusing" effect associated with the backpropagation process that is absent in backprojection [22]. Note that since γ_ρ is zero for the phantom, the generalized object profile reduces to the object profile so that the reconstructions obtained with the filtered backpropagation algorithm are of the object profile.

Shown in Fig. 8 are reconstructions for 101 views and wavelengths of 0.1, 0.5, 1 and 5 units. Comparing these to the reconstructions obtained with the filtered backprojection algorithm shown in Fig. 5 we see that except for the shortest wavelength, filtered backpropagation yields considerably higher image quality than filtered backprojection. This is, of course, to be expected since the filtered backpropagation algorithm is the generalization of the filtered backprojection algorithm to the finite wavelength case.

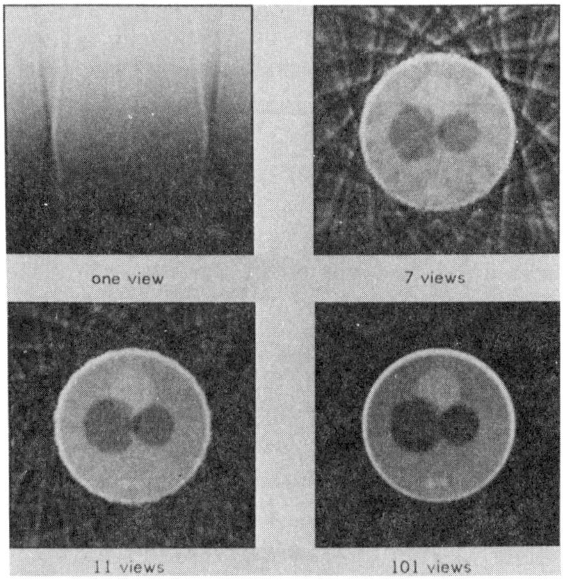

Figure 7: Reconstructions from filtered backpropagation for a wavelength of unity. (This figure is reprinted from Ref. 22).

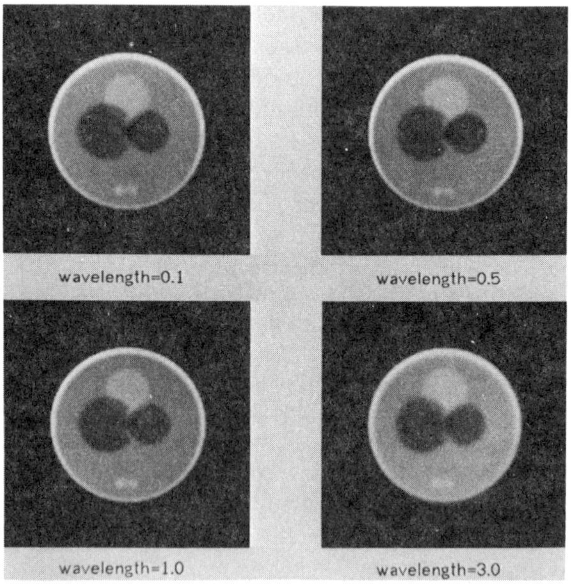

Figure 8: Reconstructions from filtered backpropagation employing 101 views. (This figure is reprinted from Ref. 22).

4.3 Recovering density and Compressibility from Acoustic Tomography.

The filtered backpropagation algorithm reconstructs the generalized object profile defined in Eq. (4.9). In the constant density case this quantity reduces to the object profile which, in the constant density case, is proportional to the normalized fluctuation in compressibility (cf., Eq. (3.6)). Thus, in this case acoustic tomography yields the compressibility (or, equivalently, the velocity) of the object. The question arises of the possibility of separately recovering the compressibility and density fluctuations when the density is not constant.

We found in Section 3.0 that this is not possible in the short wavelength limit since, in this limit, the transmitted field depends on the object profile and, hence, on the *sum* of density and compressibility fluctuations. However, in the finite wavelength case the transmitted field depends on the generalized object profile which, in turn, is *not* simply a sum of the density and compressibility fluctuations. In particular, we see on referring to Eq. (4.9) that the deviation of the generalized object profile from the object profile is inversely proportional to the square of the wavenumber (or, equivalently, frequency) of the acoustic field. Thus, by performing reconstructions at different wavenumbers it may be possible separately determine the object profile and the density fluctuation from which it is a simple matter to determine the compressibility and density. This possibility was investigated in a recent study of variable density acoustic tomography [29]. It was found that separate determination of density and compressibility is indeed possibility so long as the compressibility is non-dispersive over the frequency band of the insonifying wave. Moreover, filtered backpropagation algorithms were presented in [29] that yield density and compressibility from the Fourier amplitude of the transmitted acoustic field evaluated at two distinct frequencies. We refer the interested reader to Ref. 29 for details of the method.

REFERENCES

1. W.S. Swindell and H.H. Barrett, *Computerized tomography: taking sectional x-rays*, Physics Today **30** (1977), pp.32-41.

2. F.F. Hopkins, I.L. Morgan, H.D. Ellinger, R.V. Klinksiek, G.A. Meyer, and J.N. Thompson, *Industrial tomography applications*, IEEE Trans. Nuc. Sci. **NS-28** (1981), pp. 1717-1720.

3. M. Onoe, J. Wen Tsao, H. Yamada and M. Yoshimatsu, *Computed tomography for use on live trees*, Materials Evaluation **41** (1983), pp. 748-749.

4. M. Onoe, J.W. Tsao, H. Yamada, H. Nakamura, J. Kogure, J. Kawamura, and M. Yoshimatsu, *Computed tomography for measuring annual rings of a live tree*, Proceedings of the IEEE **71** (1983), pp. 907-908.

5. J. F. Greenleaf, S. A. Johnson, S. L. Lee, G. T. Herman, and E. H. Wood, *Algebraic reconstruction of spatial distributions of acoustic absorption with tissues from their two-dimensional acoustic projections*, in *Acoustical Holography*, Vol.5, P. S. Green, Ed. New York: Plenum (1974), pp. 591-603.

6. J. F. Greenleaf, S. A. Johnson, W. F. Samayoa, and F. A. Duck, *Algebraic reconstruction of spatial distributions of acoustic velocities in tissue from their time-of-flight profiles*, in *Acoustical Holography*, Vol.6, N. Booth, Ed. New York: Plenum (1975), pp. 71-90.

7. J. F. Greenleaf, S. A. Johnson, and A. H. Lent, *Measurement of spatial distribution of refractive index in tissues by ultrasonic computer assisted tomography*, Ultrasound Med. Biol. **3** (1978), pp. 327-339.

8. A. C. Kak, *Computerized Tomography with X-ray Emission and Ultrasound Sources*, Proc. IEEE **67** (1979), pp. 1245-1272.

9. L.A. Shepp and J.B. Kruskal, *Computerized tomography: the new medical X-ray technology*, Amer. Math. Soc. Monthly, June-July (1978), pp .420-439.

10. A.J. Devaney, *Inverse scattering as a form of computed tomography*, Proc. of the SPIE **358**, ed. W. Carter (SPIE, Bellingham, Wash., 1982), pp. 10-16.

11. R. K. Mueller, M. Kaveh, and G. Wade, *Reconstructive tomography and applications to ultrasonics*, Proc. IEEE **67** (1979), pp. 567-587.

12. R. K. Mueller, M. Kaveh, and R. D. Inverson, *A new approach to acoustic tomography using diffraction techniques*, in *Acoustical Imaging*, Vol. 8, A. F. Metherell, Ed. New York: Plenum, 1980, pp. 615-628.

13. F. Stenger and S. A. Johnson, *Ultrasonic transmission tomography based on the inversion of the Helmholtz equation for plane and spherical wave insonification*, Appl. Math. Notes **4** (1979), pp. 102-127.

14. K. Iwata and R. Nagata, *Calculation of refractive index distribution from interferograms using the Born and Rytov's approximation*, Japan J. Appl. Phys. **14** Suppl. 14-1 (1974), pp. 379-383.

15. E. Wolf, *Three-dimensional structure determination of semi-transparent objects from holographic data*, Opt. Commun. **1** (1969), pp .153-156.

16. L. Schiff, *Quantum Mechanics*, 3rd ed. New York: McGraw-Hill (1968), Ch. 8.

17. V. T. Tatarski, *Wave Propagation in a Turbulent Medium*, New York: McGraw-Hill (1961), Chap. 7.

18. S.X. Pan, and A.C. Kak, *A computational study of reconstruction algorithms for diffraction tomography*, IEEE Trans. Acoustics, Speech and Signal Processing **ASSP-31** (1982), pp. 1262-1275.

19. M. Soumekh and M. Kaveh, *Image reconstruction from frequency domain data on arbitrary contours*, in *Proc. ICASSP 84* (1984), pp.12A.2.1-12A.2.4.

20. R.M. Mersereau and A.V. Oppenheim, *Digital reconstruction of multidimensional signals from their projections*, Proc. IEEE **62** (1974), pp. 1319-1338.

21. A.J. Devaney, *A filtered backpropagation algorithm for diffraction tomography*, Ultrasonic Imaging **4** (1982), pp. 336-350.

22. A.J. Devaney, *A computer simulation study of diffraction tomography*, IEEE Trans. Biomed. Eng.**BME-30** (1983), pp. 377-386.

23. A.J. Devaney, *Inverse source and scattering problems in ultrasonics*, IEEE Trans. Sonics and Ultrasonics **SU-30** (1983), pp. 355-364.

24. A.J. Devaney and G. Beylkin, *Diffraction tomography using arbitrary transmitter and receiver surfaces*, Ultrasonic Imaging **6** (1984), pp. 181-193.

25. J.B. Keller, *Accuracy and validity of the Born and Rytov approximations*, J. Opt. Soc. Am. **59** (1969), pp. 1003-1004.

26. A.J. Devaney, *Inverse scattering theory within the Rytov approximation*, Opt. Letts. **6** (1981), pp. 374-376.

27. J.M. Tribolet, *A new phase unwrapping algorithm*, IEEE Trans. Acoustics, Speech and Signal Processing **ASSP-26** (1977), pp. 170-177.

28. P.M. Morse and H. feshbach, *Methods of Mathematical Physics*, New York: McGraw-Hill (1953), p.823.

29. A.J. Devaney, *Variable density acoustic tomography*, J. Acoustic Soc. Am. (submitted).

INVERSE PROBLEMS OF ACOUSTIC AND ELASTIC WAVES

YIH-HSING PAO*, FADIL SANTOSA* AND WILLIAM W. SYMES**

Abstract. We summarize results obtained by our research team on inverse source problems and inverse medium problems in wave propagation. We pay particular attention to the ill-posed nature of the problems, to the characterization of regularizing sets that restore continuous dependence of solutions on data, and to the prescription of *a priori* knowledge concerning physically consistent solutions as constraints used in constructing such regularizing sets. We give the results of numerical simulations and laboratory experiments which illustrate our theoretical results.

0. Introduction. The objective of our collaborative research effort is to study inverse problems that arise in acoustic and elastic wave propagation. In this report, we describe some results of our work, details of which have appeared or will appear in papers published elsewhere.

Inverse problems in wave propagation can be separated into two categories: inverse medium problems and inverse source problems. In the former, we seek to determine the mechanical properties of an inhomogeneous medium from knowledge of its response at the boundary due to a known excitation. The inverse source problems are ones in which we determine the nature of a radiating wave source from boundary data.

* Department of Theoretical and Applied Mechanics, Cornell University, Ithaca, NY 14853, USA.

** Department of Mathematical Sciences, Rice University, Houston, TX 77001, USA.

This work is sponsored by the Office of Naval Research under SRO-III grant number N000-14-83-K0051.

Our research activities were directed at studying several model problems arising in applications. The results presented here are concerned primarily with the inverse medium problem for acoustic waves under several simplifying assumptions, and the inverse point source problem in elasticity. Most inverse problems that arise in physical situations are in fact illposed, that is, the solution does not depend continuously on the data. Our interest lies in understanding precisely the characteristics of such problems, and in determining the least stringent conditions needed to be imposed to restore the continuity of the solution on the data. These facts are essential in designing constructive inversion schemes that are feasible for real data processing.

Before proceeding with the description of the illposedness and their sources, we discuss the continuum equations under consideration. The propagation of waves in an acoustic medium may be modelled by the equation for pressure $u=u(t,x,y,z)$,

$$\underline{\nabla} \cdot (\frac{1}{\rho} \underline{\nabla} u) - \frac{1}{\rho c^2} \partial_t{}^2 u = f . \tag{0.1}$$

Here, $\rho(x,y,z)$ and $c(x,y,z)$ are density and wave speed, and $f(t,x,y,z)$ is the forcing function. Many geometrical configurations can be envisioned; in particular, we use the half-space geometry to model the ocean or the Earth. If the wave speed is constant, say $c=1$, then we have the straight ray model, and setting $A=\frac{1}{\rho}$, we have

$$\underline{\nabla} \cdot (A \underline{\nabla} u) - A\partial_t{}^2 u = f . \tag{0.2}$$

Multi-dimensional inverse problems for systems governed by (0.2) are considered here, as well as problems for which ρ and c are functions of z only.

The equations for the propagation of elastic waves in an isotropic medium are given by

$$\begin{aligned} &\underline{\nabla} \cdot \underline{\tau} - \rho\partial_t{}^2 \underline{u} = \underline{f}, \\ &\underline{\tau} = \lambda (\underline{\nabla} \cdot \underline{u}) + \mu \left(\underline{\nabla}\, \underline{u} + (\underline{\nabla}\, \underline{u})^t\right) . \end{aligned} \tag{0.3}$$

Eliminating the stress tensor $\underline{\tau}$ which satisfies Hooke's law with Lame constant λ and μ, we obtain an equation governing the displacement field $\underline{u} = (u_x, u_y, u_z)$. Equations (0.1) - (0.3) are further subjected to boundary and initial conditions.

We define the model space **M** for the medium problem as the space of parameters such as (ρ, c) in (0.1), A in (0.2) and (ρ, λ, μ) in (0.3). The data space **D** is the space of measured or measurable data such as $u(t,x,y,z)$ or $u_t(t,x,y,z)$ on $\Sigma x[0,T]$ for (0.1)-(0.2), and $\underline{u}$ or $\underline{u}_t$ on $\Sigma x[0,T]$ for (0.3). Naturally we have assumed some known source of energy generated the measured pressures or displacements. The set Σ is the measurement set, usually a set of points on the surface of a half-space, and $[0,T]$ is the time duration of the collection of data.

The inverse source problem can also be cast in precisely the same manner. Identify the model space as the space of sources $f[\underline{f}]$ in (0.1) [(0.3)], the data space is as described above. For inverse source problems, the coefficients in the partial differential equation (0.1)-(0.3) are assumed known.

We describe the forward map as a map taking a point in **M** to a point in **D**. Naturally, this map is characterized by the partial differential equation which models the physical phenomena, e.g., (0.1)-(0.3). The forward map may be symbolically expressed as

(0.4) $$F(m) = d, \text{ where } m \in \mathbf{M} \text{ and } d \in \mathbf{D}.$$

In this notation, the inverse problem amounts to solving the functional equation (0.4).

Since all measured data are contaminated by noise and so they may not lie in the range of the operator F, a more practical statement of the problem, instead of (0.4), is

(0.5) $$\| F(m) - d \| \leq \varepsilon ,$$

where the norm is some suitable measure of fit and ε is a tolerance related to the assumed noise level in the data. Of course, the inequality (0.5) has many solutions, if it has any at all. Worse, the set of solutions of (0.5) may contain vastly disparate elements even when (0.4) yields unique solutions. In that case we say that the problem is illposed; the solution of (0.4) or (0.5) is inordinately sensitive to noise in the data, and assessment of the information content of the data becomes a major difficulty.

The illposedness or lack of continuity can be described alternatively as follows. Suppose we have two models m_1 and m_2, and

(0.6) $$\| F(m_1) - F(m_2) \| = O(\varepsilon),$$

This means that the pair of models fit the data given to an order of ε. Lack of continuity implies that

$$(0.7) \qquad \|m_1 - m_2\| = O(1).$$

It means that we have a set of models which fit the data to the same level of tolerance, but two elements in this set may be vastly different. Since the information content of data is understood as the amount of information about the model that can be inferred from the data, (0.6) - (0.7) state that the data reveals very little information about the model.

Suppose we are given a point $d \in \mathbf{D}$, and the set of solutions in the model space $\mathbf{M}$ that satisfy (0.5) is $\mathbf{M^*}$, that is

$$(0.8) \qquad \|F(m) - d\| \leq \varepsilon \text{ for all } m \in \mathbf{M^*} .$$

One way to represent the information content of the data of an illposed inverse problem is to choose from the set $\mathbf{M^*}$, which may be large, a subset of models which varies uniformly continuously with the data d, by imposing additional <u>regularizing constraints</u> on the solution m. That is, identify the set R such that

$$(0.9) \qquad \|F(m_1) - F(m_2)\| = O(\varepsilon) \text{ for } m_1 \text{ and } m_2 \in \mathbf{R} ,$$

imply

$$(0.10) \qquad \|m_1 - m_2\| = O(\varepsilon).$$

Once this is done, we can, for instance, pose the following as the inverse problem.

$$(0.11) \qquad \underset{m \in \mathbf{R}}{\text{minimize}} \ \|F(m) - d\|.$$

Regularization is necessary also for compelling computational reasons. All inverse medium problems under consideration by our group, with the partial exception of impedance profile inversion, are optimization (best-fit) problems, <u>by necessity</u>: i.e. the nature of the problems prevents exact matches of noisy data by <u>any</u> model (examples of this feature will be discussed below).

Algorithms for optimization typically depend for their convergence on the <u>condition</u> of (some approximation to) the derivative of the relevant map. When the problems are ill-posed, their derivatives are typically ill-conditioned. Thus, regularization is necessary not only to ensure the stability of the (theoretical) parameter estimates, but also to allow convergence of optimization algorithms, i.e. prevent unstable iterative processes.

Such a remedy for the lack of continuity is well known in the field of illposed problems. We wish to characterize the largest set **R** for each problem that would render (0.9) - (0.10) valid. Generally, smaller **R** leads to more continuous constrained optima in (0.11), whereas larger **R** allows better fit, that is smaller residual in (0.11).

For the inverse problems arising in wave propagation, we have found several sources of illposedness. These fall roughly into four categories:

(i) phase-shifting due to change of coordinates
(ii) band-limited (incomplete) data
(iii) insensitivity of the data to out-of-aperture plane wave components
(iv) near dependence of data in multi-parameter estimation

We shall discuss in some detail example problems in which each of these phenomena occur.

The report is organized into five sections. In the next section, we discuss the one-dimensional inverse problem for recovering wave speed profiles. The problems of impedance profile inversion from band limited data is presented in Section 2. Section 3 is devoted to the study of multidimensional inverse problem for waves modeled by equation (0.2). The illposed problem of complete reconstruction of density and velocity profiles in a layered acoustic medium is described in Section 4. In the final section, we present some results for an inverse point source problem of elastic waves based on synthetic and experimental data.

The other participants in our research team are Lawrence Payne, Paul Sacks, Timothy Clarke, Gabriel Raggio, and Jennifer Michaels.

In this report, we have only selected a few topics from the rather diverse scope of our collaborative research activity, which are consistent with the theme of the conference. We regret that we are not able to report all their contributions in this paper.

1. One-dimensional Velocity Profile Inversion. The inverse problem for one-dimensional profile inversion is:

Given that $u = u(t,z)$ solves

(1.1) $(\partial_t{}^2 - c^2(z)\partial_z{}^2)u=0, \quad z\geq 0, \quad t>0,$

(1.2) $u(t,z) = 0, \quad \text{for } t<0,$

(1.3) $\partial_z u \ (t,0) = \delta(t).$

Determine $c(z)$ from $G(t)= u(t,0)$, $t>0$.

Gerver (1970) showed results which imply that the map from $G(t)$ in L^2 to $c(z)$ in L^2 is not uniformly continuous. The results are best phrased in terms of the reflectivity, defined as

(1.4) $r(z) = c'(z).$

We have shown that $r(z)$ is in L^2 if and only if the velocity response

(1.5) $g(t) = G'(t), \quad t>0,$

is in L^2. However, small L^2-errors in $g(t)$ can result in arbitrarily large L^2-errors in $r(z)$. On the other hand, if r is converted to a function of travel time s defined by

(1.6) $$s(z) = \int_0^z \frac{1}{c(\zeta)}\, d\zeta ,$$

(1.7) $$\tilde{r}\,(s) = r\bigl(z(s)\bigr),$$

we have shown (e.q. Symes 1983) that the map from $r \in L^2$ to $g \in L^2$ is continuous. Thus, the illposed nature of the r to g problem is due entirely to the phase shift inherent in the coordinate change from s to z. This difficulty is present in any problem involving non-constant signal propagation speeds, that is, in nearly all practical inverse problems. A constructive example of this instability is presented in Coen and Symes (1984).

We have found (Symes 1984a) that the one-dimensional velocity inversion can be regularized by requiring the derivative of the reflectivity to be bounded _a priori_ in L^2. Define the regularizing set as

(1.8) $$\mathbf{R} = \left\{ r : \int \left| r'(z)\right|^2 dz \leq K \right\}$$

The solution r of the least squares problem

$$(1.9) \qquad \min_{r \varepsilon R} \|F(r) - g\|$$

then depends (uniformly) continuously with g in the sense of L^2. We have identified the map F as the "forward" map taking r to $\partial_t u(t,0)$ via (1.1) - (1.3).

Note that the data g is an impulse response, and hence is band unlimited in Fourier components. Nonetheless, the set R in (1.8) amounts to a band limitation on the reflectivity estimate, as the derivative of r has been constrained. Thus, we have traded resolution for stability, which is a common practice in many better known illposed problems such as backwards diffusion and analytic continuation. The loss of resolution is unavoidable if stability is to be maintained. The quantitative extent and practical importance of this resolution/stability trade-off await further investigation.

2. Impedance Profile Inversion from Band Limited Data. We have reexamined the problem of inverting bandlimited response data. The displacements u=u(t,s) satisfy

$$(2.1) \qquad \left(A(s)\partial_t u - \partial_s A(s)\partial_s u\right) = 0, \quad s>0.$$

The system which is initially quiescent, is excited at the boundary by

$$(2.2) \qquad \partial_s u(t,s=0) = f(t),$$

The response to f(t) is

$$(2.3) \qquad \partial_t u(t,s=0) = g(t).$$

The problem is to find the impedance A(s) from $\{f(t), g(t)\}$.

Much is known about this inverse problem for the case when f(t) is impulsive, i.e. $f(t)=\delta(t)$, and g(t) is the impulse response. Consider now the situation where f(t) is a band limited function. Deconvolution of f(t) from g(t) by Fourier division within the passband produces a band limited impulse response. To this band limited impulse response, we can add Fourier components whose frequencies are outside the passband, and still satisfy the observed data pair $\{f(t), g(t)\}$ to within a small error. This implies that there are infinitely many profiles A(s) which match the observation to within a small tolerance.

Perturbations of the band limited impulse response in various regions outside the passband leads to several types of ambiguity in inversion.

High-frequency error in data seems to lead only to high-frequency error in the impedance, at least for impedances of modest variation. For instance, suppression of high-frequency data components (smoothing) before inversion yields a smoothed impedance profile, in all computational instances of which we are aware. This effect is displayed in Figures 1 and 2. The calculated impulse response of the medium with impedance profile shown in dots in Figure 2 is filtered using the low pass filter shown in Figure 1. The filtered data when inverted yields the profile in Figure 2 (shown in solid line). We conjecture that smoothing the data before inversion is always equivalent to smoothing the impedance. We can prove this statement for impedance with square-integrable reflectivities. Granted the uniqueness of the inversion problem, we can also prove the statement for media of bounded variation.

It is generally believed that low-frequency error in data results only in low-frequency error in the impedances, i.e. in loss of trend information. This band-separation hypothesis seems not to have been tested in numerical experiments, and one of our current projects is to test it. In fact, we have observed that low-frequency error can also lead to spurious interfaces in the profile, provided that a high-frequency band-limit is also present. The narrow-band phenomenon occurs for quite realistic passbands. For instance, we exhibit two profiles in Figure 3 which give almost identical responses (Figure 4). The source wavelet is the upside-down version of the first segment of the time trace in Figure 4. The traces are 4 seconds long, the impedance profiles 2 (one-way travel time) seconds long, and the wavelet has most of its energy in teh 10-40 HZ band, all quite realistic numbers for reflection seismology. Observe that a false "interface" is introduced near s=1, where a multiple reflection has been misinterpreted.

Clearly the bandlimited reflection data do not determine even the gross features of the profile. Thus additional information must be supplied to select a single profile from the large set of profiles fitting the data. Levy and Fullagar (1981) and Oldenburg (1981 and 1984) have suggested selecting the profile which fits the data and simultaneously has least total variation i.e. minimizes the L^1-norm of the reflectivity. They observed, and we proved (Santosa and Symes, 1983), that the least-variation profile is associated with a sparse spike-train reflectivity (if one exists which fits the data) at least under certain circumstances. Such a profile thus matches the data with clearly defined layers and interfaces, which might perhaps aid in geological interpretation of the data.

Levy and Fullagar (1981) and Oldenburg (1981 and 1984) use the convolutional model, that is, the linearization of the impedance/data relation about constant impedance. This approach introduces inaccuracy by ignoring multiple reflections, so we approached the nonlinear problem directly (see Santosa and Symes, 1983 for a comparison). In the (nonlinear) inversion problem, the L^1-norm of the reflectivity can be minimized by a restricted gradient algorithm. The results of these simulations are displayed in Figures 5 and 6. The band limited source is the upside down version of the first wavelet in the time trace in Figure 5. The response, g(t) is shown in Figure 5. The target profile has L^1-norm in reflectivity $\|r\|=2.30$. Our algorithm began with $\|r\|=4.18$ where the missing components in the deconvolved impulse response were set to zero, and found an optimum at $\|r\|=2.43$ (discrepancies are partly due to discretization errors). The iterative process in seeking the optimum is displayed in Figure 6. Further work is being carried out for tackling noisy data problems. In the case when the data is noisy, a penalty method, such as that used by Santosa (1984) for deconvolution, is preferred.

3. Multi-Dimensional Inverse Medium Problem. We have studied the following multi-dimensional inverse problem:

Find $A(x,z)$, $x \in R^{n-1}$, $z>0$, from the data

(3.1) $$u(t,x,0) = G(t,x),\ t>0,\ x \in R^{n-1}$$

where $u = u(t,x,z)$ solves

(3.2) $$(A\partial_t^{\,2} - \underline{\nabla} \cdot A\ \underline{\nabla})u = 0,\ z>0,\ t>0,\ x \in R^{n-1}$$

(3.3) $$u(t,x,z) = 0, \text{ for } t<0,$$

(3.4) $$\partial_z u(t,x,0) = \delta(t).$$

The origin of equation (3.2) has been discussed in the Introduction. The problem (3.1) - (3.4) is the closest multi-dimensional analogue to the impedance profile inversion (for m=1 above). In particular, the signal velocity has been normalized to 1 which means that a) the phase-shift instability (Section 1) does not occur, and b) the rays of geometric optics are straight lines, the incident wave-front is the hyperplane $\{t=z\}$, and no caustics develop.

To place the multi-dimensional results in context, we first summarize the results for m=1 (impedance profile inversion) as given by Symes (1984a). As in Section 1, we work with reflectivities $r = (\log A)'$, and velocity impulse response, $g(t) = u(t,0)$ (observe that z plays the role of travel time coordinate, denoted by s elsewhere in this report). The equations for m=1 are, $u=u(t,z)$

(3.5) $\quad (\partial_t^2 - \partial_z^2 - r\partial_z)u = 0,$

(3.6) $\quad u(t,z) = 0, \; t<0,$

(3.7) $\quad \partial_z u(t,0) = \delta(t),$

and the data

(3.8) $\quad \partial_t u(t,0) = g(t).$

Denote the map from reflectivities r(z) to impulse response g(t) by F (forward map). The following facts hold:

i) F is a map from $L^2\,[0,T]$ to $L^2\,[0,2T]$
ii) F is a differentiable. This means that the formal linearization DF of the map actually approximates F locally. This is a rigorous and precise statement on linearization, or the validity of Born approximations about arbitrary reference reflectivity
iii) F is a C^2-diffeomorphism (F and its inverse are twice-differentiable maps).

In particular, F is very well conditioned:

$$\|F(r_1)-F(r_2)\|_2 < \varepsilon, \; r_1, \; r_2 \; \varepsilon L^2[0,T]$$

imply

$$\|r_1 - r_2\|_2 < K\varepsilon$$

where K is a function of $\max(\|r_1\|, \; \|r_2\|)$.

The results for n>1 are not so nice. Define the "reflectivity" by

$$r = \underline{\nabla} \log A$$

We should regard the surface values A(x,0) as data for this problem. Thus knowledge of r is equivalent to knowledge of A.

As before, set the forward map

(3.9) $F(r) = \partial_t u(t,x,0)$

We examine the formal linearization: $A = A_0 + A_1$,

(3.10) $DF(r_0) \cdot r_1 = \partial_t u_1(t,x,0)$,

where u_1 solves the perturbational BVP:

(3.11) $(\partial_t^2 - \nabla^2 - r_0 \cdot \underline{\nabla})\, u_1 = r_1 \cdot \underline{\nabla}\, u_0$,

(3.12) $u_1 = 0, t<0$,

(3.13) $\partial_z u_1(t,x,0) = 0$,

and r_1, A_1 are related by

$$A_1(x,0) = 0,$$

$$r_1 = \underline{\nabla}\left(\frac{A_1}{A_0}\right).$$

We have produced simple examples which show the illconditioned nature of DF(0) ("Born approximation") (Symes 1984b):

(i) A sequence $\{r_1^j\}$ for which

$$\|DF(0)\cdot r_1^j\| \to 0$$

but $\|r_1^j\| = 1$

(ii) A sequence $\{r_1^j\}$ for which

$$\|DF(0)\cdot r_1^j\| = 1$$

but $\|r_1^j\| \to 0.$

Thus, neither DF nor DF^{-1} are bounded maps on L^2. Consequently, any Newton-like optimization strategy for minimizing

$$(3.14) \qquad \|F(r) - g\|_2$$

is doomed to failure.

The reason for this instability is simple: the $\{r_1^j\}$ are concentrated in steeply dipping Fourier components of increasing (spatial) frequency as $j \to \infty$. These components effectively reflect high-frequency energy along horizontal rays, which either do not influence the trace at all, or have enormous boundary values, depending upon the construction. Such examples model steeply dipping interfaces, which are well-known to be more difficult to discern on seismic records than horizontal interfaces.

Although the technical details are tedious, and will not be repeated here, the regularization strategy is, in essence equally simple: Constrain r to have a priori bounded Fourier components of high horizontal wave number, in one way or another.

To obtain differentiability of the forward map, we can, for instance, require that some number of horizontal (x-) derivatives of r be square-integrable:

$$(3.15) \qquad \|r\|^2_k := \int \left| \nabla_x^k r \right|^2 < \infty.$$

On the space of r defined by this condition, F is differentiable, that is,

$$(3.16) \qquad \left\| \frac{1}{\varepsilon} \left(F(r_0 + \varepsilon r_1) - F(r_0)\right) - DF(r_0)\cdot r_1 \right\| \to 0$$

as $\qquad \varepsilon \to 0 \quad$ if $\quad \|r_0\|_k, \ \|r_1\|_k < \infty.$

Also, the derivative is a bounded operator: that is,

$$(3.17) \qquad \|DF(r_0) \cdot r_1\| \leq C(\|r_1\|^2_k + \|r_1\|^2_o)^{1/2},$$

where C depends on $\|r_0\|_k$.

For the inverse problem, we need a lower bound for DF. The examples cited above show that we cannot expect a lower bound to hold without some extra constraint on r_1. The best results of our investigation (Sacks and Symes 1984), concern the case in which r_0 depends on z alone (stratified reference medium). Then:

if $\|r_1\|_1 \leq \gamma \|r_1\|_0$

then $\|DF(r_0) \cdot r_1\| \geq c \|r_1\|_0$

where C depends on r_0, γ.

If r_0 is not stratified, our results are much harder to state and to interpret except in special cases. In general, if we demand less horizontal (x-) smoothness of the reference r_0, we must require more horizontal smoothness of r_1 to obtain lower bounds. The most stringent requirement of horizontal smoothness is that $\nabla_x r_0=0$, i.e. r_0 is stratified. Then we obtain the result stated above, merely under the assumption that $\|r_0\|_1 \leq \|r_1\|_0$. If we do not force r_0 to be stratified, then we must require more horizontal smoothness of both r_0 and r_1, and moreover consider the set of <u>horizontally bandlimited</u> r:

$$\Sigma(\Omega) = \{r : \hat{r}(\xi,z) = 0 \text{ if } |\xi|>\Omega\}$$

where $r(\xi,z)$ denotes the (partial) Fourier transform of $r(x,z)$ in x. For r_0, $r_1 \in\Sigma(\Omega)$, we have

$$\|DF(r_0)\cdot r_1\| \geq C \|r_1\|$$

where C depends on Ω.

4. Determination of a layered medium from plane wave responses. Recall the equation (0.1) which describes the propagation of acoustic pressure waves. Assume $p=p(z)$ and $c=c(z)$, hence

$$\partial_t{}^2 u = \rho(z)c^2(z)\, \underline{\nabla}\cdot \frac{1}{\rho(z)} \nabla u, \quad z>0, \quad |x|, \ |y|<\infty. \tag{4.1}$$

The problem is to find the wave speed $c(z)$ and the density $\rho(z)$ from the response at $z=0$: $\partial_t u\ (t,x,y,z=0) = g(t,x,y)$ for $0<t<T$, caused by a point excitation: $\partial_z u\ (t,x,y,z=0) = \delta(x)\delta(g)\delta(t)$.

Coen (1981) suggested a technique for recovering $c(z)$ and $\rho(z)$ from plane wave components at the surface. To obtain the plane wave components, we take the Radon transform of $u(t,x,y,z)$ defined as $\underline{x} = (x,y)$

$$P(\tau,z;\underline{\xi}) = \int_{\infty}^{\infty} d\underline{x}\ u(\tau + \underline{\xi} \cdot \underline{x}\ ,\ \underline{x}), \tag{4.2}$$

where $\underline{\xi}$ is the transform parameter and is related to the angle of incidence. Under this transformation, each plane wave component satisfy

$$\frac{1}{\rho(z)c(z)} \left(1-c^2(z)|\underline{\xi}|^2\right) \partial_\tau^2 P = \partial_z\left(\frac{1}{\rho(z)} \partial_z P\right), \quad z>0, \tag{4.3}$$

with boundary conditions

$$(4.4) \qquad \partial_z P(\tau, z=0; \underline{\xi}) = \delta(\tau)$$

$$(4.5) \qquad \partial_\tau P(\tau, z=0; \underline{\xi}) = \int_{-\infty}^{\infty} d\underline{x}\ g(\tau + \underline{\xi} \cdot \underline{x}, \underline{x}) \equiv G(\tau; \underline{\xi}).$$

The system is quiescent for $t<0$.

Notice that (4.3)-(4.5) for each $\underline{\xi}$ is a 1-D inverse problem. Coen's idea is that if two such inverse problems are solved to recover two "impedances":

$$(4.6) \qquad A(z(s), \underline{\xi}) = \frac{(1-c^2(z(s))|\underline{\xi}|^2)^{1/2}}{\rho(z(s))c(z(s))},$$

where the travel time s is given by:

$$(4.7) \qquad s(z;\underline{\xi}) = \int_0^z \frac{(1-c^2(\sigma)|\underline{\xi}|^2)^{1/2}}{c(\sigma)} d\sigma,$$

it is possible to reconstruct $c(z)$ and $\rho(z)$ uniquely.

We investigate the feasibility of such a scheme for inverting real data (Santosa and Symes 1984a) and addressed the following aspects of the problem:

(i) the nature of the data and the noise
(ii) sources of instability and error analysis
(iii) consequence of smoothing
(iv) use of multiple plane wave data

Since the impedance profile inversion, the first step in this two step method, is well understood (see e.g. Section 2), we limit the study to the effect of noise on the second part of this reconstruction procedure. This is the step of taking two "impedances" and solving for the density $\rho(z)$ and the wavespeed $c(z)$. We attempted to supply amounts and types of data noise which model the errors that arise in the measurement of the point source response at the surface. We find that the method proposed by Coen is prohibitively unstable for realistic choices of plane wave components. This type of instability is an example of the instability caused by the near dependence of data in multiparameter estimation mentioned in the Introduction. To be more specific, the two impedances for two values of $\underline{\xi}$, where $|\underline{\xi}_1 - \underline{\xi}_2|$ is small, are nearly dependent on each other.

Typically, errors in the impedances are amplified by a factor of $(|\underline{\xi}_1|^2 - |\underline{\xi}_2|^2)^{-1}$.

In the following numerical simulation (Figure 6) we took two synthetically generated $A(z(s),\underline{\xi})$ for $\underline{\xi}_1=0$, and $\underline{\xi}_2=0.2$, and added about 0.7% RMS-noise to each impedance. The reconstruction produces density and velocity estimates of 8% and 6% relative errors; a magnification of about 10-fold.

If ρ and c are smooth, then smoothing prior to reconstruction improves the result. The smoothing process only effects the high frequency noise in the data, and does not effect the profiles. However, smoothing is only valid if the target profiles ρ and c are smooth. Therefore, smoothing of the data does not stabilize Coen's procedure if the data are from non-smooth profiles.

We also consider the case where multiple plane wave data are given, that is, we have $A(z(s);\underline{\xi})$ for many values of $\underline{\xi}$. In this case, multiple use of the above procedure produces many estimates of c(z) and $\rho(z)$, which we fit in a least-square sense in a systematic way to cancel out the effect of correlated noise in the data. This leads to an improvement. We took 78 noisy planewave impedances ranging from $\underline{\xi}=0$ to $\underline{\xi}=0.2$. The noise level is as before, about 0.7% in each impedance $A(z(s);\underline{\xi})$. The many estimates of c and ρ were fitted, and the result is shown in Figure 7. The relative errors are now 3.4% in ρ and 2.3% in c.

Our work leads to the conclusion that the rich data available in such an experiment must be used in a more effective way. In particular, we can pose the problem of finding a pair of coefficients $\rho(z)$, c(z) such that the surface response in (4.5) is fitted in a least squares sense for a range of parameter values $\underline{\xi}$. In this instance, we have shown (Santosa and Symes 1984b) that such a problem is well conditioned.

5. Inverse Source Problem. The inverse source problem is the determination of properties of a source of waves from the recorded response, given knowledge of the propagating medium. This problem arises in seismologoy, passive sonar, and acoustic emission. We consider here the inverse problem for a point source; that is, a source whose extent is small compared to the typical dimensions of the medium and predominant wavelengths present in the measured data.

A source of elastic waves confined to a finite volume can be described by the moments of the body force about a point within its support. Let $f_j(t,x)$ be a force vector in a volume v_0, $\text{supp}(f_j(t,x)) \subset V_0$, for each t. Then the zeroth moment, F_j, or the total force is

$$(5.1) \qquad F_j(t) = \int_{V_0} f_j(t,\xi)d\xi.$$

With reference to $x_0 \varepsilon V_0$, the first moment, known as the moment tensor, is defined as

$$(5.2) \qquad M_{jk}(t,x_0) = \int_{V_0} (\xi - x_0)\, f_j(t,\xi)d\xi.$$

Higher order moments are similarly defined.

Our study (Michaels 1984, Michaels and Pao 1984) is directed at finding F_j and M_{jk} in an elastic medium from measurements at the boundary. In particular, we consider F_j and M_{jk} which are separable in the sense that

$$(5.3) \qquad F_j(t) = s(t)F_j$$

$$(5.4) \qquad M_{jk}(t) = s(t)M_{jk}$$

To be determined then are the components F_j of a vector and M_{jk} of a tensor, and a source time function s(t). The "location" of the point source, x_0, is presumed known.

The sources (5.3) - (5.4) generate displacement fields which can be represented in terms of the Green's function of the propagating medium via convolution. Let $u_i(t,x)$ be the displacement field caused by the source corresponding to (5.3) - (5.4), then

$$(5.5) \quad u_i(t,x) = \int_0^t \sum_{j=1}^{3} F_j\, G_{ij}(t-\tau,x,x_0)s(\tau)d\tau$$

$$+ \int_0^t \sum_{j=1}^{3} \sum_{k=1}^{3} M_{jk}\frac{\partial}{\partial x_k} G_{ij}(t-\tau,x,x_0)s(\tau)d\tau.$$

If for a given geometry, medium and source location x_0, the Green's function G_{ij} and its gradient $\partial_k G_{ij}$ are known, we can find the response $u_i(t,x)$ for any t and x generated by the source described in (5.3) - (5.4).

In the inverse problem, we are given the normal (to the surface) component of the displacement at several points $\{x_i\}_{i=1,N}$. The objective is to characterize F_j, M_{jk} and s(t). We specialize our discussion here for the determination of $M_{jk}s(t)$. Physically, this problem corresponds to finding the orientation and time signature of a microcrack in an elastic medium. We further suppose that $s(t) = 0$ after some $t > T_S$, and that the geometry to be an elastic plate, and the measured displacement be the particle displacement normal to the surface of the plate. It is convenient to write the normal displacement response at x_i as

$$(5.6) \qquad u(t;x_i) = \left\{ \sum_{m=1}^{M} C_m G_m(t;x_i) \right\} * s(t).$$

This is possible because the location x_0 is known, the Green's function $G_m(t; x_i)$ is for normal displacements only, and c_m correspond to the components of M_{jk}.

The inversion procedure is an iterative deconvolution algorithm for simultaneously finding c_m and $s(t)$. The iteration steps are

0) set c_m to some initial guess
1) find $s(t)$ by least squares estimation in (5.6)
2) use $s(t)$ in (5.6) to improve estimates of c_m
3) return to step (1) with updated c_m

This process is repeated until a stable fixed point is reached. An important condition for convergence is the linear independence of G_m for $m=1,M$. For the plate geometry, several measurement points $\{x_i\}$ are needed to guarantee uniqueness.

This technique for finding the source characteristics is illustrated by a numerical simulation with noise. We begin with an assumed M_{jk} in (5.4),

$$[M_{jk}] = \begin{bmatrix} 3 & 0 & 0 \\ 0 & 1 & 0 \\ 0 & 0 & 1 \end{bmatrix}$$

which models a tensile crack. The source time function $s(t)$ is displayed in Figure 8a. Displacements, caused by this forcing term in an elastic plate is calculated at equidistant points from the source for several angles, and those for $\theta=135°$ and $210°$ are shown in Figure 8b. Noise is then added to each trace. Our algorithm produced after 30 iterations the matrix

$$[M_{jk}] = \begin{bmatrix} 2.996 & -0.006 & 0.003 \\ -0.006 & 1.020 & 0.002 \\ 0.003 & 0.002 & 0.992 \end{bmatrix}$$

and the function s(t) displayed in Figure 8c.

The inversion algorithm was also tested against real data. To generate the data, a glass capillary is pushed with a pointer at some orientation until it breaks (See Figure 9). This experiment is modelled by a source term of the type (5.3). The problem is to find F_j, hence the orientation of the pointer, and the time signature s(t) corresponding to the breaking of a capillary.

We omit the discussion of relating the true displacements from the measured voltages in the transducers. The measured data at equidistance from the source at various angles, given in voltage displayed in Figure 10 (thin lines). Our algorithm produced a source time function s(t) in Figure 11. The recovered and true orientation vectors F_j are

true: $F_j = (0, -0.6088, -0.7934)$

recovered: $F_j = (0.0270, -0.5495, -0.8351)$.

To check how well the reconstructed s(t) and F_j fit the data, we overplot the voltage that such a force would produce at the measurement locations on the measured data in Figure 10 (thick lines). The plots show very good agreement with the measurements.

6. Summary. We have presented in this paper some results of our research effort on inverse problems in wave propagation. In particular, we discuss various strategies for handling the illposedness of the problem. Further work is being carried out to put the results in quantitative context by designing conputational schemes and performing simulations. We have alluded to the directions we will pursue. Other areas of study being explored at present include impedance profile inversion from noisy bandlimited data using a priori knowledge as constraints, and functional analysis of maps from medium models to data. In particular, we are studying media of nonconstant wave speeds and caustics.

7. Acknowledgement. The authors express their grateful appreciation to Ms. M.J. Michaels for her skillful assistance in putting this report together.

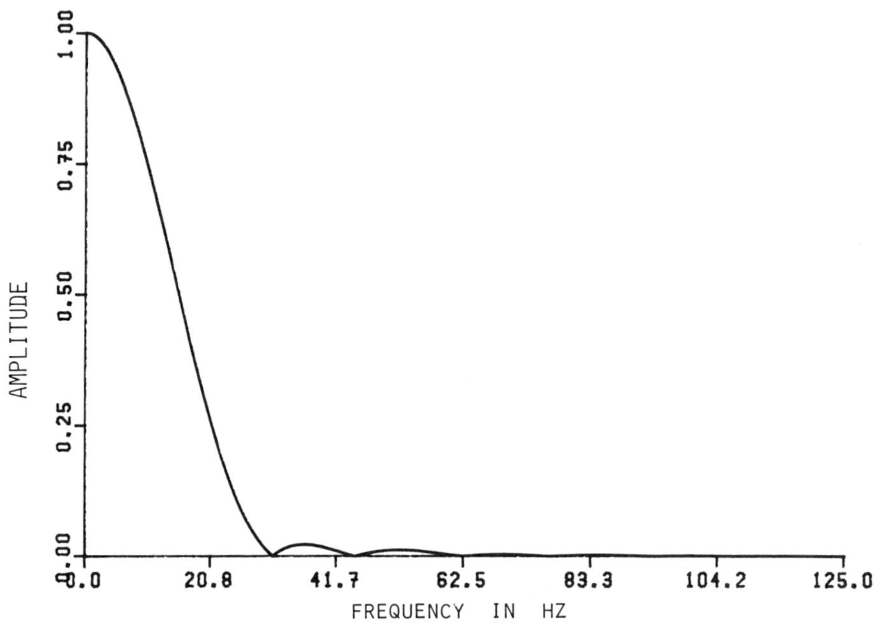

Figure 1. Low-pass filter. Impulse response of the medium whose impedance is shown in dots in Figure 2 is low-pass filtered.

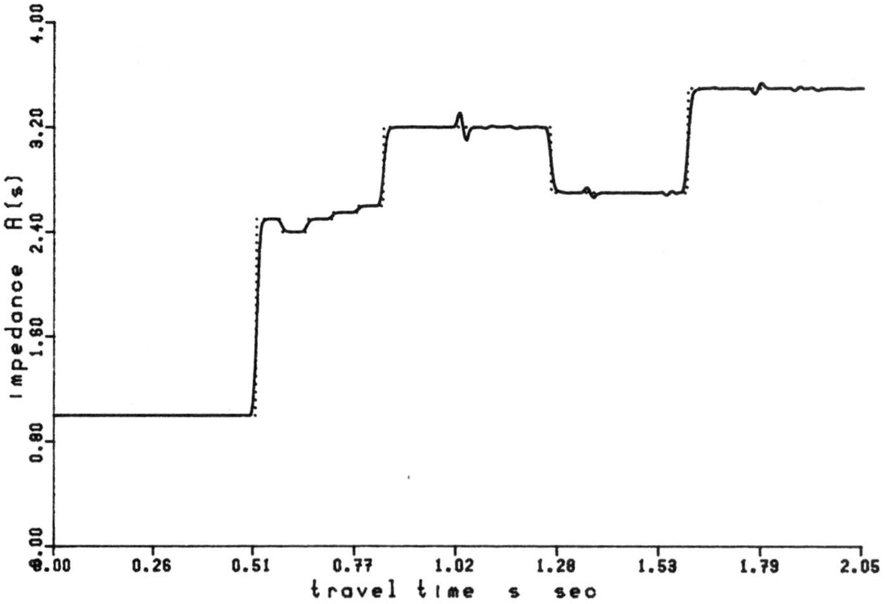

Figure 2. Inversion of low-pass filtered data. Shown in dots is the target impedance profile, and in solid line, the inverted profile using filtered data. Filter function is shown in Figure 1.

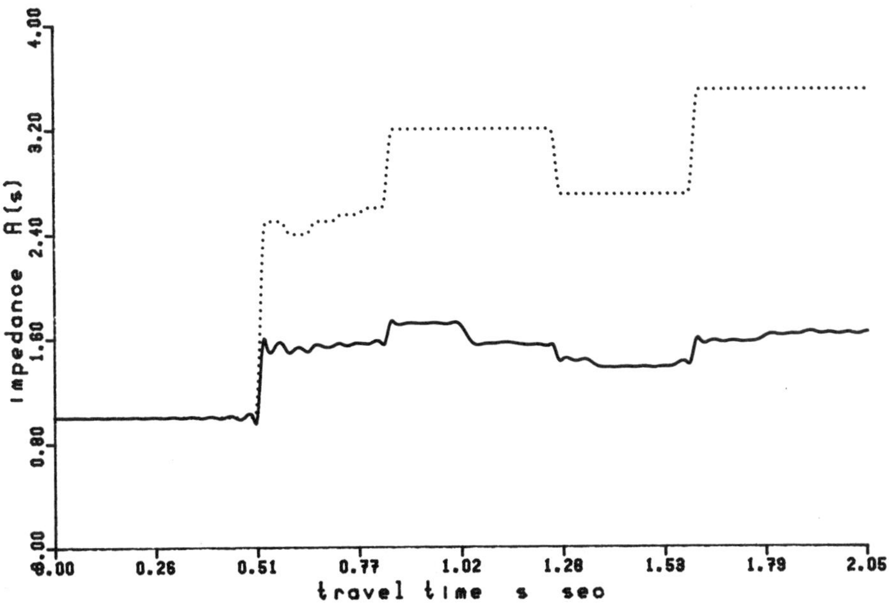

Figure 3. Example of two profiles that produce identical impulse responses in the pass-band 10-40 Hz. Observe that profile (b) has additional "interfaces".

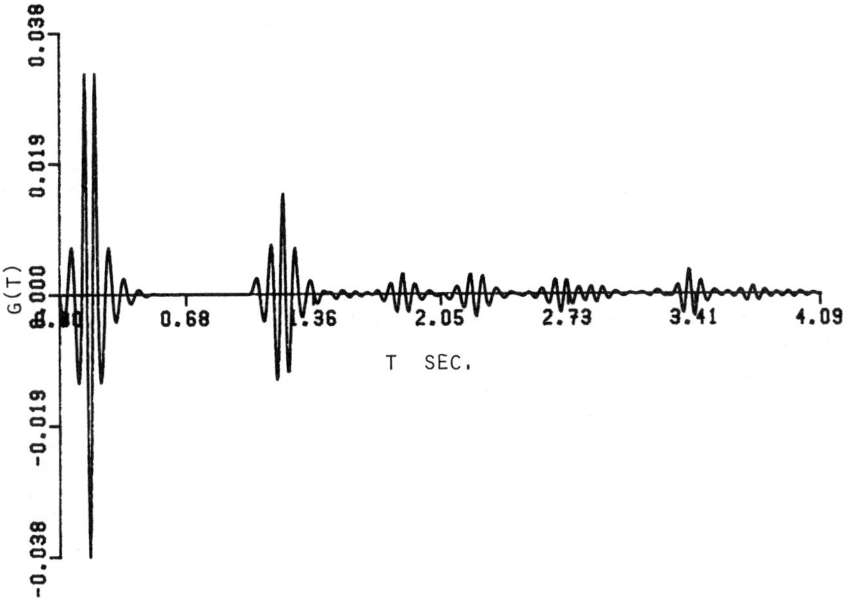

Figure 4. Response of the profiles (a) and (b) in Figure 3. Source wavelet is the upside-down version of the first wavelet in the trace. The wavelet was energy deposited mostly in 10-40 Hz range.

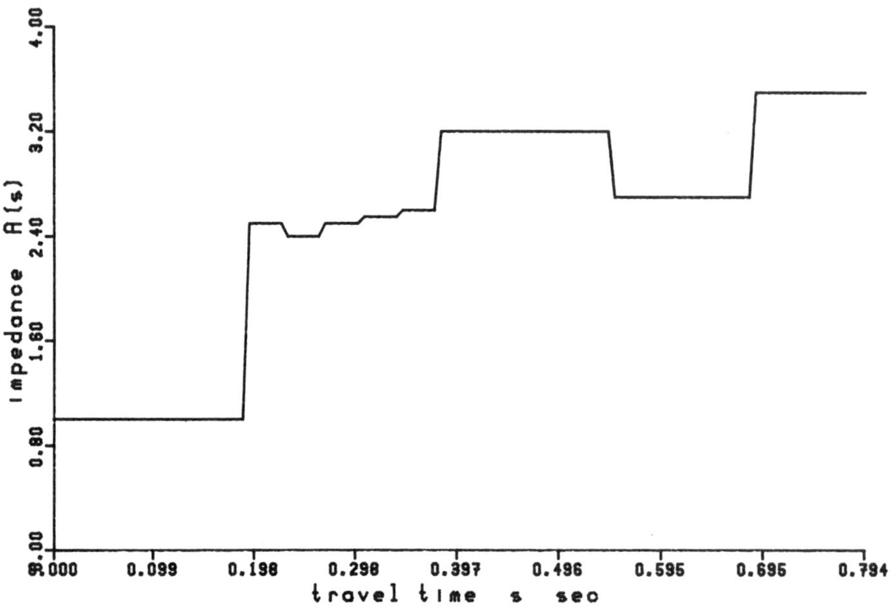

Figure 5a. Target impedance profile for band-limited data inversion. The L^1-norm of reflectivity (Log A)' is 2.30.

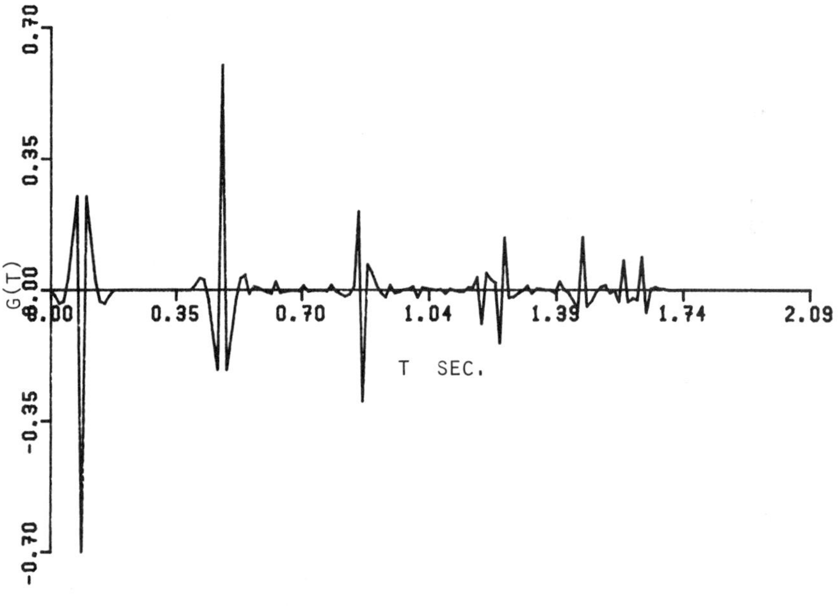

Figure 5b. Response of target profile in 5a. The source wavelet is the upside-down version of the first wavelet in the trace. The wavelet has energy mostly in the 7.5-40 Hz range (sampling frequency is 40 Hz).

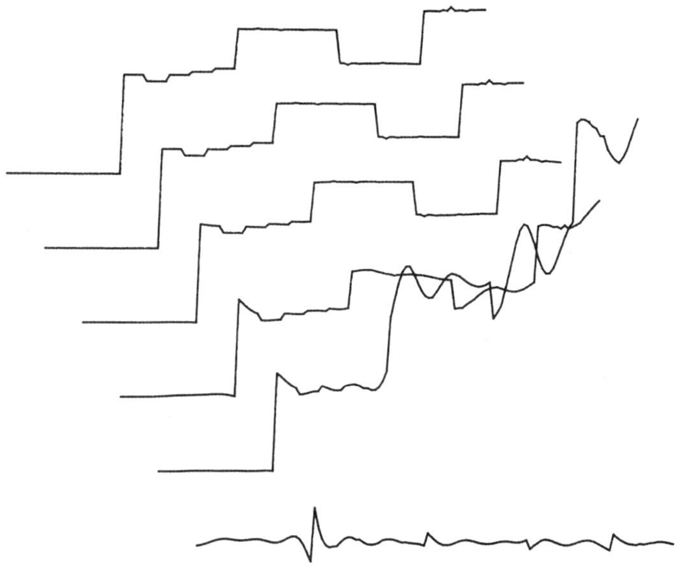

Figure 5c. The refinement process in the search for profile with least L^1-norm. Initially, all missing Fourier components are set to zero, iteration 0, the L^1-norm is 4.18. The optimization produces the minimum after 5 Gauss-Newton steps. The L^1-norm at minimum is 2.43. Discrepancies are mostly due to discretization errors.

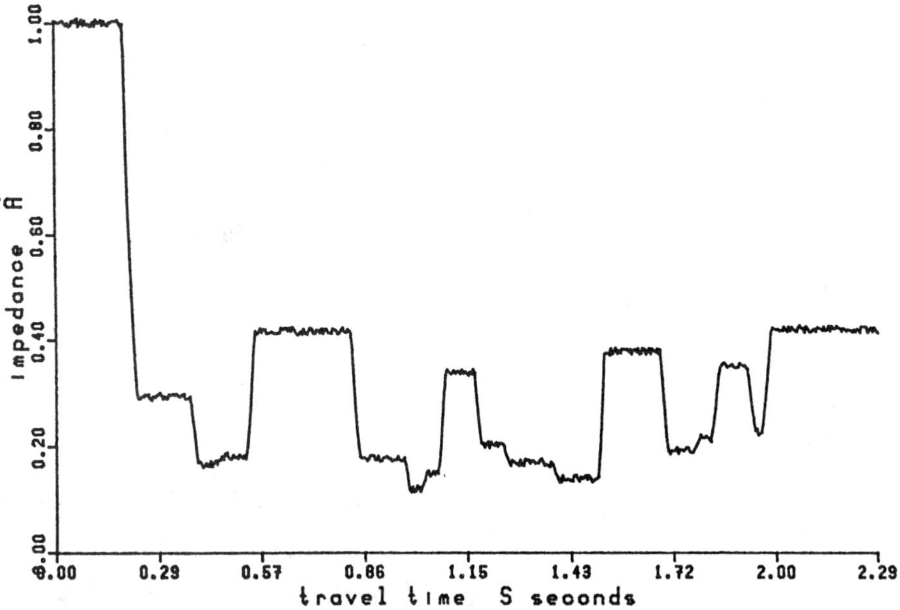

Figure 6a. The noise corrupted normal incidence "impedance" profile. Noise is 0.7% RMS.

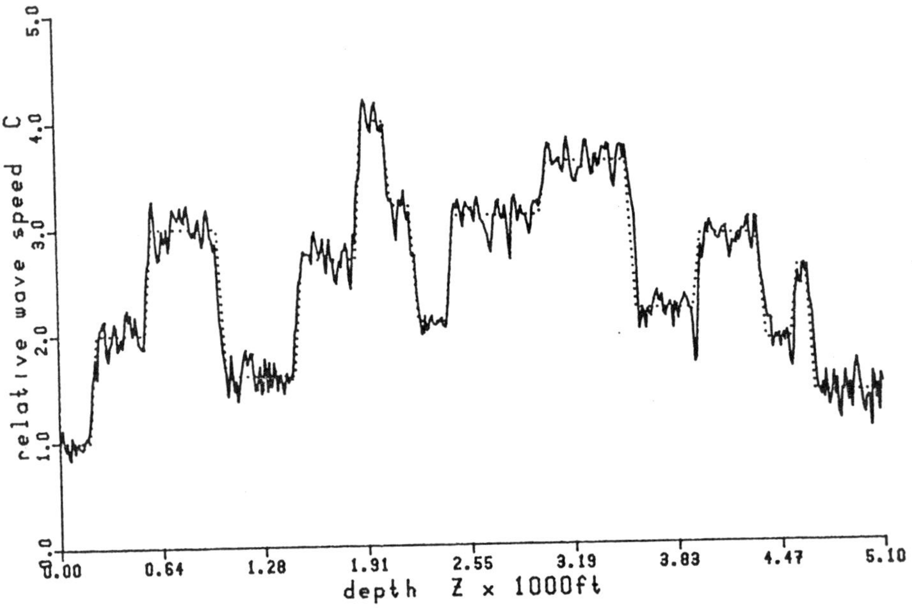

Figure 6b. Reconstructed velocity profile from 2 noisy impedances. Shown in dotted line is the largest profile. Relative error in the reconstruction is 6% RMS.

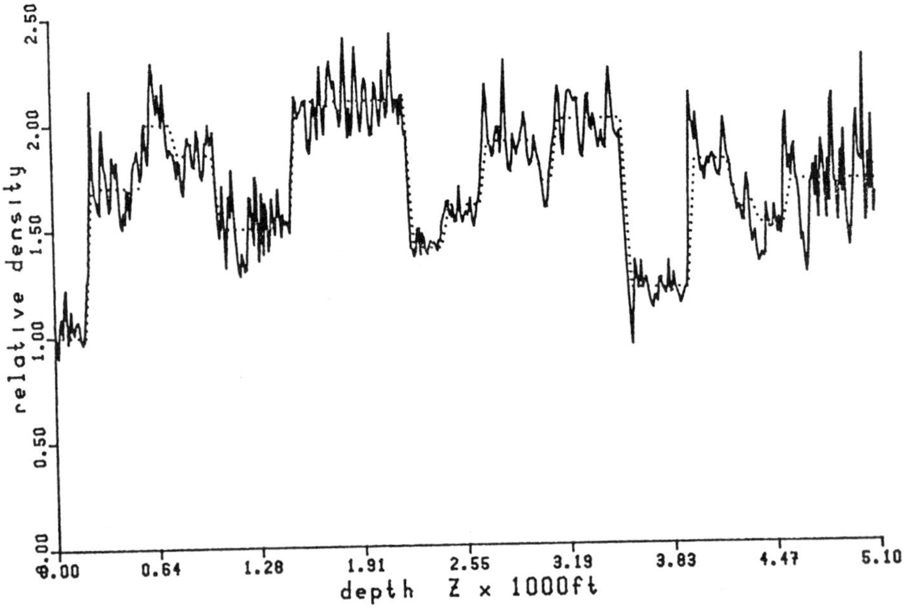

Figure 6c. Reconstructed density profile (companion to Figure 6b). Relative error in the reconstruction is 8%.

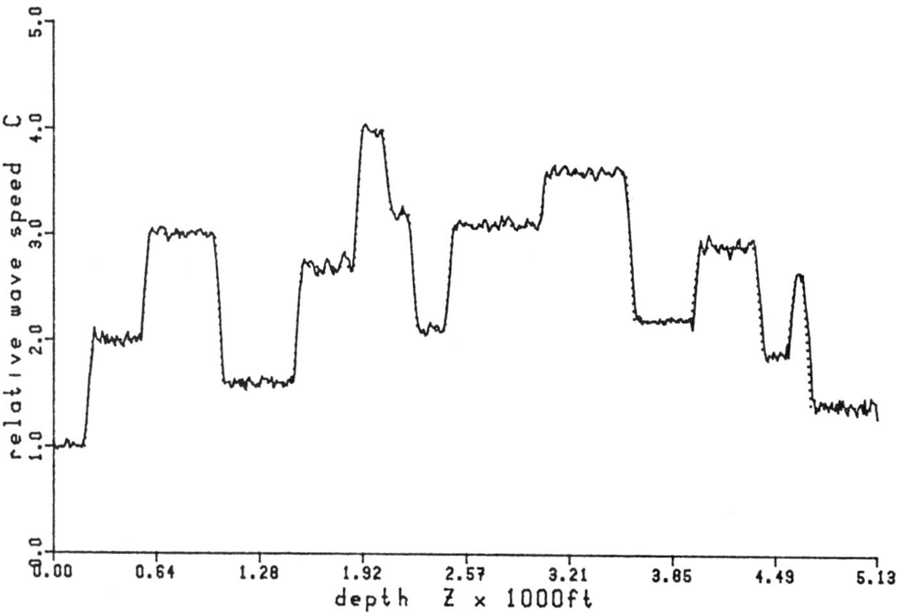

Figure 7a. Reconstructed velocity profile from 80 noisy impedances. Noise level in data is 0.7% RMS. Relative error in the reconstruction is 2.3%.

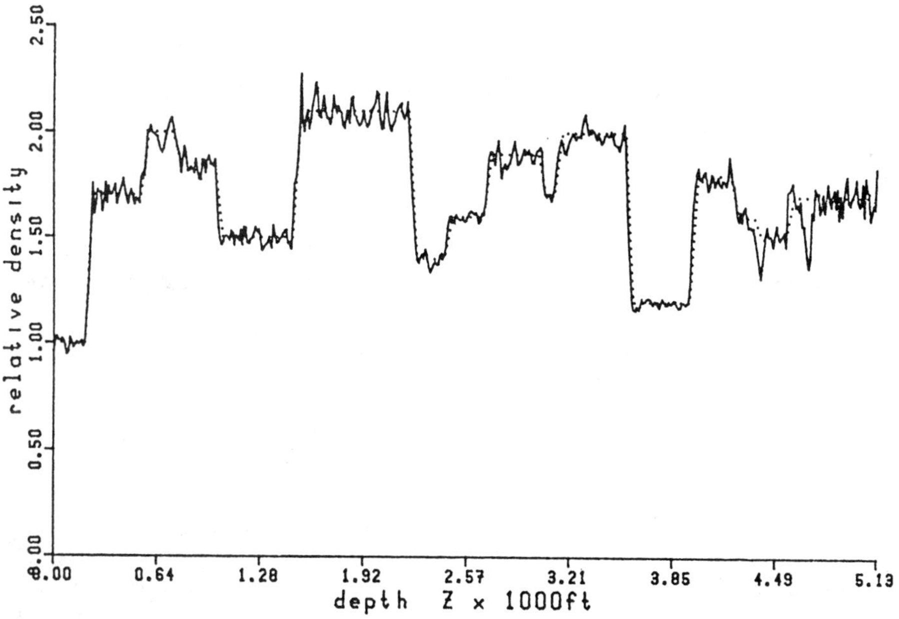

Figure 7b. Reconstructed density profile from 80 noisy impedances (companion to Figure 7a). Relative error in the reconstruction is 3.4%.

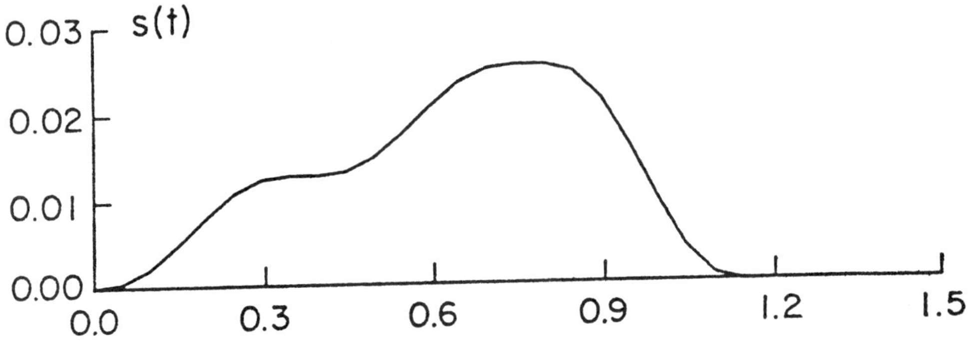

Figure 8a. Source time function s(t) for numerical simulation.

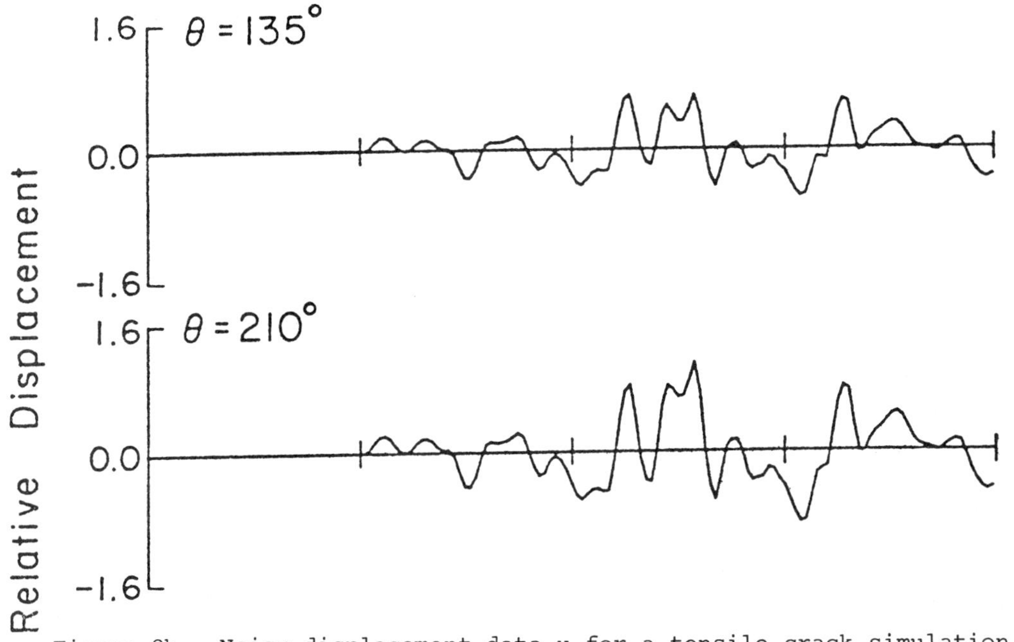

Figure 8b. Noisy displacement data u for a tensile crack simulation at various angles equidistant from the source.

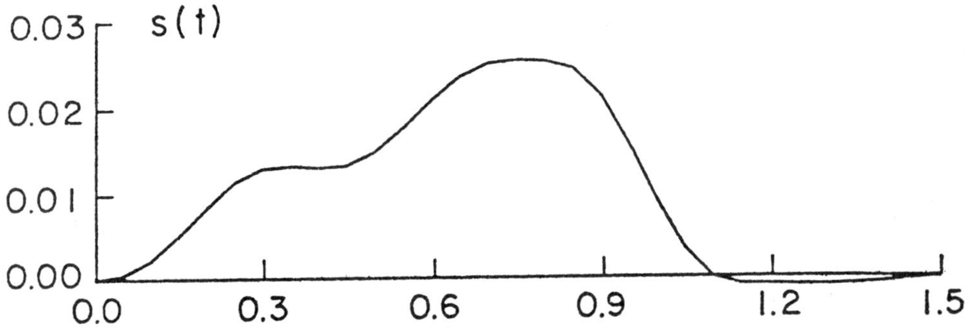

Figure 8c. The recovered s(t) from noisy data.

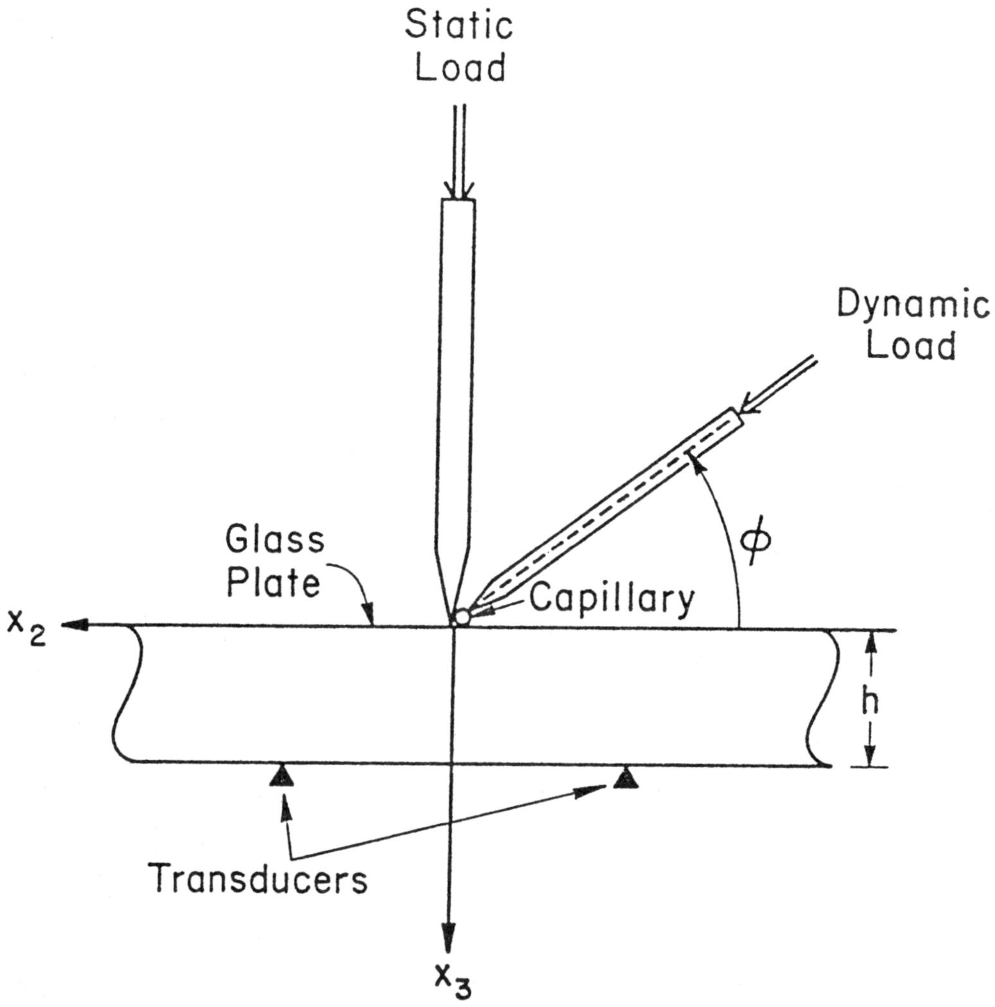

Figure 9. Experimental setup for generating oblique forces.

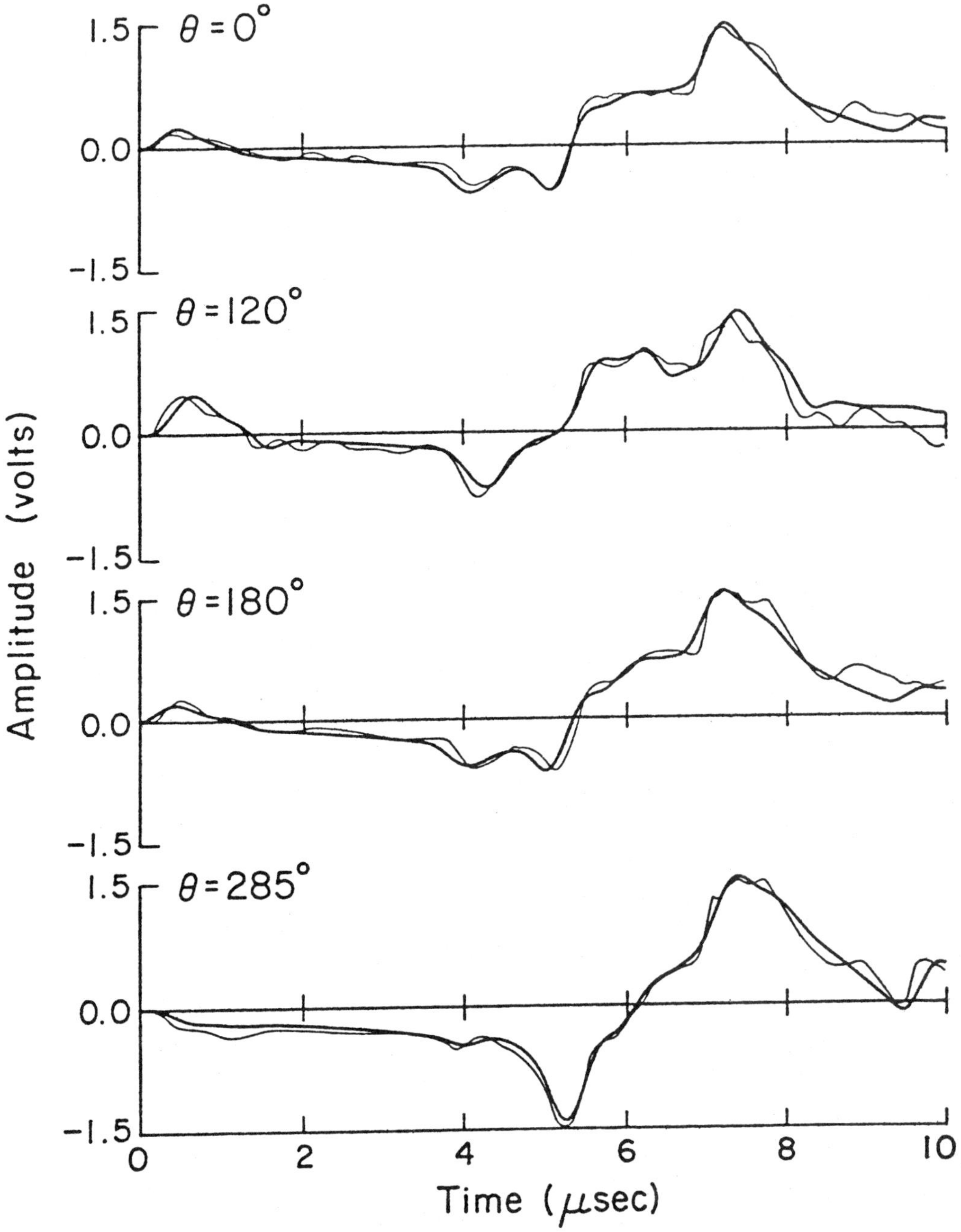

Figure 10. Experimental data for oblique force in direction (0.6088, -0.7934). The fitted data generated by the inversion algorithm is shown in thick lines.

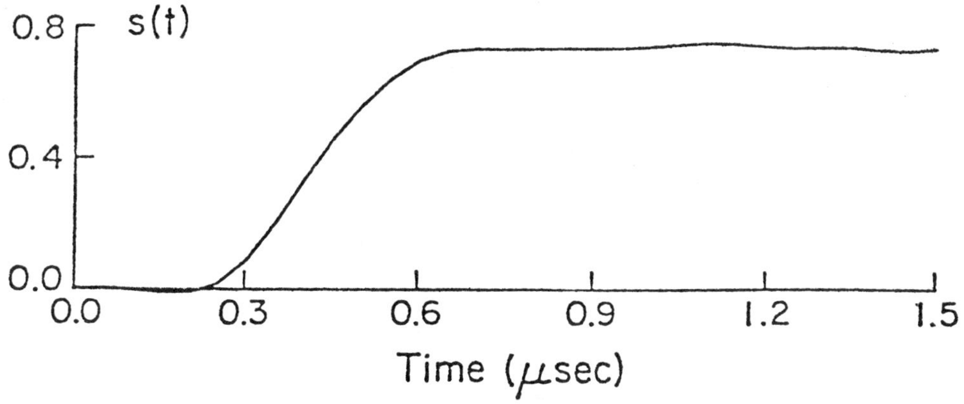

Figure 11. The recovered source function s(t) from experimental data.

References

[1] S. COEN, Density and compressibility profiles of a layered acoustic medium from precritical incidence data, Geophysics 46 (1981), 1244-1246.

[2] S. COEN and W.W. SYMES, Complete inversion of reflection data for layered acoustic media, Rice University, preprint (1984).

[3] M. GERVER, The inverse problem for the vibrating string equation, Geophys. J.R. Astr. Soc. 21 (1970), 337-357.

[4] S. LEVY and P.K. FULLAGAR, Reconstruction of a sparse spike train from a portion of its spectrum and application to high resolution deconvolution, Geophysics 46 (1981), 1235-1243.

[5] J.E. MICHAELS, An inverse source problem for elastic waves, Cornell University, Ph.D. Thesis, Department of Theoretical and Applied Mechanics (1984).

[6] J.E. MICHAELS and Y.H. PAO, Deconvolution of source line functions and the moment density tensor, in Review of Progress in Quantitative Nondestructive evaluations, D.O. Thompson and D.E. Chimanti, eds., Plenum Press, New York.

[7] D.W. OLDENBURG, T. SCHEUER and S. LEVY, Recovery of the acoustic impedance from reflection seismograms, Geophysics 48 (1983), 1318-1337.

[8] D.W. OLDENBURG, Inversion of bandlimited reflection seismograms, these proceedings.

[9] P. SACKS and W.W. SYMES, Uniqueness and continuous dependence for a multi-dimensional hyperbolic inverse problem, Cornell University preprint (1984).

[10] F. SANTOSA and W.W. SYMES, Inversion of impedance profile from band-limited data, in Digest, 1983 International Geoscience and Remote Sensing Symposium, San Francisco, CA, 1983.

[11] F. SANTOSA and W.W. SYMES, The determination of a layered acoustic medium via multiple impedance profile inversions from plane wave responses, to appear in Geophys. J.R. Astr. Soc. (1984).

[12] F. SANTOSA and W.W. SYMES, On the relation of the point source response to the coefficients of a layered medium, Cornell University preprint (1984).

[13] F. SANTOSA, Deconvolution from band-limited and noisy data using L^1-optimization, in preparation (1984).

[14] W.W. SYMES, Impedance profile inversion via the first transport equation, J. Math. Anal. Appl., 94 (1983) 435-453.

[15] W.W. SYMES, On the relation between coefficients and boundary values for Webster's Horn Equation, Cornell University preprint (1984).

[16] W.W. SYMES, Linearization stability for an inverse problem via several dimensional wave propagations, Cornell University preprint (1984).

[17] W.W. SYMES, Stability of sound speed profile inversion, in preparation (1984).

FINITE ELEMENT METHODS WITH ANISOTROPIC DIFFUSION FOR SINGULARLY PERTURBED CONVECTION DIFFUSION PROBLEMS

LARS B. WAHLBIN*

Abstract. The understanding of approximations of solutions to singularly perturbed convection diffusion problems is at present in a satisfatory state for one-dimensional problems, if those are linear and without turning points. The behavior of both finite difference methods and spline based methods is by and large known. In two dimensions, methods for problems in simple geometries can sometimes be analyzed via dimension-splitting, allowing the use of one-dimensional ideas. In curved domains, finite element methods may be of particular use since they easily incorporate the geometry. The behavior of finite element methods in singularly perturbed two-dimensional convection diffusion problems is just beginning to be known.

In this note we briefly describe the motivation for and understanding of the so-called streamline diffusion method, which incorporates artificial diffusion in the streamline direction in quite an automatic way in the finite element situation while preserving accuracy. We also consider a variation of it in which a smaller amount of artificial crosswind diffusion is added in order to gain additional control.

The local behavior of these methods is of particular interest. The original problem may have rough layers at outflow boundaries, at characteristic parts of the boundary and, typically, internal roughness following characteristics. For a method to be practical it is important that such singularities do not pollute.

*Department of Mathematics and Center for Applied Mathematics, Cornell University, Ithaca, NY 14853, USA.
Supported by the National Science Foundation, USA.

At present understanding does not explain experience with those methods in two dimensions. Considering the time it has taken to gain the present enlightenment in one-dimensional situations we may expect progress to be slow.

1. Introduction. Consider the model problem of finding $u(x,y)$ for (x,y) in Ω, a convex bounded domain, such that

$$(1.1a) \quad -\delta u_{xx} - \varepsilon u_{yy} + u_x + u = f \quad \text{in} \quad \Omega ,$$

$$(1.1b) \quad u = 0 \quad \text{on} \quad \Gamma = \partial\Omega .$$

Here δ and ε are small positive numbers and $f(x,y)$ is a given function.

In order to solve this problem by a finite element method, let $T_h = \cup_{i=1}^{I(h)} \tau_i^h$, with parameters h in a subset of $(0,\frac{1}{2}]$, be a family of partitions of Ω into disjoint elements τ_i^h. Assume that the family is of edge-to-edge type and that the family is quasi-uniform in the sense that there exist positive constants c and C such that with ρ_i the radius of the largest inscribed disc of τ_i^h,

$ch \le \rho_i \le \operatorname{diam} \tau_i^h \le Ch$, cf. [5]. To avoid certain technical points in this brief exposition assume that $\Omega = \cup_{i=1}^{I(h)} \tau_i^h$. Let finite element spaces S_h, $0<h\le\frac{1}{2}$, be associated with the partitions T_h. A typical case is as follows: The elements τ_i^h are triangular, possibly with a curved side if abutting on Γ, and

$$(1.2) \quad S_h = \{\chi : \chi \text{ continuous on } \Omega ,\ \chi = 0 \text{ on } \Gamma ,\ \chi\big|_{\tau_i^h} \text{ polynomial of degree } r-1\} .$$

A first attempt at solving (1.1) might be the <u>ordinary Galerkin</u> finite element method: Find $u^h \in S_h$ such that

$$(1.3) \quad \delta(u_x^h,\chi_x) + \varepsilon(u_y^h,\chi_y) + (u_x^h,\chi) + (u^h,\chi) = (f,\chi)$$

for all $\chi \in S_h$. Here $(f,g) = \int_\Omega fg\, dxdy$ and the equation (1.3) comes from multiplying (1.1a) by χ and integrating by parts.

In practice δ and ε might be much smaller than the quantity h which determines the resolving power of S_h . We shall concentrate on this case,

(1.4) $0 < \delta, \varepsilon < h$.

The problem (1.1) is then singularly perturbed as δ, ε tend to zero, [9, 10, 20]. Boundary layers where u undergoes sharp changes develop in general at the outflow, a.k.a. downwind, portions of the boundary and at characteristic parts. The wind direction in our model problem is the positive x-direction and the following sketch briefly indicates the situation.

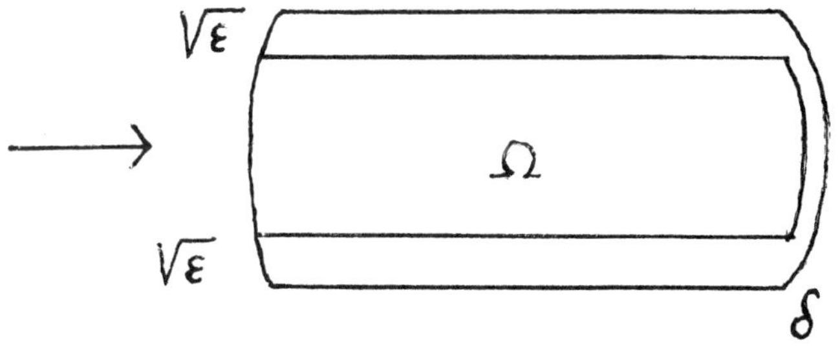

Near the characteristic parts the typical behavior of $u(x,y)$ is $\exp(-d/\sqrt{\varepsilon})$ where $d = \mathrm{dist}((x,y),\Gamma)$ and at the downwind boundary, $\exp(-d/\delta)$.

It is well known that for the ordinary Galerkin method the unresolved boundary layers pollute in an oscillatory manner all through the domain and render the approximation u^h useless for $\delta, \varepsilon < h$. The typical case is the classical one of centered finite differences in one-dimensional problems.

We point out that the aim here is not actually to resolve the boundary layers. This would be impossible without refining the mesh in those layers or combining with analytic techniques such as boundary layer expansions or a posteriori processing based on such analytical knowledge; these matters will not be considered here. The aim is rather to obtain a decent approximate solution away from boundary layers (or internal layers where u is rough) while not letting the inexact regions grow too much in the numerical approximations.

In one-dimensional finite difference theory a traditional fixup is to introduce artifical damping, e.g., via upwinding. In two-dimensional finite elements the following method, which we shall call the (h,h) - diffusion method, is well known: Seek $u^h \in S_h$ such that

$$(1.5) \quad h(u_x^h,\chi_x) + h(u_y^h,\chi_y) + (u_x^h,\chi) + (u^h,\chi) = (f,\chi) .$$

This method does indeed dampen the oscillatory pollution. However, it is also known to have the following two drawbacks. i) If a sharp interior front is present in the crosswind direction, that front (and characteristic boundary layers) are smeared too much, cf. [12, Figure 6]. ii) Away from singularities, the scheme is at most first order accurate in h . Clearly, perturbations of order h have been made to the problem (1.1) for $\delta,\varepsilon << h$.

Remark 1.1 A possible remedy for preserving accuracy in smooth parts has been suggested, e.g., in [1]. Rewrite (1.1a) as an iteration equation,

$$-hu_{xx}^{n+1} - hu_{yy}^{n+1} + u_x^{n+1} + u^{n+1} = f - (h-\delta)u_{xx}^n - (h-\varepsilon)u_{yy}^n$$

and starting with $u^0 \equiv 0$, iterate a few times in the finite element setting. Note that if many iterations are taken and if the process is convergent, the limit would be the ordinary Galerkin solution (1.3).

Having noted that the (h,h)-diffusion method smears too much in the crosswind direction, layers originally of width $\sqrt{\varepsilon}$ are smeared to width $\sqrt{h}$, the object must then be to find in a systematic way, i.e., without parameter-tuning in front of the terminal, cf. [18], methods that appear to have the following features: i) Forgetting about actually resolving boundary or interior layers, singularities should be reasonably confined and not smeared out. ii) Away from singularities, the methods should be accurate.

The streamline diffusion method is a candidate for such a method, [12, 14, 21, 22]. To derive the method, multiply (1.1a) by

(1.6) $\chi + h\chi_x$

where $\partial/\partial x$ is to be thought of as differentiating in the streamline direction. Integrate the result over Ω and obtain, via some integrations by parts,

$$-\delta h(u_{xx},\chi_x) - \varepsilon h(u_{yy},\chi_x) + (\delta+h)(u_x,\chi_x)$$

$$+\varepsilon(u_y,\chi_y) + (u_x,\chi) + (u,\chi+h\chi_x) = (f,\chi+h\chi_x) .$$

Now interpret this equaiton on S_h . In general, the terms $-\delta h(u_{xx},\chi_x)$ and $-\varepsilon h(u_{yy},\chi_x)$ do not make sense for u,χ in S_h and for simplicity we discard them. The method then consists of finding $u^h \in S_h$ such that

$$(1.7) \quad (\delta+h)(u_x^h,\chi_x) + \varepsilon(u_y,\chi_y) + (u_x,\chi) + (u,\chi+h\chi_x)$$

$$= (f,\chi+h\chi_x) \quad \text{for} \quad \chi \in S_h .$$

Considering the first two terms we now have diffusion of strength $(\delta+h)$ in the stream direction, unchanged crosswind diffusion and we have made perturbations to the original problem of order $\delta h + \varepsilon h$.

Remark 1.2 Even if the discarded terms do not make sense on S_h , one may "partially" keep them, interpreting them in the piecewise sense as

$$(f,g)_{pv} = \sum_{i=1}^{I(h)} \iint_{\tau_i^h} fg \, dxdy .$$

The streamline diffusion method has been analyzed by Nävert [21], cf. Section 4 below. While his results show that the method is quite accurate in regions where the solution is smooth, for the numerical crosswind spread they give a $\sqrt{h}$ estimate, also for $\varepsilon \ll h$. This is the same crosswind smearing as in the (h,h)-method. Numerical experiments suggest that the crosswind spread in the streamline diffusion method is much smaller than in the (h,h)-method, cf. [12, Figure 6]. What is actually true for general nonuniform meshes and complicated singularities is at present unknown. Thus, the current understanding of the methods is incomplete in two-dimensional problems.

This paper is a loose sketch of ideas and results connected with the streamline diffusion method. Proofs, or even precise statements, will not be given. An outline of the rest of the paper is as follows: In Section 2 we give a brief account of how certain ideas developed in the early-to-mid seventies in one-dimensional hyperbolic problems. While the original motivation for these ideas were quite different from the present ones, they were found, as outlined in Section 3, to tie in with classical upwinding procedures. They could thus be used in finite element methods as a very simple way of upwinding while preserving accuracy. In Section 4, we sketchily describe salient points in the fundamental theoretical work of Nävert on the understanding of the streamline diffusion method and in Section 5 we put forward some recent thoughts, [16], on how introduction of a judicious amount of crosswind diffusion leads to less crosswind smearing than that which occurs in (the theoretical results for) the streamline diffusion method.

We finally note that we have not considered the limiting case of a purely hyperbolic problem. Many results carry over and in the hyperbolic case one also has recourse to the promising discontinuous Galerkin finite element method, [15, 19, 23].

2. The Question of Optimal Accuracy in Simple Hyperbolic Problems. Consider the model problem of finding $u(t,x)$, one-periodic in the x-variable, such that

$$(2.1a)\quad u_t + u_x = 0 \;,\quad t > 0 \;,\quad x \in I = (0,1) \;,$$

$$(2.1b)\quad u(0,x) = v(x)$$

where $v(x)$ is a given smooth one-periodic function. For the numerical solution of (2.1), employ spline spaces $S^r_{h,k}$ of the following type. Partition $(0,1)$ according to the following picture

Set $h := \max h_i$. Let $r-1 > k \geq 0$ and

$$S^r_{h,k} = \{\chi : \chi \in C^k_{per} \ , \ \chi|_{I_i} \text{ polynomial of degree } r-1\} \ .$$

It is well known that for w periodic and smooth ($w \in H^r_{per}$),

$$(2.2) \qquad \min_{\chi \in S^r_{h,k}} ||w - \chi||_{L_2} \leq Ch^r \ ,$$

but higher rate of convergence is not possible in general. We call h^r <u>optimal order</u> (in L_2).

The ordinary Galerkin method in its time-continuous version consists of finding $U(t): [0,T] \to S^r_{h,k}$ such that

$$(2.3) \qquad (U_t + U_x, \chi) \equiv \int_0^1 (U_t + U_x)\chi = 0 \ , \quad \text{for } \chi \in S^r_{h,k}$$

and, say, $(U(0) - v, \chi) = 0$.

If now v and hence u are smooth in the periodic sense, will U be an optimal order approximation, i.e.,

$$(2.4) \qquad ||u(t) - U(t)||_{L_2} \leq Ch^r \ ?$$

The answer is yes and no. For smoothest splines, i.e., $r = k+2$, on a uniform mesh, $h_i \equiv h$, the answer is yes, [25]. For nonuniform meshes the answer is no, [8], and also for Hermite cubics, $r=4$, $k=1$, on uniform meshes the answer is no, [7]; the error in the Hermite case is merely $O(h^3)$ even for smooth solutions.

Note the "paradoxical" nature of the result from [7]. If we use smooth cubics, $r=4$, $k=2$, we get $O(h^4)$ convergence but if we use <u>more</u> degrees of freedom, $r=4$, $k=1$, the error <u>deteriorates</u> to $O(h^3)$.

Dendy [6] found that the method

$$(U_t + U_x, \chi + \chi_x) = 0 , \quad \chi \in S^r_{h,k}$$
(2.5)
$$U(0) \quad \text{suitable}$$

leads to optimal order error for all spline spaces.

In the ordinary Galerkin method we see upon setting $\chi = U$ that

(2.6) $$||U||^2_{L_2} \equiv \text{constant}$$

whereas for Dendy's method, taking $\chi = U$ or U_t, respectively, and rearranging suitably,

(2.7) $$\frac{1}{2}\frac{d}{dt}(||U||^2 + ||U_x||^2) + ||U_t + U_x||^2 = 0 .$$

Since in general $U_t + U_x \neq 0$, Dendy's method introduces dissipation.

A slight perturbation of Dendy's method, [26, 27],

(2.8) $$(U_t + U_x, \chi + h\chi_x) = 0$$

also leads to optimal order error for all spaces $S^r_{h,k}$. The dissipation in this method is described by

(2.9) $$\frac{1}{2}\frac{d}{dt}(||U||^2 + h^2||U_x||^2) + h||U_t + U_x||^2 = 0 .$$

Note that the method (2.8) comes about by multiplying the given equation by the test functions $\chi + h\chi_x$, cf. (1.6).

At the time technology permitted one in the case of smoothest splines on a uniform mesh to consider the methods above as finite difference schemes at knots and apply Fourier analysis, [24]. It turned out that we have the following where dissipation is in the traditional finite difference sense. The ordinary Galerkin method (2.3) is accurate of order $2r$ and has no dissipation. The method (2.8) is accurate of order $2r-1$ and dissipative of order $2r$.

Dendy's method (2.5) does not quite fit classical simple finite difference theory. Its symbol is $\exp(p(\theta))$ where $\mathrm{Re}\, p(\theta) = -h\theta^{2r}/(h^2+\theta^2)$ and $\mathrm{Im}\, p(\theta) = \theta(1+O(\theta^{2r}/(h^2+\theta^2))$. Note that all methods exhibit superconvergence at knots in the uniform smoothest spline situation.

The condition of being accurate of order $2r-1$ and dissipative of order $2r$ is the classical Kreiss condition for L_2-stability and was known, [2], to be equivalent to stability in the maximum norm for a finite difference scheme. The condition also had the implication that spread of errors from singularities is minimized, [3].

In summary, from the points of view of i) optimal accuracy for smooth solutions and ii) stability and spread of errors from singularities, the method of testing residuals against the test functions

$$\chi + h\chi_x \ , \quad \chi \in S_h$$

appeared to have desirable properties.

3. Connections with Classical Finite Differences for One-Dimensional Singularly Perturbed Problems. Consider the model equation

(3.1a) $\quad -\delta u_{xx} + u_x = f \quad \text{on} \quad (0,1)$,

(3.1b) $\quad u(0) = u(1) = 0$.

For δ small positive, the solution exhibits in general a boundary layer of width δ at the downwind boundary at $x = 1$. Let the interval $(0,1)$ be subdivided into N intervals of length $h = N^{-1}$ and set $x_i = ih$. Assume

(3.2) $\quad \delta < h$.

For the central difference scheme, with u_i an approximation to $u(x_i)$

$$(3.3) \qquad \frac{-\delta(u_{i+1} - 2u_i + u_{i-1})}{h^2} + \frac{u_{i+1} - u_{i-1}}{2h} = f_i \ ,$$

$u_0 = u_N = 0$, the approximate solution is well known to oscillate wildly even away from the boundary layer and it is useless. Traditional upwinding,

$$(3.4) \qquad \frac{-\delta(u_{i+1} - 2u_i + u_{i-1})}{h^2} + \frac{u_i - u_{i-1}}{h} = f_i$$

alleviates these oscillations but is first order accurate only. More refined schemes, like [13], which are locally exact will experience some problems when translated to two-dimensional general domains, cf. [17]; the situation is far from being understood at present.

Consider now the ordinary Galerkin method

$$(3.5) \qquad (-\delta u^h_{xx} + u^h_x, \chi) = (f, \chi) , \quad \chi \in S_h$$

and the "streamline diffusion" method,

$$(3.6) \qquad (-\delta u^h_{xx} + u_x, \chi + h\chi_x) = (f, \chi + h\chi_x) , \quad \chi \in S_h .$$

(The term $-\delta h(u^h_{xx}, \chi_x)$ is thrown away for simplicity.) Taking piecewise linear functions on a uniform mesh, the ordinary Galerkin method reduces at meshpoints to

$$\frac{\delta(-u_{i+1} + 2u_i - u_{i-1})}{h^2} + \frac{u_{i+1} - u_{i-1}}{2h} = \ldots$$

which is central differencing. The method (3.6) reduces, for δ small, to

$$\frac{1}{h}\left(-\frac{3}{2}u_{i-1} + 2u_i - \frac{1}{2}u_{i+1}\right) = \ldots$$

and exhibits the typical upwinding feature of getting more information from upstream.

The ideas indicated above can be found, e.g., in [4].

In short, the method of using $\chi + h\chi_x$ as test functions gives an automatic way of upwinding while preserving accuracy to order δh, cf. Remark 1.2.

4. Nävert's Results. Consider first the (h,h)-diffusion method (1.5). It can be viewed as coming from the equation (1.1a) rewritten as

$$-hu_{xx} - hu_{yy} + u_x + u = f - (h-\delta)u_{xx} - (h-\varepsilon)u_{yy}$$

and proceeding as in ordinary Galerkin, throwing away the last two terms on the right. It thus introduces streamline and crosswind diffusion both of strength h. Sharp layers of size $\sqrt{\varepsilon}$ in the crosswind direction in the original problem will be smeared to size $\sqrt{h}$. The formal order of accuracy is h, cf. Remark 1.1.

In the streamline diffusion case (1.7) Nävert has shown that spread in the streamline direction from discontinuities is restricted to order h, and in the crosswind direction to $\sqrt{h}$. (The latter results appears much too pessimistic in many numerical examples where $\varepsilon << h$.) He also showed that accuracy is preserved in smooth regions. Typical results of this are as follows. Let $0 < \delta, \varepsilon < h$ and let $R \subseteq \tilde{R}$ be the regions sketched.

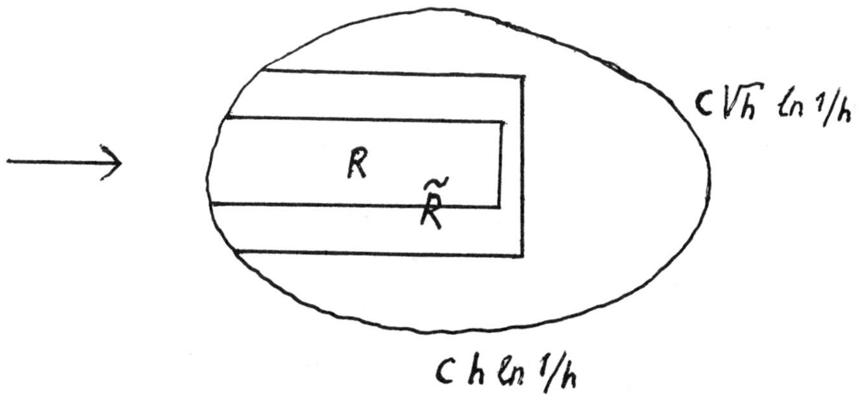

It is important, for $\tilde{R}$, that it extends all the way upstream to the inflow boundary. Then

$$(4.1) \quad ||u^h||_{L_2(R)} \leq C||f||_{L_2(\tilde{R})} + \text{NGI}$$

where NGI stands for negligible global influences, $\leq Ch^s$ for any given s . For S_h of type (1.2),

$$(4.2) \quad ||u^h-u||_{L_2(R)} \leq Ch^{r-\frac{1}{2}} ||u||_{H^r(\tilde{R})} + \text{NGI}$$

if $(\varepsilon+\delta)h < h^{r-\frac{1}{2}}$ or if Remark 1.2 is implemented. Note that the error order $h^{r-\frac{1}{2}}$ is not optimal, which would be h^r for S_h given in (1.2). Nävert shows in certain situations on almost uniform meshes and certain subspaces that the error is of optimal order. Thus the situation concerning optimality resembles very much the one in Section 2 which at one time motivated the streamline diffusion method.

As remarked two times already, the theoretical results for crosswind spread are $\sqrt{h}$, both for the (h,h)-method and the streamline diffusion method. In the final section of this paper, we shall present a method that can be shown to have less crosswind spread.

5. A Method with Some Crosswind Diffusion and Small Crosswind Smear. Let $\varepsilon_0 = \varepsilon_0(h)$ be given and set $\bar{\varepsilon} = \max(\varepsilon,\varepsilon_0)$. Consider the method of finding $u^h \in S_h$ such that, cf. (1.7),

$$(5.1) \quad (\delta+h)(u_x^h, \chi_x) + \bar{\varepsilon}(u_y^h, \chi_y) + (u_x, \chi) + (u, \chi+h\chi_x) = (f, \chi+h\chi_x) , \quad \text{for} \quad \chi \in S_h .$$

Thus, if $\varepsilon < \varepsilon_0$, we have introduced artificial crosswind diffusion of strength ε_0 . In additional to the usual perturbations to accuracy of order $(\delta+\varepsilon)h$ we have now introduced one of order ε_0 , if $\varepsilon < \varepsilon_0$. (Maybe the first one could be relieved by the technique in Remark 1.2 and maybe the second one by the ideas of Remark 1.1. Our numerical experience with the iterative technique of Remark 1.1 suggests that each iteration increases crosswind smear and since the object here is to minimize that, we shall not consider that technique further here.)

In [16] an analysis will be given of how the crosswind spread in the method (5.1) depends on ε_0 . It is found that the choice

$$\varepsilon_0 \simeq h^{3/2}$$

minimizes the spread which is then

$$h^{3/4} \ , \quad \text{for} \quad \varepsilon < h^{3/2} \ ,$$

$$\sqrt{\varepsilon} \ , \text{ for } \quad \varepsilon \geq h^{3/2} \ .$$

The latter spread is the same as in the continuous problem (1.1), of course. For small ε the crosswind spread has thus been reduced from $\sqrt{h}$ (in the streamline diffusion and (h,h)-methods) to $h^{3/4}$.

In [16] there will also be given pointwise error estimates. At a point (x_0, y_0) and with R_1 the region depicted,

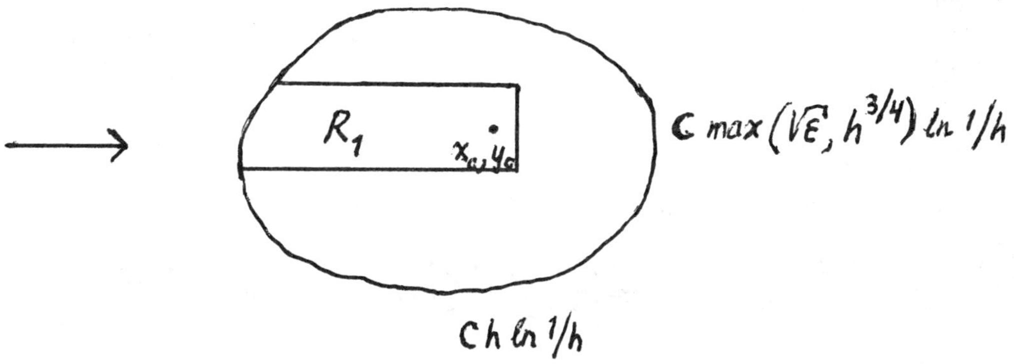

one has for the method (5.1) with $\varepsilon_0 = h^{3/2}$,

$$(5.2) \quad |(u^h-u)(x_0,y_0)| \le Ch^{5/4} \; ||u||_{C^2(R_1)} + \text{NGI}$$

whereas for the streamline diffusion method (1.7) the best presently known is

$$(5.3) \quad |(u^h-u)(x_0,y_0)| \le Ch||u||_{C^2(R_2)} + \text{NGI}$$

where now R_2 has crosswind thickness $\sqrt{h}\ \ell n\ 1/h$. Of course, if $u \in H^r(R_2)$ the estimate (4.2) combined with an inverse estimate will lead to a pointwise estimate of order $h^{r-3/2}$. For the (h,h)-method (1.5),

$$(5.4) \quad |(u^h-u)(x_0)| \le Ch^{3/4} \; ||u||_{C^2(R_2)} + \text{NGI}$$

seems to be the best known estimate although this ought to be improvable to $O(h)$.

The proofs of the pointwise estimates above involve, as is usual in present finite element technology, a careful analysis of an approximate Green's function (or some similar object such as a weighted norm). For an early example of the Green's function and its exponential decay properties, see [11, p. 574-575].

References

[1] O. AXELSSON and W. LAYTON, to appear.

[2] PH. BRENNER and V. THOMÉE, Stability and convergence rates in L_p for certain difference schemes, Math. Scand. 27, 1970, pp. 5-23.

[3] PH. BRENNER and V. THOMÉE, Estimates near discontinuities for some difference schemes, Math. Scand. 28, 1971, pp. 329-340.

[4] I. CHRISTIE, D.F. GRIFFITHS, A.R. MITCHELL, and O.C. ZIENKIEWICZ, Finite element methods for second order

differential equations with significant first derivatives, Internat. J. Numer. Meths. Engrg. 10, 1976, pp. 1389-1396.

[5] PH.G. CIARLET, The Finite Element Method for Elliptic Problems, North Holland, Amsterdam, 1978.

[6] J.E. DENDY, Two methods of Galerkin type achieving optimum L^2 accuracy for first order hyperbolics, SIAM J. Numer. Anal. 11, 1974, pp. 637-653.

[7] T. DUPONT, Galerkin methods for first order hyperbolics: an example, SIAM J. Numer. Anal. 10, 1973, pp. 890-899.

[8] T. DUPONT, The rate of convergence of Galerkin's method for first-order hyperbolic equations: two examples, to appear.

[9] W. ECKHAUS, Boundary layers in linear elliptic singular perturbation problems, SIAM Review, 14, 1972, pp. 225-270.

[10] W. ECKHAUS and E.M. deJAGER, Asymptotic solutions of singular perturbation problems for linear differential equations of elliptic type, Arch. Rat. Mech. Anal. 23, 1966, pp. 26-86.

[11] R. GORE, The dead do tell tales at Vesuvius, National Geographic 165, no. 5, May 1984, pp. 557-613.

[12] T.J.R. HUGHES and A. BROOKS, A multidimensional upwind scheme with no crosswind diffusion, in: Finite Element Methods for Convection Dominated Flows, T.J.R. Hughes, Ed., AMD, Vol. 34, ASME, New York, 1979, pp. 19-35.

[13] A.M. IL'IN, A differencing scheme for a differential equation with a small parameter affecting the highest derivative, Mat. Zametki 6, 1969, pp. 237-248 (Math. Notes 6, 1969, pp. 596-602).

[14] C. JOHNSON, U. NÄVERT, and J. PITKÄRANTA, Finite element methods for linear hyperbolic problems, Comput. Meths. Appl. Mech. Engrg., 1984, to appear.

[15] C. JOHNSON and J. PITKÄRANTA, An analysis of the discontinuous Galerkin method for a scalar hyperbolic equation, to appear.

[16] C. JOHNSON, A.H. SCHATZ, and L.B. WAHLBIN, to appear.

[17] R.B. KELLOGG, Analysis of a difference approximation for a singular perturbation problem in two dimensions, in: Boundary and Interior Layers - Computational and Asymptotic Methods, J.J.H. Miller, Ed., Boole Press, Dublin, 1980, pp. 113-117.

[18] B.P. LEONARD, A survey of finite differences of opinion on numerical muddling of the imcomprehensible defective confusion equation, in: Finite Element Methods for Convection Dominated Flows, T.J.R. Hughes, Ed., AMD, Vol. 34, ASME, New York, 1979, pp. 1-17.

[19] P. LESAINT and P-A. RAVIART, On a finite element method for solving the neutron transport equation, in: Mathematical Aspects of Finite Element in Partial Differential Equations, C. deBoor, Ed., Academic Press, New York, 1974, pp. 89-124.

[20] J.L. LIONS, Perturbations Singulières dans les Problèmes aux Limites et en Controle Optimal, Springer, Lecture Notes in Math, 323, New York, 1973.

[21] U. NÄVERT, A finite element method for convection-diffusion problems, Department of Computer Sciences, Chalmers University of Technology, Göteborg, Sweden, 1982.

[22] G.D. RAITHBY and K.E. TORRANCE, Upstream-weighted differencing schemes and their application to elliptic problems involving fluid flow, Computer and Fluids, 2, 1974, pp. 191-206.

[23] G.R. RICHTER and R.S. FALK, An analysis of a finite element method for hyperbolic equations, in: Advances in Computer Methods for Partial Differential Equations - VI , Vichnevetsky and Stepleman, Eds., IMACS, New Brunswick, Brussels, 1984, pp. 297-300.

[24] V. THOMÉE, Spline approximation and difference schemes for the heat equation, in : The Mathematical Foundations of the Finite Element Method with Applications to Partial Differential Equations, A.K. Aziz, Ed., Academic Press, New York, 1972, pp. 711-746.

[25] V. THOMÉE and B. WENDROFF, Convergence estimates for Galerkin methods for variable coefficient initial value problems, SIAM J. Numer. Anal., 11, 1974, pp. 1059-1068.

[26] L.B. WAHLBIN, A dissipative Galerkin method for the numerical solution of first order hyperbolic equations,

in : Mathematical Aspects of Finite Elements in Partial Differential Equations, C. deBoor, Ed., Academic Press, New York, 1974, pp. 147-169.

[27] L.B. WAHLBIN, A dissipative Galerkin method applied to some quasi-linear hyperbolic equations, RAIRO Anal. Numer. 8, 1974, pp. 109-117.

ADAPTIVE GRID METHODS FOR HYPERBOLIC PARTIAL DIFFERENTIAL EQUATIONS

JOSEPH OLIGER*

Abstract. Adaptive grid methods developed by the author and several coworkers for hyperbolic partial differential equations are described. The design principles underlying these methods are discussed and briefly compared with other related work. The resulting methods use piecewise uniform composite grids with different mesh intervals in both space and time constructed to explicitly control the resulting error. These grids are constructed to yield adequate approximations efficiently, but there is no attempt to optimize the placement of grid points. The system developed communicates with a user through a user supplied program to solve the equations on a uniform grid on a line segment, rectangle, or rectangular parallelopiped. A description of the grid structures generated and the computational algorithm used is followed by a discussion of the required form of stability estimates that the difference equations must satisfy to obtain an *a posteriori* error estimates and of the classes of problems that can be treated using this approach. Computational experience with these methods is discussed briefly with accompanying references.

* Department of Computer Science, Stanford University, Stanford, CA 94305.
Partially supported by ONR contract N00014-75-C-1132.

1. Background and Design Principles. Adaptive numerical methods have long been the standard for many problems such as the solution of initial and boundary value problems for ordinary differential equations and for quadrature. The solutions of hyperbolic or incompletely parabolic partial differential equations often vary greatly in character over their domains and it is natural to consider using methods which take advantage of this to achieve better results or reduce costs. Over the past ten years or so several lines of development of such methods have been pursued by various researchers. These methods are just beginning to emerge from the research environment as useful computational tools. A recent survey paper by Hedstrom and Rodrigue [15] compares several of the approaches as they apply to hyperbolic equations and a recent conference proceedings [1] is an excellent source of further information on most of these ideas. It is probably true that several of these different ideas will prove to be advantageous for different classes of problems, and that it will prove useful to combine several of the ideas in some instances. These different approaches will only be discussed here briefly to clarify and illustrate several design issues. The systems designed at Stanford by Berger, Bolstad, Gropp and Oliger ([2], [6], [9], [12], [13]) will then be described and discussed in these terms.

For a given computational environment one wants to achieve *adequate results* (desired accuracy) at minimal cost. This cost will include both *computer expense* and *program development expense* which need to be given appropriate weights. The computer expense is a function of both the number of *degrees of freedom* used to represent the approximation and of the *complexity of the algorithm* used to compute them.

Several moving mesh algorithms use a fixed number of points and try to find optimal placement for them, see [15]. Error estimates have been obtained by Dupont [10] for the moving finite element method of Miller *et al*, see [15] for references, but none of these methods explicitly confront the issue of whether or not the results are adequate (sufficiently accurate). In contrast, the adaptive mesh methods discussed here and the adaptive finite element methods of Bieterman and Babuška [7,8] explicitly relate the grid to desired accuracy, drop the quest for an optimal grid and instead focus upon generating an adequate *good* grid.

It is clear that one can achieve an adequate approximation of a solution using fewer degrees of freedom if one allows grid points or finite elements to be distributed in an arbitrary manner. However, this can only be accomplished if the positions of the points are computed and stored. Miller formulates and solves an optimization problem which incorporates the local error for nodal placement. Most of the other moving mesh methods solve equations based upon heuristic measures of difficulty or upon known or expected characteristics of the solution such as jumps or large derivatives. Our approach utilizes piecewise regular grid structures so that the overhead associated with the refinements is proportional to the number of refined zones rather than the total number of points in the refinements or grid. We are using more grid points but have less overhead (storage and computational effort) per point. It is clear that cost functions associated with these different methods are complicated and can be expected to intersect. We need considerably more experience to make judgements about which approach one should choose for a given problem. However, we have always had an underlying model situation in mind as we have developed our adaptive grid methods: problems whose solutions are rather uniformly difficult to approximate over most of their domain with a few more difficult zones which one wishes to approximate well. Prototypical examples are found in weather prediction and oceanographic computations where the difficult features are fronts and currents.

It was clear from the outset that these were going to be difficult programs to write and that one would need to provide potential users of these methods with software aids if these methods were to be used for all but a very few problems. The program organization and data structures we have used are described by Berger and Oliger [6] and Berger [3]. We have not tried to provide a monolithic program package which implements these methods and might be used as a typical numerical subroutine or applications package. We have instead focused upon developing a system which takes care of data management, error control and refinement generation. The user is required to program a standard method on a simple region, a line segment with uniform grid in one space dimension, a uniform grid on a rectangle in two space dimensions, *etc.* Of course, the domain and appropriate boundary approximations are required from the user in another program as well. These user provided programs are then used by our system on adaptive grids generated by the system. We have only used our methods on problems in two space dimensions which have been mapped to domains which are unions of rectangles to date. We plan to continue the development of these systems to make them easier for others to use and to handle more complicated geometries without excessive demands on the user. Component grid methods like those used by Starius [18] and B. Kreiss [17] are currently being combined with our method to handle more complicated geometries.

This program structure allows one to easily use different methods on different portions of the domain, *e.g.*, one might want to use a higher order method where the solution is smooth and a special method designed for shocks where they occur. Preliminary computations of this sort were done by Bolstad [9]. One can also simultaneously execute the algorithm on different subregions using this same structure to utilize parallelism. W. Gropp is currently investigating this possibility.

2. Description of the Algorithm. We will describe our method as it might be used for a hyperbolic problem in two space dimensions. The restriction to one space dimension and extension to three space dimensions are both obvious. In one space dimension there are several obvious simplifications that can be made. We will also restrict our discussion to the implementation of explicit methods. At this point one can think of using almost any method of choice but minor restrictions will arise later when we discuss error estimation and the method will generally need to be dissipative to obtain the necessary form of stability estimate which is discussed in the next section.

We begin by describing the spacial grid structure. At the start of a computation the coarsest or base grid is specified by the user. this base grid denoted by G_0 will remain fixed for the duration. We generally use the term grid to refer to the convex hull of a point set rather than the point set. This base grid may be composed of several possibly overlapping component grids, $G_{0,j}$. Each component grid is locally uniform in some coordinate system. For simplicity of exposition we assume that we have equal mesh spacing in both coordinates x and y, say of h_0 on all of the components $G_{0,j}$. During a computation refined subgrids will be created adaptively to control the error. We create subgrids which are rectangles of arbitrary rotation. This allows us to reduce the anisotropic errors associated with numerical methods. These subgrids are not patched into the coarse grid but should be thought of as overlying it. We do not *move* these subgrids during the computation but create and destroy them periodically as required so that they do *follow* the phenonema for which they are required. These restrictions allow us to use a simple data structure [3,6].

We will generate even finer subgrids within the boundaries of our initial refinement if they are required. We label the subgrids of G_0 as G_1 and those of G_1 as G_2, *etc.* In this way we generate a nested sequence of grids with finer and finer discretizations. A point in the problem domain may thus lie under several grids in which case we will define the solution there in terms of the finest grid which covers it.

In practice we assume a possible set of grid increments $\{h_0, h_1, \ldots, h_{\max}\}$ has been specified in advance where each h_l is an integer multiple of h_{l+1}. We have found a factor of 4 to be a generally good choice through experimentation. We maintain all intermediate level grids and require that every point of a grid at level l be interior to a grid of level $l-1$ unless it is on the boundary of the problem domain. However, not all points in a fine grid are necessarily interior to the same coarse grid. Figure 2.1 illustrates a possible grid configuration.

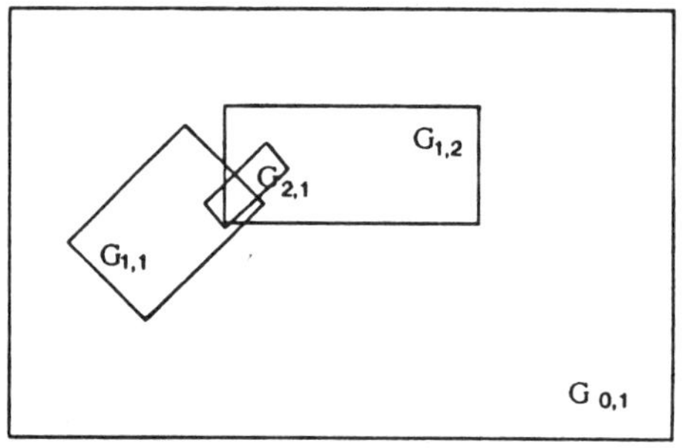

Figure 2.1. A Sample Grid Structure

The complete grid structure is denoted by

$$G = \cup_l G_l$$

where the grid at level l is the union of rectangular components

$$G_l = \cup_j G_{l,j}.$$

We will now describe the integration algorithm on a grid structure like that described above. There are three main components: (1) the time integration using finite differences that is done on each grid, (2) the error estimation and grid generation, and (3) the grid-to-grid operations done every time step that arise because of the mesh refinement.

Since each grid is defined separately the solution on each grid can be integrated in time independently of the other grids except for its boundary values. We use the same mesh ratio $\lambda = k_l/h_l$ on all grids, k_l is the time-step used on

G_l. Our mesh refinement is thus one of both space and time. This contributes greatly to the efficiency of the algorithm since the restrictive smallest time step on the finest grid is only used there. The grids are updated in order from coarsest to finest. One coarse grid time-step is the basic unit of the algorithm. At every coarse grid time-step all grids are defined. See Figure 2.2 which illustrates the space-time grid structure.

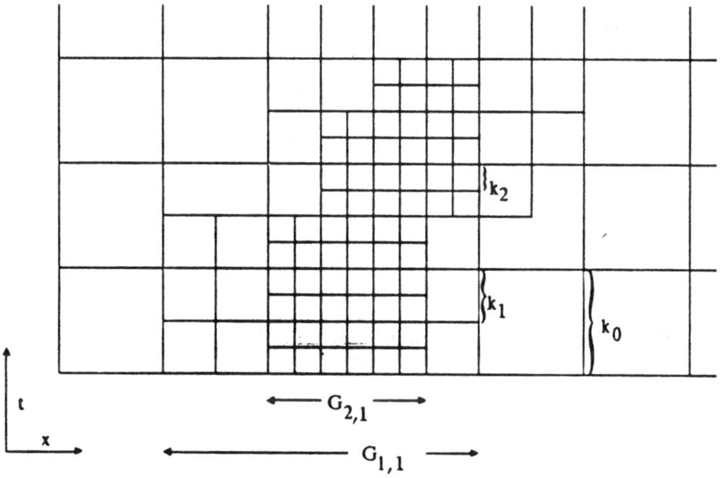

Figure 2.2. Space Time Grid Structure

Since we use rotated rectangles the difference equations will, in general, need to be transformed into the rotated coordinate system. This can be done in an automatic way so that the user supplied integrator can be separated from the adaptive mesh routines, see [6]. However, if the physical problem is invariant to translation and rotation we can use the same difference equations on each grid.

Error estimation and subsequent regridding is the second major task. Every several time steps we estimate the error at all grid points and adjust the grid structure. If a new grid is created, its initial values are interpolated from the finest existing grids underlying it. If a grid is no longer needed, it is discarded. For hyperbolic problems one can estimate the speed of propagation to determine how fast a phenomenon needing refinement can move. If a buffer zone is added to the fine grid we can lengthen the time interval over which the grid will be adequate and avoid regridding so often, but this does introduce extra points and require more work at each step. Typical calculations have optimal regridding frequency of every 3-4 steps. If there are several levels of refinement we apply this procedure at each level. Finer grids are regridded more often than coarser ones.

We use estimates of the local truncation error to determine where we need to refine. We generally use a variant of Richardson extrapolation, which we call step doubling, to obtain these estimates. The following result from [12] both explains this procedure and justifies its use.

Theorem: Let $u(x,t)$ be the solution to the partial differential equation and $v(x,t)$ be the approximation given by $v(x,t+k) = Q(h)v(x,t)$, with $v(x,0) = u(x,0)$ and Q consistent ($Q = I + O(h)$). Assume that $v(x,t)$ is continuous in t. Further, let $u(x,t+k) - Q(h)u(x,t) = h^{p+1}T(x,t) + O(h^{p+2})$, with $T(x,t)$ continuous in t. Finally, let

$$\tilde{v}(x,t+2k) = Q(2h)v(x,t)$$
$$v(x,t+2k) = Q^2(h)v(x,t)$$

Then

$$\frac{v(x,t+2k) - \tilde{v}(x,t+2k)}{2(2^p - 1)} = h^{p+1}T(x,t) + O(h^{p+2})$$

Having so obtained $T(x,t)$ above, we require a refinement with grid interval h_l which satisfies $|T|h_l^{p+1} \leq \varepsilon h_l$ where ε is a user supplied parameter – its choice is discussed in [12]. We successively refine until this inequality is satisfied. We can then obtain global error estimates as discussed in Section 3. Step doubling as described here can only be used when the spacial and temporal orders of accuracy of the method are equal. Other methods of estimation are discussed in [12].

We finally need to describe the required interaction between grids to complete the specification of the algorithm. There are three such tasks. First, we need to specify the boundary values on the refined grids. If a boundary value is in the interior of a different grid at the same level, we can interpolate its value form that grid. If there are no intersecting fine grids we compute the required boundary values form the underlying coarse grids where the error estimate is satisfied, otherwise the refinement would extend further. We might simply interpolate the needed values to the same accuracy as that of the method being used. The stability of such procedures and other alternatives are discussed by Berger [4]. Interpolation is satisfactory for use with the Lax-Wendroff and Mac Cormack methods, *e.g.*

The second intergrid operation is updating. When a fine grid is nested in a course grid, the coarse grid values underlying the fine grid are updated by interpolating them from the fine grid. If this is not done the solution on the coarse grid can smear out so that it will appear that refinement is unnecessary and/or possibly pollute the boundary values of the refined grid.

The final communication operation is averaging. This arises when two subgrids at the same level of refinement overlap. To make the approximation well-defined we use averaged values to represent the solution in the overlapped zone.

There are still several issues remaining but we have completed our description of the essential features of the algorithm. One remaining issue is the construction of the rectangles once we have determined that given points need to be included in refined grids. It is accomplished in three stages. First the points needing refinement are flagged, then the flagged points are grouped into clusters, and finally the clusters are covered by rectangles which best fit the cluster in a certain sense. This is discussed in Berger and Oliger [6] and we refer there for details.

Another important issue is the choice of data structures for which we refer to Berger and Oliger [6] and Berger [3].

3. Stability, Error Control and Limitations. In this section we will discuss the form of stability estimate which the method must satisfy in order to obtain *a posteriori* error estimates and limitations on the class of problems which can be approximated with these methods.

Suppose that we want to approximate the equation

$$u_t + \sum_{\nu=1}^{d} A_\nu u_{x_\nu} + Cu = f \tag{3.1}$$

on a domain $\Omega \times [0,T]$ where $\Omega \subset \mathbf{R}^d$, with initial data

$$u(x,0) = u_0(x), \quad x \in \Omega,$$

and boundary conditions and data given by

$$Bu(x,t) = g(t) \quad (x,t) \in \partial\Omega \times [0,T].$$

We will say that this problem is well-posed if

$$(3.2)\quad \|u(x,T)\|_\Omega + \|u(x,T)\|_{\partial\Omega\times[0,T]} \le K_T\left(\|u_0\|_\Omega + \|g\|_{\partial\Omega\times[0,T]} + \|F\|_{\Omega\times[0,T]}\right)$$

where the norms are L_p norms over the domains indicated.

If we consider the adaptive grid method we have described here which uses piecewise uniform grids on slabs $\Omega^j = [t_{j-1} \le t \le t_j]$ and is reinitialized for each such slab, we immediately see that our method must satisfy a stability estimate of the same form as (3.2) with discrete ℓ_p norms instead of the L_p norms. We must be able to bound $\|v(x,t_j)\|_\Omega$ to guarantee that we can restart our procedure and $\|v(x,t)\|_{\partial\Omega\times[t_{j-1},t_j]}$ to provide boundary data where v is our difference approximation. This is a stronger stability estimate than that used by Gustafsson, Kreiss and Sundström [14] and unfortunately there is no general stability theory to guarantee such estimates. One can only resort to using the energy method, summation by parts.

One can establish an ℓ_p error estimate by applying the stability estimate of the form (3.2) to the error function $e = u - v$ repeatedly on each regular subgrid of each slab and then on each slab. This introduces a factor of K_t^m where m counts the slabs. We cannot bound m as h_l, $k_l \to 0$, so we must have $K_T = 1 + O(T)$ if we are to obtain an estimate. Therefore, we can only use methods like those discussed here for problems where $K_T = 1 + O(T)$ in the well-posedness estimate (3.2). Similarly, the interpolation which is used to reinitialize the procedure must be done with the same order of accuracy as that of the interior difference approximation if an optimal order of convergence is to be obtained. For some problems, methods and strategies, the interior boundary approximations can be one order less accurate – one must show that the number of refined zeros in each slab remains bounded. This is discussed in detail in [12].

The detailed error estimate following from (3.2) as indicated above is contained in Gropp and Oliger [12]. As mentioned in the previous section, a global error bound which is $O(\varepsilon)$ can be obtained if the local errors are pointwise bounded by $h\varepsilon$. However, this is stronger than needed and a modified algorithm given in [12] computes the h_l required for a given rectangle covering a given cluster, which is a function of the ℓ_p norm of the local truncation error over the rectangle rather than a pointwise bound and is more appropriate for the ℓ_p estimate.

4. Computational Experience. In this final short section we will summarize our experience with these methods. In Bolstad [9] results are given using a method which is similar, but not exactly the same as that described here, for both linear and nonlinear problems in one space dimension. Gropp [11] has done a sample computation for a scalar nonlinear conservation law in two space dimensions with another method that is similar, but different, than that described here – the grid refinements were all aligned with the base grid. Berger [2] and Berger and Oliger [6] include sample calculations in both one and two space dimensions for both linear and nonlinear problems. Gropp [13] has used this method for the solution of a transport equation in one space dimension. Results for gas dynamics in one space dimension are given in Hedstrom *et al* [16]. Berger and Jameson [5] have reported results with the Euler equations in two space dimensions.

These experiments have confirmed that one can obtain the optimal order of convergence and that the associated overhead is surprisingly low. We have generally measured performance by comparing computer time for adaptive mesh calculations with that of uniform fine grid calculations which achieve the same accuracy. This ratio is very problem dependent, as one must expect, but we have found that we can often get speed-ups of from 5 to 10 and sometimes as much as 20. We were very encouraged to find that the overhead was small enough that we were able to do a calculation with between 2/3 and 3/4 of the domain refined without running slower than a uniform fine grid code without the adaptive overhead. We expect that our efficiency will be even greater in three space dimensions – but have not yet done any such calculations.

We are currently developing an adaptive numerical weather prediction code and extending the method to handle coupled elliptic equations for applications such as incompressible flows.

References.

[1] I. BABUŠKA, J. CHANDRA and J.E. FLAHERTY eds., *Adaptive Computational Methods for Partial Differential Equations*, SIAM, Philadelphia, 1983.

[2] M.J. BERGER, *Adaptive mesh refinement for hyperbolic partial differential equations*, Ph.D. Dissertation, Computer Science Dept. Stanford University, 1982.

[3] M.J. BERGER, *Data structures for adaptive mesh refinement* in *Adaptive Computational Methods for Partial Differential Equations*, SIAM, Philadelphia, 1983, pp. 237-251.

[4] M.J. BERGER, *Stability of interfaces with mesh refinement*, Report 83-42, ICASE, NASA Langley Research Center, 1983.

[5] M.J. BERGER, and A. JAMESON, *Automatic Adaptive Grid Refinement for the Euler Equations*, Report DOE/ER/03077-202, Courant Mathematics and Computing Laboratory, New York University, 1983.

[6] M.J. BERGER and J. OLIGER, *Adaptive mesh refinement for hyperbolic partial differential equations*, J. Comp. Phys., **53** (1984), pp. 484-512.

[7] M. BIETERMAN and I. BABUŠKA, *The finite element method for parabolic equations, I. A posteriori error estimation*, Numer. Math., **40** (1982), pp. 339-371.

[8] M. BIETERMAN and I. BABUŠKA, *The finite element method for parabolic equations, II. A posteriori error estimation and adaptive approach*, Numer. Math., **40** (1982), pp. 373-406.

[9] J. BOLSTAD, *An adaptive finite difference method for hyperbolic systems in one space dimension*, Ph.D. Dissertation, Computer Science Dept., Stanford University, 1982.

[10] T. DUPONT, *Mesh modification for evolution equations*, Math. Comp., **39** (1982), pp. 85-108.

[11] W.D. GROPP, *A test of moving mesh refinement for 2-D scalar hyperbolic problems*, SIAM J. Sci. and Stat. Comp., **1** (1980), pp 191-197.

[12] W.D. GROPP and J. OLIGER, *Error estimation, stability and convergence for adaptive difference methods for time dependent problems*, to appear.

[13] W.D. GROPP, *Numerical solution of transport equations*, Ph.D. Dissertation, Computer Science Dept., Stanford University, 1981.

[14] B. GUSTAFSSON, H.-O. KREISS and A. SUNDSTRÖM, *Stability theory of difference approximations for mixed initial boundary value problems. II*, Math. Comp., **26** (1972), pp. 649-686.

[15] G.W. HEDSTROM and G.H. RODRIGUE, *Adaptive-grid methods for time- dependent partial differential equations*, in *Multigrid Methods*, W. HACKBUSCH and U. TROTTENBERG eds., Springer-Verlag, New York, 1982, pp. 474-484.

[16] G.W. HEDSTROM, G.H. RODRIGUE, M. BERGER and J. OLIGER, *Adaptive mesh refinement for 1-dimensional gas dynamics*, in *Scientific Computing*, R. Stephenson *et al* eds., IMACS North Holland, New York, 1983, pp. 43-47.

[17] B. KREISS, *Construction of curvilinear grids*, SIAM J. Sci and Stat. Comp., **4** (1983), pp. 270-279.

[18] G. STARIUS, *On composite mesh difference methods for hyperbolic differential equations*, Numer. Math., **35** (1980), pp. 241-255.

SOME SIMPLE STABILITY RESULTS FOR INVERSE SCATTERING PROBLEMS

JOHN FAWCETT*

The initial boundary value problem central to this abstract is:

(1a) $R_{tt} - R_{yy} + q(y)R = \delta(t)\delta(y)$

(1b) $R_y + hR = 0 \qquad y = 0,\ t \geq 0$

(1c) $R(y,t=0) = R_t(y,t=0) = 0$

The impulse function $F(t)$, defined as the regular part of $R_t(y=0,t)$, is considered known from observation. The inverse problem is to determine $(q(y),h)$ from $F(t)$. Symes [4] has shown the stability of the mapping $F \to (q,h)$ with respect to various Sobolev norms. We give analytical and numerical examples which show that mapping $F \to (q,h)(t\varepsilon[0,2\tau], y\varepsilon[0,\tau])$ is not stable with respect to the L_∞ norm ($||F_1(t)-F_2(t)||_\infty \to 0$ does not imply $||q_1(y)-q_2(y)||_\infty + |h_1-h_2| \to 0$; $t\varepsilon\ [0,2\tau]$, $y\varepsilon[0,\tau]$). However, we prove that the mapping $F \to (\int_0^y q(u)du, h)$ is stable; there exists a $\delta(\varepsilon)$ such that $||F_1(t)-F_2(t)||_\infty < \delta$ implies $||\int_0^y (q_1-q_2)du||_\infty + |h_1-h_2| < \varepsilon$, for ε an arbitrary positive number, and $F_1(t)$, $F_2(t)$ two impulse response functions. The proof of this result is based upon the analysis of the linear Gelfand-Levitan [3] integral equation.

As an application of the above stability result, we prove a stability result for the one-dimensional acoustic equation.

---

*Department of Mathematics, Stanford University, Stanford, California U.S.A. 94305

(2a) $\rho(y)U_{tt} - (\rho(y)c^2(y)U_y)_y = \delta(y)\delta(t)$

(2b) $U_y = 0 \quad y = 0,\ t \geq 0$

(2c) $U(y,t = 0) = U_t(y,t=0) = 0$

The Liouville transformation changes (2) into the form of (1). By proving an additional stability result for ordinary differential equations, we can then establish the stability of the mapping $F(t) \to (\eta(x), \eta'(x);\ t \in [0,2\tau],\ x \in [0,\tau])$; there exists a $\delta(\varepsilon)$ such that $||F_2(t) - F_1(t)||_\infty < \delta(\varepsilon)$ implies $||\eta_1(x) - \eta_2(x)||_\infty < \varepsilon$, $||\eta_1'(x) - \eta_2'(x)||_\infty < \varepsilon$, for ε an arbitrary positive number and F_1, F_2 two impulse response functions corresponding to (2). Here $\eta(x)$ is the impedance $\eta(x) \equiv (\rho(y(x))c(y(x)))^{1/2}$, and x is the travel time variable. A similar stability result, derived in a different fashion can be found in Symes [5].

Analytical and numerical examples illustrate all the results. Further detail can be found in Fawcett [1],[2].

REFERENCES

[1] J. FAWCETT, I. Three Dimensional Ray-Tracing and Ray-Inversion in Layered Media, II. Inverse Scattering and Curved Ray Tomography with Applications to Seismology, Ph.D. Thesis, California Institute of Technology, Pasadena, 1983.

[2] J. FAWCETT, On the stability of inverse scattering problems, to appear in Wave Motion.

[3] I.M. GELFAND and B.M. LEVITAN, On the determination of a differential equation from its spectral function, AMS Translation Series 2, 1, pp. 253-304.

[4] W. SYMES, Inverse boundary value problems and a theorem of Gelfand and Levitan, Journal of Mathematical Analysis and Applications, 71(1979), pp. 379-402.

[5] W. SYMES, Stable solution of the inverse problem for a smoothly stratified medium, SIAM Journal of Mathematical Analysis, 12(1981), pp. 421-453.

INVERSE SCATTERING FOR STRATIFIED, ISOTROPIC ELASTIC MEDIA USING THE TRACE METHOD

D. C. STICKLER*

Trace formula methods were introduced by Trubowitz [1] and Deift and Trubowitz [2], and in the latter paper it was used to study one-dimensional inverse quantum mechanical scattering. Using these ideas Stickler [3] has derived trace formulas for the matrix coefficients of a one-dimensional wave equation which supports N distinct types of waves. In the present analysis using the techniques developed in [3] the trace method is used to obtain trace formulas for the Lame' parameters and density in a stratified, isotropic elastic medium.

The trace formulas are functional relationships between λ, μ, ρ and a reflection coefficient $R(k)$, obtained from measured data, and two solutions of elastic equation with known initial values; these two solutions are denoted by $u_2(z,k)$ and $u_3(z,k)$. The reflection coefficient data is obtained from a measurement of the displacement field, described below, due to a point harmonic source at a single fixed frequency.

References to other applications of the trace method are given in reference [3].

The Lame' parameters are denoted by $\lambda(z)$, $\mu(z)$ and the density $\rho(z)$. They are assumed known for $z \leq 0$, but unknown for $z > 0$, and are assumed to approach constant values λ_1, μ_1, ρ_1, as $z \to \infty$ and unknown but constant values λ_2, μ_2, ρ_2 as $z \to -\infty$. The condition

$$(1) \quad c_{s_1} \equiv \sqrt{\mu_1/\rho_1} \leq \sqrt{\mu_2/\rho_2} \equiv c_{s_2}$$

is assumed to hold.

*Courant Institute, New York University, New York, New York 10012

A cylindrically symmetric point harmonic body force with no angular component is located at $z_s<0$, and is assumed to excite no proper modes. The source thus has a radial and a z-coordinate and each component of the source excites both a radial and a z-component of the displacement field, which is represented by a 2x2 matrix reflection coefficient R(k), where the parameter k is the vertical wave number, and is defined explicitly below. An alternate measurement with some advantages is to measure both $g(r,z_r,z_s)$ and $g_{z_r}(r,z_r,z_s)$, $0\geq z_r\geq z_s$ for all $r\geq 0$.

For the source described above, the displacement field s(r,z) has z and r-components, but no ϕ component and does not depend on the angular coordinate. For a vector source of this type the displacement field is defined by

$$\text{(2a)}\qquad s(r,z) = \begin{pmatrix} s_r(r,z) \\ s_z(r,z) \end{pmatrix}$$

and the vector body force by

$$\text{(2b)}\qquad f(r,z) = \begin{pmatrix} f_r(r,z) \\ f_z(r,z) \end{pmatrix} .$$

The Hankel transform exploits the cylindrical symmetry of the problem and is defined by

$$\text{(2c)}\qquad u(z,\beta) \equiv \int_{r=0}^{\infty} J(\beta r)s(r,z)rdr = \begin{pmatrix} u_r(z,\beta) \\ u_z(z,\beta) \end{pmatrix}$$

where

$$\text{(2d)}\qquad J(\beta r) = \begin{pmatrix} J_1(\beta r) & 0 \\ 0 & J_0(\beta r) \end{pmatrix}$$

with $J_1(x)$ and $J_0(x)$ Bessel functions of the first kind. The self-adjoint differential equation satisifed by u(z,b) is given by

$$\text{(3)}\quad Mu = \frac{d}{dz} P \frac{d}{dz} u + \beta \left(N \frac{du}{dz} - \frac{d}{dz} \tilde{N}u\right) + Qu = F$$

where

$$P = \begin{pmatrix} \mu & 0 \\ 0 & \lambda+2\mu \end{pmatrix}, \quad N = \begin{pmatrix} 0 & -\lambda \\ \mu & 0 \end{pmatrix}$$

$$Q = \omega^2 \rho I - \beta^2 \begin{pmatrix} \lambda+2\mu & 0 \\ 0 & \mu \end{pmatrix}$$

and $F(z,B)$ is the Hankel transform of the bodyface.

Let $u_1(z,k)$, $u_2(z,k)$ and $u_3(z,k)$ denote the solutions of Eq. 3 which represent an outgoing wave as $z\to+\infty$, and outgoing wave as $z\to-\infty$ and an incoming wave as $z\to-\infty$, respectively. Note, that $u_2(z,k)$ and $u_3(z,k)$ satisfy known initial conditions as $z\to-\infty$ and thus can be calculated for $z\leq 0$. The Hankel transform of $g(r,z,z')$ is defined by $G(\bar{z},z',k)$ and satisfies

$$(4)\quad MG = \delta(z - z')I$$

with appropriate outgoing conditions as $|z|\to\infty$. The vertical wave number k is defined by

$$(5)\quad k = \sqrt{\left(\frac{\omega}{c_{s1}}\right)^2 - \beta^2},\ \mathrm{Im}\ k\geq 0$$

and in the k- plane $\beta(k)$ is chosen such that $\mathrm{Re}\beta\geq 0$, and thus $\beta(k)$ is analytic in the upper half k- plane.

The reflection coefficient $R(k)$ is obtained from a measurement of $g(r,z_r,z_s)$, $z_s\leq 0$, and $z_r\leq 0$ for all $r\geq 0$ or from a measurement of $\partial g/\partial z_r\ (r,z_r,z_s)$ and $g(r,z_r,z_s)$ for $z_s\leq z_r\leq 0$ for al $r\geq 0$.

The latter method offers some advantages. The Hankel transform of this data is taken to obtain in the second case, to obtain an impedance

$$(6)\quad Z(z_r,z_s,k) \equiv G(z_r,Z_s,k)\ \frac{dG^{-1}}{dz_r}\ (z_r,z_s,k)$$

The Green function $G(z_r,z_s,k)$ can be expressed in terms of $u_2(z,k)$, $u_3(z,k)$ and $R(k)$, and since $Z(z_r,z_s,k)$ is measured, u_3 and u_2 can be calculated at z_r and z_s, $R(k)$ can be recovered. The necessary relationship is provided by the condition that there are two constant 2x2 matrices, which are denoted by $R(k)$ and $T(k)$, such that

$$(7)\quad u_3(z,k) + u_2(z,k)\ R(k) = u_1(z,k)\ T(k)$$

The functional relationship $\tilde{\lambda}(u_2,u_3,R)$, $\tilde{u}(u_2,u_3,R)$, $\tilde{\rho}(u_2,u_3,R)$ are derived by exploiting the behavior of $G(z,z',k)$ for large k. To obtain the Lame' parameters and the density for $z>0$, these functional relations as substituted in the differential equations for u_2 and u_3 and the resulting non-linear equations solved; the initial values of u_2 and u_3 are known at $z=0$. The solution for each $z\geq 0$ is then substituted back into the functional relations $\tilde{\lambda}$, $\tilde{m}$, $\tilde{\rho}$ to obtain $\lambda(z)$, $\mu(z)$ and $\rho(z)$.

REFERENCES

[1] E. TRUBOWITZ, The inverse problem for periodic potentials, Comm. Pure and Appl. Math., Vol. XXX, pp. 321-328, 1977.

[2] P. DEIFT and E. TRUBOWITZ, Inverse scattering on the line, Comm. Pure Appl. Math., Vol. XXXII, pp. 121-151, 1979.

[3] D.C. STICKLER, Application of the trace method for inverse scattering to a medium which supports N types of waves, SIAM Journal of Appl. Math., accepted for publication.

A LAYER-STRIPPING SOLUTION OF THE INVERSE PROBLEM FOR A ONE-DIMENSIONAL ELASTIC MEDIUM

ANDREW E. YAGLE* AND BERNARD C. LEVY*

This paper presents a fast algorithm for recovering profiles of the density $\rho(z)$ and Lamé parameters $\lambda(z)$ and $\mu(z)$ of a layered elastic medium as functions of depth. It is assumed that a homogeneous half-space is located above the inhomogeneous medium that we seek to reconstruct, and that the surface of the medium acts like a free surface. Due to the vast difference in material parameters between the ground and the air, this assumption is quite reasonable.

The following experiment is performed. An impulsive planar P wave is obliquely incident on the medium. The horizontal and vertical velocities are measured at the surface $z=0$ as functions of time. The experiment is then repeated with impulsive planar SV waves. The angles of incidence of the P and SV plane waves with respect to the vertical are chosen (i.e. the point source data is stacked) so that the horizontal ray parameter p is the same for both experiments. Note that the desired responses for P and SV excitations could be obtained by an appropriate superposition of the responses to a P-wave source and to a mixed P- and SV-wave source.

Since the medium is elastic, there is continuous conversion between P-type and SV-type seismic waves as the inhomogeneous medium is penetrated. This makes the problem far more complex than the acoustic problem, for which several efficient reconstruction methods have been proposed. There may also be propagation of SH waves, if the impulsive SV-wave source is not completely polarized. Since SH waves are decoupled from P and SV waves, these waves will not be considered here. The interconversion of P and SV waves defines reflection coefficients from which the medium parameter profiles are reconstructed. It is the complexity of wave propa-

*Department of Electrical Engineering and Computer Science, Massachusetts Institute of Technology, Cambridge, Massachusetts 02139 gation in the elastic medium that allows the recovery of all three medium parameter profiles as functions of depth instead of travel time.

It is assumed that the density $\rho(z)$ and the Lamé parameters $\lambda(z)$ and $\mu(z)$ are differentiable functions. Then, the differential functions satisfied by the horizontal and vertical components u_x and u_z of the displacement, and by the horizontal and vertical stresses τ_{zx} and τ_{zz} on an element perpendicular to the z axis, are transformed into an equivalent system for the downgoing and upgoing P and SV waves propagating in the medium. This differential system is parametrized entirely by the reflectivity functions $r_P(z)$, $r_S(z)$, $r_C(z)$ which describe the local reflection at depth z of P to P, SV to SV and P to SV or SV to P waves, and by the transmission function $t_C(z)$ which describes the local transmission of P to SV or SV to P waves. This system can also be viewed as a vector Zakharov-Shabat wave system.

Our reconstruction algorithm works on a <u>layer-stripping principle</u>. At depth z, it is assumed that the downgoing and upgoing P and SV waves in both experiments have been computed, and that $\rho(z)$, $\lambda(z)$ and $\mu(z)$ are known. Then, to reconstruct the medium parameters over the segment $[z, z+\Delta)$, we note that the probing waves contain impulses, which by the method of characteristics can be used to obtain $r_P(z)$, $r_S(z)$ and $r_C(z)$. These reflectivity functions given in turn $\frac{d}{dz}\ln\rho(z)$, $\frac{d}{dz}\ln\alpha(z)$, and $\frac{d}{dz}\ln\beta(z)$, where $\alpha(z)$ and $\beta(z)$ denote the local P and SV wave speeds at depth z, which can be used to obtain the medium parameters at depth $z+\Delta$. In addition, $t_C(z)$ can be expressed in function of $\frac{d}{dz}\ln\rho(z)$, $\frac{d}{dz}\ln\alpha(z)$, and $\frac{d}{dz}\ln\beta(z)$, and the waves in both experiments can be computed at depth z+W. The effect of our layer stripping technique is therefore to identify and then strip away the layer $[z, z+\Delta)$. The reconstruction problem at depth $z+\Delta$ is then identical to what it was at depth z.

A physical interpretation of this procedure can be given in terms of a lattice filter showing how the first reflection of various wave types at each depth yields the medium reflectivities at that depth. A computer run of the algorithm on the synthetic impulsive plane wave responses of a twenty layer medium shows that the algorithm works satisfactorily. More details on this procedure can be found in [1].

<u>Acknowledgements</u>: The work of the first author was supported by the Exxon Education Foundation, and that of the second author was supported by the Air Force Office of Scientific Research under Grant AFOSR-82-0135A.

REFERENCES

[1] YAGLE, A.E. and LEVY, B.C., A layer-stripping solution of the inverse problem for a one-dimensional elastic medium, Report LIDS-P-1329, Laboratory for Information and Decision Systems, MIT, Cambridge, MA., September 1983, to appear in Geophysics.

ON CONSTRUCTING SOLUTIONS TO AN INVERSE EULER-BERNOULLI BEAM PROBLEM

JOYCE R. McLAUGHLIN*

The problem at hand is to determine a set of spectral data which will (uniquely) determine physical characteristics of a rod or beam. A method of determining these characteristics is then sought using the spectral data. The model which will be used is the Euler-Bernoulli model for the beam. Of primary importance is the following eigenvalue problem associated with this model

(1) $$\left(EIu_{xx}\right)_{xx} - \lambda \rho u = 0$$

(2) $$u_{xx}(0) = 0 , \left(EIu_{xx}\right)_{x}(0) = 0 , u(1) = 0 , u'(1) = 0.$$

The given spectral data will all be associated with this problem. It is desired to determine the flexural rigidity, EI, and the density, ρ, from the given data.

The motivation for the solution of this problem comes from, [1], where a Gel'fand-Levitan, [2], type integral equation approach succeeds for determining the coefficient A in

(3) $$\left(Aw'\right)' + \lambda A w = 0$$

(4) $$w'(0) = 0 = w(1)$$

from the spectral data contained in a spectral function. It is important to note that the boundary conditions may be fixed in advance and are not determined by the spectral data.

*Rensselaer Polytechnic Institute, Troy, New York 12181

The integral equation approach is best described after making the independent variable change $s = \int_0^x \left(\frac{\rho(v)}{EI(v)}\right)^{1/4} dv$. Defining $c^2 = (EI)^{3/4}\rho^{1/4}$ and $a^2 = \left(\frac{\rho}{EI}\right)^{1/2}$, we now consider consider the eigenvalue problem

$$(5) \qquad (a(c^2(au_s)_s)_s)_s - \lambda\, a^2 c^2 u = 0 \quad , \qquad 0 < s < L$$

$$(6) \qquad u(L) = 0 = u_s(L) \ ,$$

$$(7) \qquad (au_s)_s(0) = 0 = (c^2(au_s)_s)_s(0) \ ,$$

where now

$$L = \int_0^1 [\rho(v)/EI(v)]^{1/4}\, dv \ .$$

We seek $a > 0$, $c > 0$, $a \in C^3[0,L]$, $c \in C^2[0,L]$. The given data will be $a(0) = 1$, $c(0) = 1$, L and the spectral data (that contained in a spectral matrix, see [3]). The spectral data is the set of eigenvalues $0 < \lambda_1 < \lambda_2 < \ldots$ and two additional sequences $\gamma_i > 0$, $\xi_i < 0$, $i = 1,2,\ldots$ defined as follows. Let $\hat{u}(s,\lambda_i)$ be the eigenfunction associated with λ_i, with weighted L^2 norm, $\int_0^1 a^2c^2\hat{u}^2 dx = 1$, then $\hat{u}(0,\lambda_i) = \gamma_i$ and $\hat{u}_s(0,\lambda_i) = \xi_i$. Existence is not assumed, a priori, but it is determined through the integral equation approach. Asymptotic forms <u>are</u> assumed for λ_i, γ_i, ξ_i as well as conditions on a determinant.

The approach is to first develop a relationship between solutions of (5) and solutions of a known problem

$$(8) \qquad \{a_0[(c_0)^2(a_0 z_s)_s]_s\}_s - \lambda\, a_0^2\, c_0^2\, z = 0$$

$$(9) \qquad z(L) = 0 = z_s(L)$$

$$(10) \qquad (a_0 z_s)_s(0) = [(c_0)^2\,(a_0 z_s)_s]_s = 0$$

where a_0, c_0 are known and $a_0(0) = 1 = c_0(0)$. This relationship is defined in terms of a Volterra integral operator. This is, if $u(s,\lambda)$ satisfies (5) with initial conditions

$$u(0) = \gamma \ , \ u_s(0) = \xi \ , \ (a\ u_s)_s(0) = 0 = (c^2(a\ u_s)_s)_s(0)$$

and $z(s,\lambda)$ satisfies (8) with initial conditions

$$z(0) = \gamma \ , \ z_s(0) = \xi \ , \ (a_0 z_s)_s(0) = 0$$
$$= (c_0{}^2(a_0 z_s)_s)_s(0)$$

then u and z are related by

$$\text{(11)} \qquad a\ c\ u(s,\lambda) = a_0\ c_0\ z(s,\lambda) + \int_0^s k(s,t)\ a_0^2\ c_0^2\ z(t,\lambda)dt.$$

The kernel $k(x,t)$ is determined by requiring that $\{a\ c\ \hat{u}(s,\lambda_i)\}_{i=1}^{\infty}$ is a complete orthonormal set in L^2. As such the kernel satisfies an integral equation

$$\text{(12)} \qquad a_0c_0(s)f(s,t) + \int_0^s [a_0c_0(u)]^2k(s,u)f(u,t)du + k(s,t) = 0$$

where $f(s,t)$ is defined as follows. Let

$$\text{(13)} \qquad f(s,t) = \int_0^{\infty} \sum_{i,j=1}^{2} z_i(t,\lambda)z_j(s,\lambda)d(\rho_{ij} - \rho_{ij}^0)(\lambda) \ ,$$

where z_1 and z_2 satisfy (8) and the initial conditions $z_1(0) = 1$, $z_{1,s}(0) = 0$, $(a_0z_{1,s})_s(0) = 0$, $[(c_0)^2(a_0z_{1,s})_s]_s(0) = 0$ and $z_2(0) = 0$, $z_{2,s}(0) = 1$, $(a_0,z_{2,s})_s(0) = 0$, $[(c_0)^2(a_0z_{1,s})_s]_s(0) = 0$, respectively.

Further $\rho(\lambda) = \left(\rho_{ij}(\lambda)\right)$ is the 2×2 (spectral) matrix which is constant as a function of λ except for jump discontinuities at each eigenvalue $\lambda = \lambda_k$. The "jump" at $\lambda = \lambda_k$, see [3], can be shown to be

$$\begin{pmatrix} (\gamma_k)^2 & \gamma_k\ \xi_k \\ \gamma_k\ \xi_k & (\xi_k)^2 \end{pmatrix}. \tag{14}$$

The matrix $\rho^0(\lambda) = \left(\rho^0_{ij}(\lambda)\right)$ is the corresponding spectral matrix for (8), (9), (10).

From the integral relationship for solutions it is seen that

$$ac = a_0\ c_0 + \int_0^s k(s,t)a_0^2\ c_0^2\ (t)dt\ ,$$

$$ac(s) \int_0^s \frac{dw}{a(w)} = a_0\ c_0(s) \int_0^s \frac{dw}{a_0(w)}$$

$$+ \int_0^s k(s,t)a_0^2\ c_0^2(t)\ \left[\int_0^t \frac{dw}{a_0(w)}\ \right] dt.$$

An important feature of this method is that when all but a finite number of pieces of data for (5), (6), (7) and (8), (9), (10) are the same, the integral equation (12) has finite rank. The function k can thus be determined by solving a set of linear equations provided a certain determinant is non-zero. (Positivity conditions on additional determinants are also required in order to insure that $a > 0$, $c > 0$). It is important to point out that while k can be determined by solving a set of linear equations, the dependence of the solution on the data is in general nonlinear. That is a and c show an exponential-type dependence on the eigenvalues and are rational functions of the norming constants γ_i and ξ_i.

Bounds, giving a continuity result, on a and c in terms of a_0, c_0 and the differences $\lambda_i - \lambda_i^0$, $\gamma_i - \gamma_i^0$, $\xi_i - \xi_i^0$ can be determined. For example when

$$\delta = \sum_{i=1}^{\infty} \lambda_i^{-1/4} \left[|\lambda_i - \lambda_i^0| + |\gamma_i - \gamma_i^0| \lambda_i^{3/4} + \left| \frac{\xi_i}{\gamma_i} e^{\lambda_i^{1/4}} - \frac{\xi_i^0}{\gamma_i^0} e^{(\lambda_i^0)^{1/4}} \right| \lambda_i^{1/2} \right]$$

is sufficiently small then there exists $M > 0$ such that

$$\left\| \frac{(ac)_s}{ac} - \frac{(a_0 c_0)_s}{a_0 c_0} \right\|_\infty \leq M\delta .$$

References

[1] Joyce R. McLaughlin, Bounds for Constructed Solutions of Second and Fourth Order Inverse Eigenvalue Problems, Proceedings International Conference on Differential Equations, Birmingham, Alabama, 1983.

[2] I. M. Gel'fand and B. M. Levitan, On the Determination of a Differential Equation From Its Spectrum, Izv. Akad Nauk SSSR Ser. Math. 15 (1951), 309-360; Amer. Math. Soc. Trans., 1 (1955), 253-304.

[3] E. A. Coddington and Norman Levison, Theory of Ordinary Differential Equations, McGraw Hill Book Company, 1955.

THE INVERSE PROBLEM FOR THE EULER-BERNOULLI BEAM

G. M. L. GLADWELL*

This paper is concerned with the problem of constructing the continuous mass and stiffness distributions of a thin untwisted beam from a knowledge of spectral data. It is known (Barcilon, 1982) that three spectra, corresponding to three different sets of end conditions, are needed to determine the mass and stiffness distributions; this paper is concerned with the conditions which must be satisifed by the spectra for them to correspond to an actual beam. The paper is a sequel to Gladwell (1984) which solved the corresponding problem for a discrete model. That model consisted of N massless rigid rods of lengths $(\ell_n)_1^N$, connected by torsional springs of stiffness $(k_n)_1^N$ and having masses $(m_n)_1^N$ at the joints. Necessary and sufficient conditions on the spectral data were established for them to relate to a system with positive values of m_n, ℓ_n, k_n, and an explicit construction procedure was constructed for the determination of these quantities.

The (natural frequencies), ω^2 and mode shapes $w(x)$ of free undamped flexural vibration of a beam are the eigenvalues and eigenfunctions of the self-adjoint integral equation

$$\int_0^\ell H(x,s)u(s)ds = \omega^2 u(s)$$

where $u(x) = A^{1/2}(x)w(x)$, $H(x,s) = A^{1/2}(x)A^{1/2}(s)G(x,s)$

$$G(x,s) = \frac{E}{\rho}\int_0^{\min(x,s)} \frac{(x-t)(s-t)dt}{I(t)},$$

and $A(x)$, $I(x)$ are the area and second moment of area of the beam. Since the kernel $H(x,s)$ is <u>oscillating</u> (Gantmakher

*Solid Mechanics Division, University of Waterloo, Waterloo, Ontario, Canada

and Krein, 1950) the eigenvalues are positive and distinct, and the eigenfunctions $\{w_i(x)\}_1^\infty$ satisfy certain properties which may be expressed in terms of the end values $w_i(\ell)$ and $\theta_i(\ell) \equiv w_i'(\ell)$. The derivative $w_i'(x)$ is also the eigenfunction of an oscillating (but non self-adjoint) kernel; this yields further inequalities which must be satisfied at the free end.

The natural frequencies for the clamped-pinned and clamped-sliding beams are the zeros of the equations

$$\sum_{i=1}^{\infty} \frac{[w_i(\ell)]^2}{\bar{m}_i(\omega_i^2-\omega^2)} = 0 , \quad \sum_{i=1}^{\infty} \frac{[\theta_i(\ell)]^2}{\bar{m}_i(\omega_i^2-{}^2)} = 0 ,$$

respectively. If these two spectra are known, then the coefficients $[w_i(\ell)]^2/\bar{m}_i$ and $[\theta_i(\ell)]^2/\bar{m}_i$ may be computed. The inequalities mentioned above may therefore be expressed in terms of the spectra corresponding to clamped-free, clamped-pinned and clamped-sliding ends. These conditions are therefore <u>necessary</u> for the existence of a beam having these three spectra. A conjecture is made concerning a set of sufficient conditions.

REFERENCES

[1] V. BARCILON, On the inverse problem for the vibrating beam, Phil. Trans. R. Soc. Lond. A., 304(1982), pp. 211-252.

[2] G.M.L. GLADWELL, The inverse problem for the vibrating beam, Proc. Roy. Soc. Lond. A., 1984, to appear.

[3] F.R. GANTMAKHER and M.G. KREIN, Ostsillyatsionnye Matritsy i Malye Kolebaniya Mekhanicheskihk Sistem, Moscow, Leningrad, 1950.

FAR FIELD PATTERNS IN ACOUSTIC AND ELECTROMAGNETIC SCATTERING THEORY

DAVID COLTON*

A basic task in the investigation of the inverse scattering problem for time-harmonic acoustic and electromagnetic waves is the study of the class of far field patterns corresponding to the scattering of entire incident fields of a given wave number by a bounded obstacle. Indeed if T denotes the operator mapping the incident field and scattering obstacle onto the far field pattern, then the inverse scattering problem is to construct T^{-1} defined on the range of T, and the determination of this range is nothing more than the description of the class of far field patterns. Unfortunately, little is known concerning this class except for the well known fact that the far field patterns are entire functions of their independent (complex) variables for each positive fixed value of the wave number ([3]), i.e. the range of T is not all of $L^2(\Omega)$ where Ω is the unit sphere. We note that this implies that the inverse scattering problem is an improperly posed problem since the far field patterns are in practice determined from inexact measurements.

Recently Colton ([1]) and Colton and Kirsch ([2]) have investigated the case of acoustic scattering and asked the question if the class of far field patterns corresponding to a fixed scattering obstacle and all entire incident fields is dense in $L^2(\Omega)$. The rather surprising answer to this question is that, if the impedance of the scattering obstacle is positive, then the far field patterns are dense in $L^2(\Omega)$, whereas if the scattering obstacle is sound-soft or sound-hard, then the far field patterns are dense in $L^2(\Omega)$ if and only if there does not exist an (interior) eigenfuction that is an entire Herglotz wave function, i.e. a solution u of the Helmholtz equation defined in all of space such that

*Department of Mathematical Sciences, University of Delaware, Newark, Delaware 19711

$$\lim_{r\to\infty} \frac{1}{r} \int\int_{|x|<r} |(u(x)|^2 dx < \infty.$$

This phenomenon is rather unusual since in a wide variety of improperly posed problems in mathematical physics the range of the operator which one wants to invert is dense in (but not equal to) the Banach space in which the measurements are being made. Hence, the inverse scattering problem is peculiar even in the class of improperly posed problems. Furthermore, not only is the property of the far field patterns being dense very sensitive on the shape of the domain, but from physical considerations the interior eigenvalues should in fact have nothing to do with the exterior scattering problem at all.

The above results for acoustic scattering have been extended by Colton and Kress to the electromagnetic case. In particular, it was shown in [4] that the far field patterns of the electric fields corresponding to the scattering of entire incident fields by a bounded perfectly conducting obstacle are dense in the space of square integrable tangential vector fields defined on Ω if and only if there does not exist a Maxwell eigenfunction that is an electromagnetic Herglotz wave function, i.e. a solution $\{E,H\}$ of Maxwell's equations defined in all of space such that

$$\lim_{r\to\infty} \frac{1}{r} \int\int_{|x|<r} (|E(x)|^2 + |H(x)|^2 dx < \infty.$$

Future research will be concerned with the following questions: For which domains are the eigenfunctions entire Herglotz wave functions (a ball is one such domain)? What are the corresponding results for wave propagation in a penetrable non-homogeneous medium (partial results are contained in [2] and [5])?

REFERENCES

[1] D. COLTON, Far field patterns for the impedance boundary value problem in acoustic scattering, Applicable Analysis, 16(1983), pp. 131-139.

[2] D. COLTON and A. KIRSCH, Dense sets and far field patterns in acoustic wave propagation, SIAM J. Math. Anal., to appear.

[3] D. COLTON and R. KRESS, Integral Equation Methods in Scattering Theory, John Wiley, New York, 1983.

[4] D. COLTON and R. KRESS, Dense sets and far field patterns in electromagnetic wave propagation, SIAM J. Math. Anal., to appear.

[5] D. COLTON and A. KIRSCH, Dense sets and far field patterns for the transmission problem, to appear.

RENAISSANCE INVERSION

R. H. T. BATES*

A fully informative but intolerably cumbersome title for this abstract would be: an approach to interpreting and processing measured data in a way permitting the Born approximation to be usefully invoked in inverse-diffraction contexts to which it is conventionally inapplicable. Contraction to "Born-again inversion" would have been acceptable perhaps, but the chosen heading avoids any possible sectarian confusion. No rediscovery of a lost manuscript by Leonardo, nor yet even Boccaccio, is implied.

In a typical inverse diffraction problem the quantity which one wishes to recover depends upon the same number of parameters as do the measured data. However, the quantity that must first be reconstructed depends upon at least one more parameter. This "dimensionality difficulty", so-called[1], emphasizes that diffracted fields must somehow be continued analytically from the parts of free space where they are measured back into the region containing the inhomogeneities which cause the diffraction.

Exact inversion is so complicated that one is forced in practice to make stringent approximations[2]. The most popular approximation, which owes its many adherents to its extreme simplicity, is the Rayleigh-Gans[3] (or Born) approximation. It rests on two assumptions: (i) the incident field swamps the diffracted field throughout the inhomogeneous region, and (ii) multiple scattering is inconsequential.

Both of the above assumptions are unfortunately invalid in most realistic situations involving macroscopic wave motion (e.g. acoustics, electromagnetism, fluidic surface waves, etc.) - which seems not to attenuate the fervor of the aforesaid adherents! The latter are of course right to opt for

*Electrical and Electronic Engineering Department, University of Canterbury, Christchurch, New Zealand.

simplicity, but they should seek rational justifications for their prejudices.

It so happens that the article of faith can be substantiated, and fairly straighforwardly, by appropriate transformations and reinterpretations.

The trouble with assumption (i) is that it requires there to be negligible incremental delay (over that experienced in free space) when the wave motion penetrates the inhomogeneities. The incremental delay is very appreciable in almost all realistic situations, so that the Born approximation (as conventionally invoked) is only valid in practice for either isolated scatterers embedded in highly homogeneous media or extremely tenuous inhomogeneous media. Nevertheless, it transpires that the incremental delay problem can be circumvented by invoking the (LSJ)WKB transformation[4].

While the difficulties conventionally experienced over assumption (i) are largely ameliorated by the WKB transformation,[1,4] it has no effect on assumption (ii). However, by appropriate processing of wideband signals - by, for instance, separating them into many narrow bands and invoking the shift-and-add technique[5] - multiple-scattered signals can be strongly rejected[4].

Inversion algorithms can thus be Born-wise interpreted - reBorn in fact - even when the Born approximation, as conventionally employed and understood, is quite unjustifiable. Although this is comforting, it has an unfortunate practical disfigurement. It breaks down when the wave motion suffers attenuation on its passage through the inhomogeneous region.

There is a partial salve for this blemish. It is conveniently explained in the context of the basic wave equation characterizing a wide range of linear wave motions representable by a scalar wave function[2], ϕ say:

(1) $$\nabla^2\phi + \nabla\alpha\cdot\nabla\phi - \nu^2\ddot{\phi} - \beta\,\dot{\phi} + \gamma\,\phi = 0$$

where ν (the refractive index), α, β (representing absorption of the wave motion by the medium) and γ are arbitrary functions of space, and the free-space wave speed has been normalized to unity. This equation must be transformed to a more standard form before one can hope to use it as the basis of a useful inversion algorithm.

On introducing the new wave function

(2) $$\psi = \phi/A$$

the absorption term can be compensated by the following strategy. One begins by writing

(3) $a = \log_e A$

and substituting (2) into (1), which then becomes

$$\nabla^2\psi - \nu^2\ddot{\psi} + (2\nabla a + \nabla\alpha)\cdot\nabla\psi - (2\nu^2\dot{a} + \beta)\dot{\psi}$$

(4) $$+ (\nabla^2 a + \nabla a\cdot\nabla a + \nabla\alpha\cdot\nabla\alpha - (\ddot{a} + \dot{a}^2)\ \nu^2 + \gamma - \beta\dot{a})\psi = 0$$

On setting

(5) $a = -\beta t/2\nu^2$,

where t denotes time, (4) transforms to

$$\nabla^2\psi - \nu^2\ddot{\psi} - (t\nabla\beta/\nu^2 - \nabla\alpha)\cdot\nabla\psi + (t2\nabla(\beta/\nu^2)\cdot\nabla\beta/\nu^2$$

(6) $$- 2t\nabla^2\beta/\nu^2 + \beta^2/\nu^2 - 2t\nabla\alpha\cdot\nabla\beta/\nu^2 + 4\gamma)\psi/4 = 0$$

Provided $|\nabla\beta/\nu^2|$ is nowhere excessive and the diffraction process is rapid enough (with the origin of time chosen roughly half-way through the process), it must often be reasonable to approximate (6) by

(7) $\nabla^2\nu - \nu^2\ddot{\psi} + \nabla\alpha\cdot\nabla\psi + (\beta^2/\nu^2 + 4\gamma)\psi/4 = 0$

which can further usefully modified by writing

(8) $\theta = \psi/B$ with $b = \log_e B = -\alpha/2$

so that

(9) $\nabla^2\theta - \nu^2\ddot{\theta} + \mu\theta = 0$

where

(10) $\mu = \beta^2/4\nu^2 + \nabla\alpha\cdot\nabla\alpha/4 - \nabla^2\alpha/2 + \gamma$

The form of (9) is such that the above-outlined procedure for validating assumptions (i) and (ii) can be immediately invoked. Since rebirth has been achieved, it is appropriate to call (2), (3), (5) and (8) the "renaissance transformations."

After the LSJWKB transformation[4] is introduced, (9) takes on the form

(11) $\nabla^2 u - \ddot{u} + \sigma u = 0$

where u is appropriately called the "renaissance wave function." The implication is that it is the spatially varying function σ which is recovered from the inversion procedure. Finally, note that the way in which ϕ, ψ, θ, and u are defined ensures that they are all identical in free space, thereby permitting inverse diffraction problems to be posed in terms of any of these wave functions.

REFERENCES

[1] BATES, R.H.T. and MILLANE, R.P., Trans. IEEE AP-29(1981), pp. 359-363.

[2] BATES, R.H.T., Full-wave computed tomography: I. fundamental theory, Proc. IEE Part A, accepted for publication.

[3] JONES, D.S., Theory of Electromagnetism, Section 6.13, MacMillan, 1964.

[4] BATES, R.H.T. and MINARD, R.A., Proc. SPIE 413(1983), pp. 56-60.

[5] BATES, R.H.T. and ROBINSON, B.S., Ultrasonic Imaging, 3(1981), pp. 378-394.

ON THE EQUILIBRIUM EQUATIONS OF POROELASTICITY

KENNETH R. DRIESSEL*

In this report, we derive equations that govern the equilibrium (static) behavior of a porous material. We assume that on a microscopic scale (e.g., "grain size" in the case of a porous rock) the material consists of an elastic solid and fluid filled pores. We "average" the equations governing the microscopic equilibrium behavior to obtain equations on a larger, "macroscopic" scale (e.g., size of a core sample in the case of a porous rock). We obtain equations that are similar to the equations of elasticity but which have an additional term involving the pressure of the pore fluid.

For a macroscopically uniform isotropic poroelastic medium these equations are

$$\nabla \cdot \tau = 0$$

$$\tau = A\nabla u + Mp$$

where τ is the effective stress tensor of the medium, p is the pore pressure of the fluid, and A and M are tensors of the form

$$A_{ijkl} = \lambda\delta_{ij}\delta_{k\ell} + \mu(\delta_{ik}\delta_{j\ell} + \delta_{i\ell}\delta_{jk})$$

$$M_{ij} = \alpha\delta_{ij} .$$

Here λ, μ, and α are scalars.

Biot (1941, 1955, and 1956) has also derived equations governing the equilibrium behavior of a porous rock. Our equations have the same form as those given by Biot. We use a procedure that is more general than his. We do not explicitly solve a ("cell") problem on the microscopic scale;

*Amoco Research, Box 591, Tulsa, Oklahoma 74102

consequently, we do not obtain explicit relationships between the parameters appearing in the microscopic equations and those appearing in the macroscopic equations.

In particular, we do not explicitly relate the parameters λ, μ, and α to the Lamé parameters of the solid and the incompressibility modulus of the fluid. Consequently, λ, μ, and α generally must be determined by experiment. We describe a procedure for the experimental determination of λ, μ, and α.

To derive the above equations, we use the two space method of homogenization (see Bensousssan-Lions-Papanicolaou 1978, and Sanchez-Palencia 1980). The method of homogenization is closely related to the classical Lindstedt-Poincare perturbation technique. (See, for example, Nayfeh 1981 -especially Chapter 5, "The Linear Damped Oscillator" - or Kevorkian-Cole 1981 - especially Chapter 3, "Multiple Variable Expansion Procedure.") We do not assume that the reader is familiar with these techniques. We closely follow the work of Burridge-Keller (1981). They use the method of homogenization to derive equations governing the dynamic behavior of a porous material. The calculations that we perform are simpler than theirs. (Generally, equilibrium problems are easier than dynamic ones.) However, when we are not able to improve their argument, we simply use it.

REFERENCES

[1] BENSOUSSAN, A., LIONS, J.L., and PAPANICOLAOU, G.C., Asymptotic Analysis for Periodic Structures, North-Holland, Amsterdam, 1978.

[2] BIOT, M.A., General theory of three-dimensional consolidation, J. Appl. Physics, 12(1941), pp. 155-164.

[3] BIOT, M.A., Theory of elasticity and consolidation for a porous anisotropic solid, J. Appl. Phys., 26(1955), pp. 182-185.

[4] BIOT, M.A., General solutions of the equations of elasticity and consolidation for porous material, J. Appl. Mech., 23(1956), pp. 91-95.

[5] BURRIDGE, R., and KELLER, J.B., Poroelasticity equations derived from microstructure, J. Acoust. Soc. Am., 70(1981), pp. 1140-1146.

[6] KEVORKIAN, J., and COLE, J.D., Perturbation Methods in Applied Mathematics, Springer-Verlag, Berlin, 1981.

[7] NAYFEH, A.H., Introduction to Perturbation Techniques, Wiley, New York, 1981.

[8] SANCHEZ-PALENCIA, E., Non-Homogeneous Media and Vibration Theory, Lecture Notes in Physics 127, Springer-Verlag, Berlin, 1980.

GPST—A VERSATILE NUMERICAL METHOD FOR SOLVING INVERSE PROBLEMS OF PARTIAL DIFFERENTIAL EQUATIONS

Y. M. CHEN*

The Generalized Pulse-Spectrum Technique (GPST) is a versatile (and efficient) iterative numerical method for solving multi-parameter inverse problems of systems of nonlinear partial differential equations of the following type,

$$\sum_{j,k,\ell=1}^{m\ 3\ 3} \partial\{A_{ijk\ell}(\underline{x})a_{ijk\ell}(\underline{u})\partial u_j/\partial x_\ell\}/\partial x_k$$

$$+ \sum_{j,k=1}^{m\ \ 3} B_{ijk}(\underline{x})b_{ijk}(\underline{u})\partial u_j/\partial x_k$$

$$+ \sum_{j=1}^{m} C_{ij}(\underline{x})c_{ij}(\underline{u})u_j - D_i(\underline{x})d_i(\underline{u})\partial u_i/\partial t$$

$$- E_i(\underline{x})e_i(\underline{u})\partial^2 u_i/\partial t^2 = R_i(\underline{x},t),$$

$$\underline{u} \equiv (u_1,u_2,u_3,\ldots,u_m), \quad i = 1,2,3,\ldots,m,$$

$$\underline{x} \equiv (x_1,x_2,x_3) \in \Omega\ , \quad 0 < t \leq T < \infty,$$

with initial conditions

$$\left.\begin{array}{l}\underline{u}(\underline{x},0) = \underline{f}(\underline{x}),\\ \partial\underline{u}(\underline{x},0)/\partial t = \underline{g}(\underline{x}),\end{array}\right\} \underline{x}\in\Omega,$$

and boundary conditions

$$\left.\begin{array}{l}\alpha(\underline{x})\underline{u} + \beta(\underline{x})\partial\underline{u}/\partial\underline{n} = \underline{h}^{in}(\underline{x},t),\ \underline{x} \in \partial\Omega^{in},\\ \gamma(\underline{x})\underline{u} + \zeta(\underline{x})\partial\underline{u}/\partial\underline{n} = \underline{h}^{ex}(\underline{x},t),\ \underline{x} \in \partial\Omega^{ex},\end{array}\right\} 0 \leq t \leq T.$$

*Department of Applied Mathematics and Statistics, State University of New York Stony Brook, New York 11794, U.S.A.

In other words, GPST is capable to determine the unknown coefficients, $A_{ijk\ell}(\underline{x})$, $B_{ijk}(\underline{x})$, $C_{ij}(\underline{x})$, $D_i(\underline{x})$ and $E_i(\underline{x})$, and the unknown domain Ω characterized by $\partial\Omega^{in}$ and $\partial\Omega^{ex}$ simultaneously and efficiently from the known initial-boundary values, $\underline{f}$, $\underline{g}$, $\underline{h}^{in}$ and $\underline{h}^{ex}$, the source functions R_i, and the additionally measured auxiliary data (independent from the initial-boundary values) either in the space-time domain,

$$\underline{u}(\underline{x},t) = \underline{p}_\lambda(\underline{x},t), \quad \underline{x} \in \Omega_\lambda, \; \lambda = 1,2,3,\ldots,\Lambda,$$
$$0 \leq t \leq T,$$
$$\text{and } (\underline{\nu}_\sigma \cdot \nabla)\underline{u}(\underline{x},t) = \underline{q}_\sigma(\underline{x},t), \; \underline{x} \in \Omega_\sigma, \; \sigma = 1,2,3,\ldots,\Sigma,$$

where Ω_λ and Ω_σ are the disjoint subsets of $\Omega \oplus \partial\Omega^{in} \oplus \partial\Omega^{ex}$ and $\underline{\nu}_\sigma$ is a given unit vector in Ω_σ, or in the space-complex frequency domain (useful only for the systems of linear partial differential equations),

$$\underline{U}(\underline{x},s) = \underline{P}_\lambda(\underline{x},s), \quad \underline{x} \in \Omega_\lambda, \quad \lambda = 1,2,3,\ldots,\Lambda,$$
$$0 \leq s \leq S,$$
$$\text{and } (\underline{\nu}_\sigma \cdot \nabla)\underline{U}(\underline{x},s) = \underline{Q}_\sigma(\underline{x},s), \quad \underline{x} \in \Omega_\sigma, \; \sigma = 1,2,3,\ldots,\Sigma,$$

where $\underline{U}(\underline{x},s)$, $\underline{P}_\lambda(\underline{x},s)$ and $\underline{Q}_\sigma(\underline{x},s)$ are the Laplace or Fourier transforms of $\underline{u}(\underline{x},t)$, $\underline{P}_\lambda(\underline{x},t)$ and $\underline{q}_\sigma(\underline{x},t)$ respectively.

In this paper, only the presentation of GPST for solving multi-parameter inverse problems of systems of linear partial differential equations is given and it also can be found in [1]. Here, typical examples in nondestructive evaluation [2],[3] and seismic exploration [4],[5] are used for the purpose of demonstration. Numerical simulations are carried out to test the feasibility and to study the general performance of GPST without the real measurement data. The real computational accuracy of GPST (not the function-theoretical error estimates), its actual efficiency (in terms of CPU time) and its robustness when the measurement data are erroneous are examined. Finally, the mathematical foundation of GPST, e.g., local uniqueness, global uniqueness, convergence, etc., are discussed.

REFERENCES

[1] Y. M. CHEN, Generalized pulse-spectrum technique (GPST) for solving problems in parameter identification, Proc, U.S.-China Workshop on Adv. in Comput. Eng. Mech., Dalian, China.

[2] Y. M. CHEN and G. Q. XIE, A numerical method for simultaneous determination of bulk modulus, shear modulus and density variations for nondestructive evaluation, to appear in Nondestructive Testing Communications.

[3] S. L. WANG and Y. M. CHEN, The determination of shapes and sizes of scatterers in the inverse scattering problem, Proc. of 5th IMACS Int. Symp. on Computer Methods for PDE, Lehigh University, Bethlehem, Pa., June 19-21, 1984.

[4] Y. M. CHEN and G. Q. XIE, An iterative method for simultaneous determination of bulk and shear moduli and density variations, submitted to J. Comput. Phys.

[5] G. Q. XIE and Y. M. CHEN, A modified pulse-spectrum technique for solving inverse problems of two-dimensional elastic wave equation, Proc. of 5th IMACS Int. Symp. on Computer Methods for PDE, Lehigh University, Bethlehem, Pa., June 19-21, 1984.

APPLICATIONS OF SEISMIC RAY-TRACING TECHNIQUES TO THE STUDY OF EARTHQUAKE FOCAL REGIONS

W. H. K. LEE*, F. LUK** AND W. D. MOONEY***

The earth is continuously being deformed by internal and external stresses. If the stresses are not too large, elastic and/or plastic deformation will take place. However, if the stresses build up sufficiently over a period of time, earthquakes may eventually occur. Earthquakes are essentially a fracture process involving the sudden release of stress and the generation of elastic waves that travel through the earth. Earthquakes have caused considerable damage throughout human history. For example, over 1 million people were killed since 1900. Recently, the town of Coalinga, California was nearly destroyed, and the San Francisco Bay region was shaken by a strong earthquake near Mount Hamilton.

To reduce earthquake hazards, we need a better understanding of the earthquake processes and the properties of the rocks in the focal region. The principal tool in seismology is the seismograph which detects and records ground motion caused by the passage of elastic waves from earthquakes. A major problem in seismology is the determination of earthquake source parameters and the seismic properties of the earth from a set of observations made near the earth's surface. This is an inverse problem and considerable advances have been made during the past decade. For example, Aki, Christoffersson, and Husebye (1976) introduced a technique to invert teleseismic observations, and Aki and Lee (1976) developed a method to invert P-wave arrival times from local earthquakes. Since then, these methods have been refined and applied to many regions of seismological interest.

*United States Department of the Interior, Geological Survey, Menlo Park, California 94015
**Department of Computer Science, Cornell University, Ithaca, New York 14853
***United States Department of the Interior, Geological Survey, Menlo Park, California 94015

In simultaneous inversion for earthquake source and path parameters, we solve the nonlinear optimization problem by taking iterative linear steps. The linear step is usually formulated to minimize the residuals between the calculated and the observed variables (e.g., P arrival times from earthquakes and explosions) in a least squares sense. To start the inversion scheme, we must have an initial earth model of seismic velocities and the ability to calculate travel times and ray paths for a given earth model (i.e., ability to trace seismic rays). The most direct way to obtain an initial earth model is to conduct a refraction experiment using artificial explosions and to interpret the data with the aid of seismic ray-tracing techniques. In this paper, we will briefly summarize the seismic ray-tracing problem and its applications to the study of earthquake focal regions.

A modern treatment of wave propagation in elastic media has been presented by Karal and Keller (1959). Although many seismological problems depend on the solution of the elastic wave equations with appropriate initial and boundary conditions, finding analytic solutions is difficult. Therefore, two practical approaches were applied: (1) to develop through specific boundary conditions a solution in terms of normal modes (see Officer, 1958), and (2) to transform the wave equation to the eikonal equation and seek solutions in terms of wavefronts and rays that are valid at high frequencies (see Cerveny, Molotkov, and Psencik, 1977). The ray approach is particularly relevant to the earthquake inversion problem. By solving the ray equation for a given earth model, we obtain travel times and derivatives as well as ray paths.

As shown by Lee and Stewart (1981), the seismic ray equation may be derived either from the wave equation or from Fermat's principle. If r is a position vector and u is the slowness (reciprocal of the velocity v), then the seismic ray equation may be written compactly as:

$$(1) \quad \frac{d}{ds}\left[u\left(\frac{d\vec{r}}{ds}\right)\right] = \vec{\nabla} u$$

where ds is an element of the ray path. Equation (1) is a set of three second-order differential equations. It can be reduced to a first-order system by following the approach of Pereyra, Lee and Keller (1980) as modified by R.P. Comer. If we introduce

$$\begin{aligned} &\omega_1 = x, \qquad &&\omega_3 = y, \qquad &&\omega_5 = z, \\ (2) \quad &\omega_2 = u(dx/ds), \qquad &&\omega_4 = u(dy/ds), \qquad &&\omega_6 = u(dz/ds) \end{aligned}$$

then Equation (1) becomes:

$$(3)\quad d\omega_1/ds = v\omega_2,\quad d\omega_3/ds = v\omega_4,\quad d\omega_5/ds = v\omega_6,$$

$$d\omega_2/ds = \partial u/\partial x \quad d\omega_4/ds = \partial u/\partial y \quad d\omega_6/ds = \partial u/\partial z$$

Since in many seismological applications, the travel time between point A and point B, i.e., $T=\int_A^B u\,ds$, is of utmost interest, we introduce an additional variable ω_7 to represent the partial travel time τ, along a segment of the ray path from point A. The corresponding differential equation is simple: $d\omega_7/ds=u$. With this addendum, the total travel time is given by $T=\tau$ (at point B)$=\tau(S)$, where S is the total path length. To determine S (which is a constant for a given ray path), we introduce one more variable $\omega_8 = s$, and its corresponding differential equation, $d\omega_8/ds = 0$. It is also computationally convenient to scale the path length s such that its value is between 0 and 1. Therefore, we introduce a new variable p for s such that $p = s/S$ and use the prime (') symbol to denote differentiation with respect to p. Thus the final set of equations to be solved is:

$$\begin{aligned}
&\omega_1' = \omega_8\,\omega_2, && \omega_3' = \omega_8\,\omega_4, && \omega_5' = \omega_8\,\omega_6,\\
(4)\quad &\omega_2' = \omega_8(\partial u/\partial x), && \omega_4' = \omega_8(\partial u/\partial y), && \omega_6' = \omega_8(\partial u/\partial z),\\
&\omega_7' = \omega_8\,u, && \omega_8' = 0, && p \in \{0,1\}
\end{aligned}$$

The variables corresponding to the solution of Equation (4) are:

$$\begin{aligned}
(5)\quad &\omega_1=x, && \omega_3=y, && \omega_5=z, && \omega_7=\tau,\\
&\omega_2=u(dx/ds), && \omega_4=u(dy/ds), && \omega_6=u(dz/ds), && \omega_8=s
\end{aligned}$$

and the boundary conditions are:

$$\begin{aligned}
&\omega_1(0) = x_a, && \omega_3(0) = y_a, && \omega_5(0) = z_a,\\
(6)\quad &\omega_1(1) = x_b, && \omega_3(1) = y_b, && \omega_5(1) = z_b,\\
&\omega_7(0) = 0, && \omega_2^2(0) + \omega_4^2(0) + \omega_6^2(0) = u^2(0)
\end{aligned}$$

where the coordinates for point A are (x_a,y_a,z_a), and for point B are (x_b,y_b,z_b). The boundary condition for ω_7 is determined by the fact that the partial travel time at the initial point A is zero. The last boundary condition simply states that the sum of squares of direction cosines is unity.

In the past decade, several accurate and efficient numerical methods for solving boundary value and initial value problems have been developed by mathematicians. We have applied the following numerical methods to solve the seismic ray equation: (1) a finite difference method (Pereyra, Lee and Keller, 1980), (2) a collocation method (Ascher and Lee, in preparation) and (3) an initial-value method (Comer, Lee and Luk, in preparation) using the numerical technique developed by Shampine and Gordon (1975). The reason to apply three different methods in solving the seismic ray-tracing problem is the fact that complex computer codes are hard to debug, and this approach offers a direct comparison of results.

Although it is straightforward to apply existing numerical techniques to solve the seismic ray equation, many practical difficulties arise. First, the earth is heterogeneous, and especially complex in earthquake fault zones (see Page, 1982). Therefore, it is difficult, if not impossible, to represent the earth adequately in a numerical model. Second, the derivation of the seismic ray equation and numerical methods to solve it are based on the assumption that the seismic velocity is continuous and differentiable. Thus we face a difficult problem in interpolating seismic velocity and its derivatives based on values specified for a simplified grid of points. After many experimentations, we finally developed a 3-dimensional hermite interpolation scheme using cubic bessel interpolants that seems to provide the desired interpolated values.

We have applied ray tracing techniques to study two earthquake focal regions: the Bear Valley and the Coyote Lake areas in California. In the Bear Valley region, we found significant seismic velocity variations across the San Andreas fault zone (see Aki and Lee, 1976; Engdahl and Lee, 1976). The Coyote Lake study was stimulated by the occurrence of a strong earthquake (magnitude 5.7) on August 6, 1979. Recently, a slightly stronger earthquake (magnitude about 6) also occurred in this region, about 20km north of the 1979 event. Both earthquakes occurred within the Calaveras fault zone, a major branch of the San Andreas fault system.

Active seismic experiments have not been carried out within the San Andreas fault zone in the past because seismologists thought that the results would be too complex to interpret. However, a series of strong earthquakes in our nearby areas prompted us to carry out the first major refraction experiment within the San Andreas fault zone. Three explosions were set off (two within the fault zone) and the permanent seismic network was supplemented by 100 temporary seismographs deployed along a main profile in the fault zone and two fan profiles across the fault zone. Analysis of the explosion data with the aid of mostly two-dimensional seismic ray tracing reveals a complicated crustal velocity structure with strong lateral variations. Details are given in the paper by Blumling, Mooney and Lee (in press).

In the past several years, our efforts have been concentrated toward developing computer programs that can trace seismic rays in 2- and 3-dimensional earth models, and to collect the necessary seismic data. We are just beginning to apply our ray tracing programs to model earthquake focal regions. It is clear that a combination of efficient seismic ray-tracing techniques, and detailed geologic and geophysical data are required to resolve the complexity of earthquake focal regions. Although some progress has been made, much work remains to be carried out.

REFERENCES

[1] AKI, K., CHRISTOFFERSSON, A., and HUSEBYE, E.S., Three-dimensional seismic structure of the lithosphere under Montana LASA., Bull. Seism. Soc. Am., 66(1976), pp. 501-524.

[2] AKI, K. and LEE, W.H.K., Determination of three-dimensional velocity anomalies under a seismic array using first P arrival times from local earthquakes, Part I. A homogeneous initial model, J. Geophys. Res., 81(1976), pp. 4381-4399.

[3] ASCHER, U. and LEE, W.H.K., Solving two-point seismic ray-tracing problems in a heterogeneous medium using a collocation method, in preparation.

[4] BLUMLING, P., MOONEY, W.D., AND LEE, W.H.K., Crustal structure of the southern Calaveras fault zone, central California, from seismic refraction investigations, Bull. Seism. Soc. Am., in press.

[5] CERVENY, V., MOLOTKOV, I.A., AND PSENCIK, I., Ray Method in Seismology, Univ. Karlova Press, Prague, 1977.

[6] COMER, R.P., LEE, W.H.K., and LUK, F., Solving seismic ray-tracing problems in a heterogeneous medium by a general shooting method, in preparation.

[7] ENGDAHL, E.R. and LEE, W.H.K., Relocation of local earthquakes by seismic ray tracing, J. Geophys. Res., 81(1976), pp. 4400-4406.

[8] KARAL, F.C. and KELLER, J.B., Elastic wave propagation in homogeneous and heterogeneous media, J. Acoust. Soc. Am., 31(1959), pp. 694-705.

[9] LEE, W.H.K. and STEWART, S.W., Principles and Applications of Microearthquake Networks, Academic Press, New York, 1981.

[10] OFFICER, C.B., Introduction to the Theory of Sound Transmission, McGraw-Hill, New York, 1958.

[11] PAGE, B.M., The Calaveras fault zone of California, an active plate boundary element, Calif. Div. Mines & Geol., Spec. Pub., 62(1982), pp. 175-184.

[12] PEREYRA, V., LEE, W.H.K., and KELLER, H.B., Solving two-point seismic ray-tracing problems in heterogeneous medium, Part 1. A general adaptive finite difference method, Seism. Soc. Am., 70(1980), pp. 79-99.

[14] SHAMPINE, L.F. and GORDON, M.K., Computer Solution of Ordinary Differential Equations: the Initial Value Problem, Freeman, San Francisco, 1975.